Applied Geography

The GeoJournal Library

Volume 77

Managing Editor: Max Barlow, Concordia University,
Montreal, Canada

Founding Series Editor:
Wolf Tietze, Helmstedt, Germany

Editorial Board: Paul Claval, France
Yehuda Gradus, Israel
Risto Laulajainen, Sweden
Sam Ock Park, South Korea
Herman van der Wusten, The Netherlands

The titles published in this series are listed at the end of this volume.

Applied Geography

A World Perspective

edited by

ANTOINE BAILLY

University of Geneva,
Switzerland

LAY JAMES GIBSON

University of Arizona,
Tucson, USA

KLUWER ACADEMIC PUBLISHERS
DORDRECHT / BOSTON / LONDON

A C.I.P. Catalogue record for this book is available from the Library of Congress

ISBN 1-4020-2441-X (HB)
ISBN 1-4020-2442-8 (e.book)

Published by Kluwer Academic Publishers,
P.O. Box 17, 3300 AA Dordrecht, The Netherlands.

Solid and distributed in North, Central and South America
by Kluwer Academic Publishers,
101 Philip Drive, Norwell, MA 02061, U.S.A.

In all other countries, sold and distributed
by Kluwer Academic Publishers,
P.O. Box 322, 3300 AH Dordrecht, The Netherlands

Printed on acid-free paper

Funded by European Union (ERBIC 18CT 970152)

Printed in the Netherlands.

TABLE OF CONTENTS

Introduction   1
Antoines Bailly and Lay James Gibson

Part I: History and epistemological foundations

Chapter 1
Managing geography after Y2K   9
Antoine Bailly and Lay James Gibson

Chapter 2
The principles and practice of applied geography   23
Michael Pacione

Chapter 3
Historical foundations of applied geography   47
Michel Phlipponneau

Chapter 4
Political geography, public policy and the rise of policy analysis   69
Kingsley E. Haynes, Qingshu Xie and Lei Ding

Chapter 5
The role of geographic information science in applied geography   95
Arthur Getis

Chapter 6
Economic base theory and applied geography   113
Lay James Gibson

Chapter 7
Retail location and consumer spatial choice behavior   133
Harry Timmermans

Part II: A world perspective

Chapter 8
Applied geography in Western and Southern Europe   151
Jorge Gaspar

Chapter 9
Applied geography in Central Europe 169
György Enyedi

Chapter 10
Applied geography in 20th century North America: A perspective 187
John W. Frazier

Part III: Case studies

Chapter 11
Disability, disadvantage, and discrimination: An overview with
special emphasis on blindness in the usa 213
Reginald G. Golledge

Chapter 12
Human Wayfinding 233
Reginald G. Golledge

Chapter 13
International trade 253
Jessie P.H. Poon and James E. McConnell

Chapter 14
Medicometry and regional development 273
Antoine S. Bailly

Chapter 15
Monitoring and benchmarking regional and local performance 287
Robert Stimson

Chapter 16
Applied geography for the future 305
Antoine Bailly and Lay James Gibson

Chapter 17
Biographies and fields 309

Index 319

ANTOINES BAILLY AND LAY JAMES GIBSON

INTRODUCTION

Applied Geography: A World Perspective makes geography's utility as an applied science explicit. It has been argued that geographic applications in corporations, government agencies and in non-governmental organizations "just happen." Many geographers, however, are less trusting. Geography's position as an academic discipline has been challenged frequently over the years. Further, geographic approaches have often been adopted by those with little or no formal training in geography per se. The authors who have contributed to this book feel that there is real value to having a strong and readily recognizable discipline of geography. We want geography to get full credit for its contributions to policy and practice and we want graduates of our university-based programs get full consideration when they apply for positions in industry, government, and in NGO's.

The book is organized around three main themes: "History and Epistemological Foundations" of applied geography; "A World Perspective" on the practice of applied geography in different regions; and finally "Case Studies" which show how applied geography can contribute to problem solving in different institutional environments.

Part 1, History and epistemological foundations has seven of the book's 15 papers. The first paper by Antoine Bailly and Lay Gibson sets the tone. The authors build the case for an aggressive and proactive approach for establishing geography as a vital problem-solving discipline and for making applied geography an explicit part of the geography curriculum.

Michael Pacione deals with the Principles and Practices of Applied Geography and more specifically the ways that the worlds of practice and theory intersect. His chapter is organized into nine main sections. These address the definition of applied geography; the concept of useful knowledge; the relationship between pure and applied research; the value of applied geography; the question of values in applied geography; types of applied geographical research; the practice of applied geography; the history of applied geography; and the prospects for applied geography.

The paper by Michel Phlipponneau favors "French geography;" nevertheless it complements Pacione's paper nicely. Whereas the latter paper is especially good at qualifying

1

A. Bailly and L. J. Gibson (eds.), Applied Geography, 1–5.
© 2004 Kluwer Academic Publishers. Printed in the Netherlands.

and positioning the special role of applied geography in the contemporary world, Phlipponneau's paper digs deep into the history of applied work including geography and early explorations, the roles of geographical societies in heightening awareness of geography, and involvement by geographers in surveying and mapping. A special treat are his frequent mentions of the International Geographical Union, its growth and development, and the place of applied work in the IGU's agenda.

The paper by Kingsley Haynes et al. (Political Geography, Public Policy and the Rise of Policy Analysis) also complements the Pacione paper and Phlipponeau's paper too in that it looks at trends in academic geography and its influence on applied geography. But there is a major difference, Pacione focuses on geography broadly defined whereas Haynes focuses on political geography and public policy. He begins with a brief review of the literature on political geography and a categorization of scale issues. He then moves onto two simple but extremely appropriate applied case studies which show how geographic factors affect public policy analysis and how applied public policy studies can be carried out with sensitivity to geographic considerations. The two case studies are regional income convergence and the rise of regional transportation management institutions.

The paper by Arthur Getis represents a change in direction – to technology driven applied geography. He points out that powerful new technologies have emerged that greatly improve our ability to collect, store, manage, view, analyze, and utilize information regarding the critical issues of our time. These technologies include geographic information systems (GIS), global positioning systems (GPS), satellite-based remote sensing, and a great variety of remarkable software that allows for the analysis of the compelling problems. The issues include globalization, global warming, pollution, security, crime, public health, transportation, energy supplies, and population growth. Geographic Information Science (GISc) is more than just GIS and it has given rise to an essentially multidisciplinary approach to applied problems. No single person is an expert in all of these areas. It is necessary to emphasize coordination and collaboration and to find bridges that reduce barriers between disciplines.

Whereas Getis featured a technique, Lay Gibson focuses on a concept – economic base theory. Gibson's paper deals explicitly with the ways that economic base theory can contribute to practical understandings of value to regional planning and development professionals who need to better understand how regions work and how they might work even better. Base theory is often employed as a research tool on the promise that it can produce a multiplier. But it can do much more if the practitioner thinks to ask or if the applied geographer bothers to offer. This article identifies seven economic development problems commonly faced by development practitioners and regional planners and illustrates how solutions can be drawn from the application of economic base theory.

The final paper in this section was prepared by Harry Timmermans; it deals with "Retail Location and Consumer Spatial Choice Behavior." He reminds us that geographers traditionally have made substantial contributions to retail management and retail policy. Retail location decisions have a major impact on the success of a store. The saying "location, location, location" suggests that bad location decisions are difficult to compensate by other elements of the marketing mix such as pricing, merchandising and promotion. Space and location are the very core of the geography discipline; hence many geographical theories, results of spatial analyses and spatial methods are potentially relevant for better informed decisions in retailing. But he also discusses the role of retail geography in land-use policy development. For example, to avoid empty downtown areas, retail policy in many countries has stimulated retail business in downtown areas and prohibited retail growth in out-of-town or peripheral location by implementing appropriate planning control. Equity issues (good accessibility and a basic provision of retail stores in every neighborhood) have also been high on the policy agenda, especially in European countries in the 1970's and 1980's.

There are three papers in Part 2, A World Perspective. The three papers cover critical regions but there is clearly a bias toward the Western World. Jorges Gaspar's paper focuses on Western and Southern Europe. Whereas the historical origins of applied geography are considered, his real emphasis is on geographers as key players in the regional planning movement and on the emergence of the European Community and it's role in stimulating the demand for applied geographers. He also discusses the shifting fortunes of geography as a primary and secondary school subject and the implications of these shifts for university based geography departments.

György Enyedi's paper focuses on East-Central Europe. In a nominal way he deals with some of the same topics that Gasper deals with but the real story in this paper is the move away from communism and to capitalism. The implications for applied geographers have been enormous; as some opportunities disapper, others come into play. In some respects applied geographers in this realm have needed to totally reinvent themselves.

The final paper in this section is a review of applied geography in North America. John Frazier approaches applied geography from all sides. He talks about the influence of individuals, about the roles of academic departments, and he talks about the emergence of applied geography both as a part of the agenda of the Association of American Geographers and as *the* agenda of the free-standing national Applied Geography Conference which is now in its 26th year. Frazier ends his chapter with a thoughtful essay on "The Future of Applied Geography in North America."

The third and final group of papers is found in Part 3, Case Studies. This collection of five papers focuses on different settings where geography has or can make a contribution to finding answers to real questions and solutions to real problems.

Reginald Golledge is the author of two papers in this section, whereas his papers deal with very different topics and they share a common attribute – both have substantial significance for the business of helping those with physical disabilities deal with the world around them. The movement of people in the aggregate, over time has been successfully modeled by migration models. But knowledge of why people move, particularly with regard to individual or household daily decision, is still in its infancy. The emphasis in Golledge's wayfinding paper is on the process of wayfinding by humans. Specifically, the paper focuses on an activity approach in general, discusses the cognitive basis of human wayfinding, and examines how this knowledge is being integrated by applied geographers into agent-based computational and simulation models.

Golledge's other contribution deals with disability, disadvantages, and discrimination which in his opinion has received too little attention by geographers. In the first part of this chapter he discusses alternative definitions of disability and he surveys the nature of disability in the United States and elsewhere. He then moves on to an examination of enabling legislation that provides an umbrella for disability research and he discusses avenues of future research for which the applied geographer is eminently suited to pursue.

In the chapter on international trade, Jessie Poon and James McConnell note that since the early 1990's, geographers have been on the forefront of the debate on globalization and the shape of the contemporary world economy. While the number of geographers researching international trade issues remains relatively small, the authors argue that these specialists have nonetheless contributed to applied geography by sensitizing trade issues, models, and interaction. By rescaling trade analyses through their works at the urban and regional levels, trade geographers potentially pave the path for a more bottom-up as opposed to the more dominant top-down approach in formulating and assessing trade and industry policies at regional and national levels.

Antoine Bailly is one of the key figures in the field of regional medicometry – the study of medical provider and facility locations and the role that location plays in meeting patient needs. His chapter presents the epistemological foundations of regional medicometry and its five criteria: ethical, economic, social, spatial, and temporal. By its global approach, medicometry, studys' the direct and indirect effects of medical infrastructure and employment on society and its quality of life. Two case studies, one dealing with an urban hospital in Geneva, Switzerland, the other with peripheral hospitals, give a step-by-step example of the multiplier effect approach. Detailed measures of economic and social impacts of hospitals are given to show that medical planning needs a careful and long-term analysis.

The final paper, by Robert Stimson, is based on the observation that measuring and benchmarking how regions and local communities perform and cope with socio-economic change is a methodological issue of interest to geographers. His chapter outlines two approaches used in projects in which he has been involved. The first involves monitoring and evaluating the performance of three of Australia's metropolitan cities on a range of indicators relating to population and employment, investment in economic activities and housing markets. The second involves developing a multivariate model to measure the socio-economic performance of local communities across Australia's cities and towns.

Applied Geography: A World Perspective is not just about geographic approaches. It is about trained geographers using geographic approaches to evaluate real-world problems. We hope that you will agree that this collection builds the case for the special place that we think geographers and geography deserve in industry, government, and in NGO's.

We are extremely grateful for the assistance that Gérard Widmer has provided throughout all the phases of this publication. His knowledge of graphical packages, ability to manipulate texts, producing tables and figures were extremely useful as his critical analysis of the texts. We thank him for all of his hard work in making this publication happen.

Part I

History and epistemological foundations

ANTOINE BAILLY AND LAY JAMES GIBSON

CHAPTER 1
MANAGING GEOGRAPHY AFTER Y2K

INTRODUCTION

All academic disciplines evolve and geography is no exception. Physics, French, or even economics change over time, but at the end of the day (or decade) they are still physics, French, or economics. Geography is different – it can be a natural science, a social science, or a humanity. Ideally, perhaps geography is all three. Many feel that the inevitable tension between these three content areas and scientific and non-scientific orientations is not only healthy but essential. Keeping these perspectives in balance while recognizing that geography's real value is tied to its role as a measurement oriented science and managing the tensions that are inherent in a discipline where both basic research and applied research are appropriate can be a challenge. Put all of this together with the fact that there is rapid change and increased competition for limited resources in the academic institutions that house geography departments and it seems worthwhile for us to revisit the ways that we manage our discipline and the ways that we position it within our universities and within the job markets that our graduates enter.

GEOGRAPHY: A VERTICALLY INTEGRATED DISCIPLINE

In brief, we will propose an applied geography model that is sensitive to both the supply side and the demand side, a model which suggests that we should be proactive in cultivating markets both inside and outside of educational institutions. We discuss the need for programs at the bachelors, masters, and Ph.D. levels that transfer existing knowledge and prepare the discipline's next generation for intellectual and technical leadership. We do not discuss primary and secondary education per se in this paper, but we recognize that teaching the teachers and being supportive of primary and secondary curriculum development is an essential and fundamental responsibility of university-based geography departments.

By building bridges to the governmental, NGO, and corporate sectors we expect to realize at least two benefits. First, we stimulate demand for our graduates and for the problem-solving capabilities that they bring to the workplace. This encourages the educational sector to expand supply. Second, it puts geographic models and techniques

A. Bailly and L. J. Gibson (eds.), Applied Geography, 9–22.
© 2004 *Kluwer Academic Publishers. Printed in the Netherlands.*

to market tests that ultimately encourage innovation and "continuous improvement." Does "size matter?" We think it does (Gibson, 2000). But we also think that the way to position geography in a competitive market place for ideas is to push it beyond the academy into the large world of application. The links between academic institutions and outside markets for information technologies, biotechnologies and dozens of other areas have certainly had beneficial effects for both applied science and the basic science that supports it. University departments of management information systems have seen extraordinary increases in demand for students. Many of these students will move on to relatively routine jobs but others will move on to positions of scientific leadership both in universities and in government and industry. As Bowen (2001) has suggested, geography wants to be known as a discipline that leads societal changes – rather than one that follows by adapting to change. We argue that the best way to do this is by increasing our access to students and by increasing our access to those outside the academy who hire our students and put our approaches to market tests.

INTRINSIC VALUE AND DEMONSTRATED VALUE

Some university-based disciplines have intrinsic value and others do not. This has more to do with perceived utility than real utility but it is a fact of academic life nevertheless. Unfortunately geography in most countries does not have intrinsic value, at least in many of the best universities. It is almost impossible to imagine a good university without an economics department or a physics department. It is all to easy to imagine a good university without a geography department.

Geographers can agree that this statement is unfortunate and regrettable but they will also agree that it is a sad but true fact of academic life. The premise that is implicit and now explicit in this paper is that academic disciplines that do not have intrinsic value can reduce their at-risk status by not just asserting their intellectual value but by demonstrating their value for solving problems and informing policy in the worlds of commerce, government and in the non-governmental organization (NGO) sector.

OVERVIEW: GEOGRAPHY AND INSTITUTIONAL RESPONSES

Education philosophers and theorists such as Kuhn (1962) and Schlanger (1983) have developed models to help us understand how academic disciplines and fields come into existence and how they change over time. The basic ideas are straight forward. Some sort of representation is "tested" by society. If acceptable, it is embraced by educational institutions (Figure 1).

There is an initial representation. For example, geographers make substantial and well-documented contributions to tasks of strategic planning for resource utilization and industrial location to support the efforts of the allies during World War II. For geographers the timing is good. The United States and Europe are "thinking regionally"

Figure 1. Representations, societal changes and educational institutions respond

and "location matters" as they go about the business of making war-time and post war regional adjustments, especially with the Marshall Plan. Subsequently, the cold war is driving programs to build capacity and to develop networks to facilitate flows between regions (e.g.: the interstate highway system in the United States and the St. Lawrence seaway). Social, cultural, economic, and political institutions find something of value and accept the new representation. It is compatible with dominant ideologies. Globalization, economic growth and emerging social disparities, and the baby boom are the sorts of things that both shape the representation and encourage acceptance. At the same time, the development of the computer, information technologies, and biotech advances have changed our lives. As innovations gain acceptance there are associated changes in the ways that we view space and in the location of goods and services. Educational institutions respond by offering new courses. Students are drawn to the new offerings and graduates are placed in appreciative firms and agencies. Over time, educational institutions respond by expanding offerings, establishing formal programs and ultimately by establishing free standing departments. Educational institutions are building capacity, demand is robust, public funding is available for research to help the nation meet its needs and to support curiosity-driven "entitlement" research.

Change in geography is constant (Bailly and Ferras, 2001), as societies change and educational institutions follow. New representations are proposed and they compete with existing ones; ways of life change, values change, and aspirations change (Figure 2a–c). Academic fields respond to advances in scientific knowledge and to changing values, aspirations, and preferences. Some fields maintain their positions or even improve them. Others are either pushed aside by new representations or they struggle to maintain themselves. Institutions find their resource bases frozen or in decline. Competition for resources among units becomes keener. Faculty who are not entrepreneurial slip backwards. Extramural resources grow in importance as do public-private partnerships and leveraging schemes. Eventually the system reaches a temporary equilibrium – and then the cycle starts again.

Geography got a good boost from the Second World War and from the events that followed. In Asia and Europe, geographers were in demand as planners and as managers of the war effort and of the post war reconstruction efforts. In America, "Washington

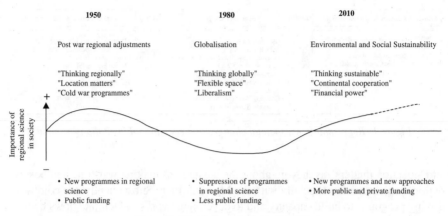

Figure 2. Three periods for geography

geographers" stayed in government agencies and at least some veterans used the G.I. Bill benefits to study geography and to train as geography teachers in departments across the country. The French language book, *La géographie, ça sert à faire la guerre* by Lacoste (1973), clearly identified geography's strategic role in war. But for geography it was not all smooth sailing. Advances such as the creation of geography programs and new universities following the war were sometimes off-set by set-backs such as departmental closures at places like Harvard, Yale, and Stanford in the 1970s.

"Thinking regionally" was replaced by "thinking globally" and "spatial planning by liberalism." Pessimists saw these closures as the beginning of the end. Optimists saw them as minor set backs that could be overcome if we maintained a positive attitude and avoided words like "crisis" (Bailly and Coffey, 1994 and Bailly, Coffey, Gibson, 1996). Chances are good that neither cohort is right. Somewhere in between the pessimists and the optimists are the realists. Whereas realists would be delighted to see departments of geography continue to spring up on six continents, they do not see this as vital or even likely event until geography moves ahead on several fronts such as the one opened by Secretary General of the United Nations, Kofi Annan, at the 2001 Association of American Geographers annual meeting (AAG, 2001). In the meantime, advances on any one of several fronts should be enough to keep geography alive as a disciplinary program focus within university departments or as a distinctive and worthwhile way to look at the world when geographers are based in other disciplinary units or in multidisciplinary units.

The realist takes comfort in the fact that geography has largely maintained its core values and that it has also responded to societal changes (Gould and Bailly, 2000). But the realist is also aware that nothing stays the same and that new initiatives will be needed simply to maintain the status quo. Geography needs to heighten its profile both within universities and outside universities.

SOME BACKGROUND

A paper by Gibson (1998) suggested that the business of all sciences is "business." It might be said about geography. It was suggested that our scientific organizations need to rethink the way that they do business and embrace new cohorts including those in the private research sector. Ultimately, (geography) needs to grow in size if it expects to have clout (Gibson, 1998). This chapter will take a somewhat different direction. We are guided by the assertion that businesses that make a profit survive; those that do not make a profit die. Specifically, we will look for a bottom-line for geography as a university-based discipline. We will explore ways for geographers to "make a profit." Whereas university departments are not typically expected to generate more revenue than they spend, they are often judged in terms of ROI (Return On Investment). Alert administrators want to know what each of their departments deliver in terms of enrollments, research dollars, and visibility externally given the university's investment in the department. We will not concern ourselves with the content of geography. We will be concerned with the several beachheads that geographers may want to consider taking and holding on to ensure that our discipline is adequately represented in the university curriculum.

The business of managing geography in Y2K and beyond raises delicate questions that are sometimes avoided for some of the same reasons that sex, religion and politics are avoided as topics for dinner table conversation – they get people riled. In other times it seems that these questions get pushed to the side by the geographers, who treat geography as a secondary or tertiary interest or worse yet, pretend to be something other than geographers and, therefore have less of a stake in the health of the field (Golledge, 2000). But the fact that people sometimes shy away from these issues does not imply that they should be down-played or moved to the back burner.

Throughout its more than 100 years of existence, geography has been actively (and fairly successfully) engaged in a balancing act involving fundamental and applied research. From almost the beginning, geographers have been committed to addressing natural science and societal problems including regional differentiation and national identities and later locational analyses for new production facilities and service activities, regional development initiatives to deal with economic and social disparities, and enhanced spatial models to improve the efficiency and effectiveness of regional planning activities. Theoretical research and applied research, especially at mid-century and particularly in Ph.D. programs were well articulated components in a geographers education. "Space Matters" became a defining slogan in the geographic community.

In more recent years it seems that different regions have conspicuously followed different paths. North America and some countries in Western Europe (e.g. Sweden and the Netherlands) were among the regions that seemed to increasingly emphasize theory and to focus on fine-tuning models, sometimes at the expense of concern with operalization of models. At the other extreme, many countries, including lesser developed ones, sometimes seem preoccupied with producing research with few if any real links to theory. Australian geographer Sorensen (1997) summarizes these issues in three points: (a) many economic models are contextually away from the world in which practitioners are acting; (b) they underestimate variables essential to development including leadership, invention, innovation, entrepreneurship, networking, social values, and confidence; and (c) there is a poor understanding of the political and bureaucratic processes which enable us to move from scientific knowledge to action. Such drifts seem to have weakened geography and diminished its usefulness as a problem-solving science. With a low number of students or a spotty record of extramural funding, geography departments risk losing credibility with academic authorities and make it increasingly easy for expanding free-market economies to neglect the spatial dimension.

Geography's competitive advantage continues to lie in its ability to bring together the several elements that characterize human and physical systems (Ehlers and Krafft, 1998) and more thematic approaches such as urban and regional research and planning. If there is a crisis in geography it is in no small measure associated with a widening gap between scientific and thematic approaches.

If geography's promise is to be realized, the balance between these complimentary approaches needs to be restored and in the process, greater attention needs to be given to geography's position within the university curriculum and to engaging (or re-engaging) non-university based constituencies in the work of professional geographers.

WHAT DRIVES THE AGENDA?

The notion that good applied research cannot exist without fundamental research has been frequently asserted (White, 1972). But perhaps even more basic is the question "who wants to know?" Geographer Wellar (1998) likes to think of two research models – curiosity-driven research and client-driven research. The trick, of course, is to assure that each model feeds off the other. We want to be able to demonstrate our value to managers and to political leaders without debasing the currency of geography or denying progress in terms of theoretical research. The following simple model (Figure 3) is offered as one way of thinking about a well articulated university-based applied geography program.

In this model, basic concepts from geography and other complimentary spatial disciplines are drawn together in applied geography core courses. Core courses establish a solid and rigorous scientific foundation and they introduce students to the institutions and corporate culture of geography. If the core courses are essentially inward looking, geography research seminars are essentially outward looking. The seminars and the research related to them are tied to external constituencies and to external funding. It is here that geography is pushed forward as a fundamental research field and it is here that geographic approaches are tried, fine-tuned and used to solve real-world problems.

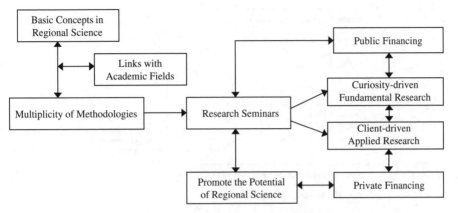

Figure 3. A university-based regional science programme

Ideally, the business of introducing students to basic concepts and methods with applied examples would start early in the student's career. The performance of geography students could be expressed with reference to four areas of achievement: knowledge and understanding, geography skills, intellectual skills and key skills useful for applied geography such as GIS (and geographical information sciences GISc) techniques (Benchmark statement for Geography, London, 2000). Figure 4 offers a generalized overview.[1]

On completion of their degree, geography students should be able to apply geographical concepts in different situations and work as a participant in, or leader of, research groups able to develop geographical ideas.

Perhaps the biggest challenges in putting a program of this sort in place are (a) designing the curriculum for the most introductory courses and (b) assuring that appropriate texts and other instructional materials are in place. The establishment of the new journal *Applied Geographic Studies* by Wiley was encouraging. It was a high quality journal in all regards but, unfortunately it never found a market and was discontinued after just a few issues. The loss of *Applied Geographic Studies* points to the fragility of the applied field, when it is separated from mainstream geography. A strategy suggested by Isserman (1993) shows special promise. It involves a follow-up of our student's

---

1. The Bologne Convention of 2001 produced an agreement among European states designed to standardize B.A. degrees at 180 credits and M.A. degrees at 120 credits. According to the Bologne Convention, European undergraduate degrees should be designed as three year degrees and M.A. programs should be two year programs. The B.A. in geography will probably include some applied geography in lecture classes and seminars. At the M.A. level applied practices could easily be one-half of the credits with a thesis or master report based on work completed within a public or private firm.

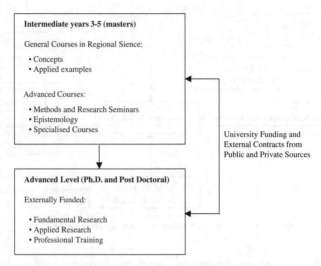

Figure 4. An overview of a regional science curriculum

careers would help to define major channels in terms of preparation of new students while opening new branches of inquiry that allow us to respond to new demands for spatial research.

THE CHANGING LANDSCAPE

A broad vision for the future which appears to have some relevance for applied geography is offered by Geenhuizen and Nijkamp (1996):

1. We need to make progress in theory and in improvement of research tools and techniques both rooted in spatial sciences and in other areas such as chaos theory, choice models, multi-criteria analysis, and GIS and GISc.
2. We must track changes in the economy such as structural shifts based on technological trajectories, increased flexibility of production and the emergence of the network economy.
3. We should raise new spatial policy questions such as questions about transportation infrastructure in Europe, the relationships between states and regional and local governments, ecological issues, and new urban problems.

Geographers Cutter, Golledge and Graf (2002) identify 10 "big questions" for the future of geography. We quote theme since it shows the changing landscape of the field and for its applications (Table 1).

The first question deals with the nature of uneven economic development and how geography can contribute to understanding this phenomenon. The second is linked to

Table 1. Big questions in Geography

1. What makes places and landscapes different from one another, and why is this important?
2. Is there a deeply held human need to organize space by creating arbitrary borders, boundaries, and districts?
3. How do we delineate space?
4. Why do people, resources, and ideas move?
5. How has the earth been transformed by human action?
6. What role will virtual systems play in learning about the world?
7. How do we measure the unmeasurable?
8. What role has geographical skill played in the evolution of human civilization, and what role can it play in predicting the future?
9. How and why do sustainablility and vulnerability change from place to place and over time?
10. What is the nature of spatial thinking, reasoning, and abilities?

the interactions between people and places, and the third how we partition space. Then geography has also to understand the flows of people, resources and ideas from place to place and how humans have altered the Earth by their actions. As a sixth question what will virtual systems allow us to do in the future that we cannot do now? And how can we measure all the previous key issues? Another question is linked to the geographic base of human history and how we can improve our ability to predict spatial events. The two last questions deal with the sustainability of our systems and the nature of spatial thinking and reasoning.

These big questions represent a way of looking at the future of a geography facing modern society instead of doing investigation on small problems more easy to solve.

When translating this vision into action, we would be wise to keep in mind Hägerstand's (1973) concern about the need to consider the effects of programs and policies on people. And we will need to consider what, if anything, paradigm shifts in the social and natural sciences mean for geography.

First, a number of geographers have moved away from the dominant positivist paradigm. In some circles, research agendas are driven by concerns with issues such as quality of life, aging populations, and "sustainable" regional development (Ives and Messerli, 1990) and regional performance, social segregation and exclusion, regional identity, awareness of new forms of citizenship and participation, quality of environment and resource utilization, and "territorial equity." It is probably fair to say that geographers have not always been leaders in addressing these questions. But it is also fair to point out that geographers have been made fully aware of the value of social perspectives (White, 1992). Should geography move more aggressively into these areas? Can geography maintain its distinctive identity and still make significant contributions to the business of dealing with, and managing, these issues? While the jury is still out it seems that the answer is either "yes" or perhaps a "possible yes", Kates (2001) would respond with a clear "yes". He opens a recent paper by saying "the central question of my scientific work has always been, what is and ought to be the human use of the earth? It is a ground query, derivative of my discipline of geography...". Geographer John Rees (1999), on the other hand, is cautious... "the recent shift away from spatial analysis to social theory [in the social sciences] has not found much support in regional science but it is the implications of postmodernism that should be of most concern. This is because postmodernism can be compared with "anything goes" as an espistemology, and can be characterized as "antitheory" and "antiscience." What does this mean for geography as a discipline and what are the implications for geography's standing as a decision science for business and public policy?

Second, there are potential payoffs associated with expanding instructional programs to ensure exposure to undergraduates, graduate students in allied fields, university based academics and researchers from the public and private research sectors who deal with regional issues. The general payoff is greater awareness that like time, space is essential if we are to more fully appreciate human activities. The historians have built the case for time; geographers should continue to be effective advocates for the role of space.

But there are more specific payoffs too. By focusing on undergraduate instruction we are in a position to create additional demand for instructions – demand for teaching assistants and eventually, more demand for professors. Further, we build awareness that can create demand for graduate instruction.

Similarly, we have so far been unable to adequately sustain programs designed to draw into geography university based academics and researchers from the public and private research sectors. There was a time in the United States when the Office of Naval Research and other such funding sources were available to support geography; but such opportunities seem to be much less frequent today. In today's funding environment entitlements are rare. Capturing extramural funds usually requires strong leadership, an entrepreneurial spirit, and political capital. Funding also requires institutional partners and institutional partners come and go. It appears that we are coming up short when it comes to initiatives designed to sell geography to academics, to the research sector, to public and private managers, and to elected officials. This is unfortunate inasmuch as these are our links to institutional partners.

Third, we should follow the good examples set by medicine and business, engineering and the natural sciences, and many other fields of study, and aggressively pursue new forms of public and private funding. This means that we must have a clearly articulated vision of the relationship between basic research and applied research, including the potential for relatively routine operations such as data collection and data bank maintenance and the delivery of training programs to support and subsidize methodological improvements. GIS (and GISc) are obvious areas to exploit but ironically they have grown so large so fast that it simply became too much for geography to handle. As producers of research, we should envisage creating laboratories that are revenue centers supporting both research and instructional programs while at the same time moving us closer to "the business of translating theoretical processes into public policy and corporate action" (Gibson, 1998). In the process of doing so, we will likely find ourselves rethinking our definition of geography and broadening the scope of our inquires along lines suggested, over the years, by a variety of thoughtful students of geography. Hägerstrand's (1973) rhetorical question "what about people . . . " has been expanded upon by others, such as geographer Thomas (1977), who talks about "space society", not just "space economy."

The real value of the approaches advocated by Thomas and others is that they remind us that we should be searching for methodologies that help us solve real problems and not just problems that fit our methodologies. The World Bank, the Centre National de la Recherche Scientifique, (a French national organization for scientific research), UNESCO, many major foundations, and some private research groups typically start with problems that have both social and economic dimensions. Geography has the potential to provide the research perspectives and geographers should further develop the capacity to manage these projects and follow them through the implementation phase.

Fourth, whereas it is essential that we continue to communicate with one another, we need to make particular efforts to reach new publics. We should, of course, continue to publish specialized books and articles but there is also merit in the publication of basic books for students beginning their training and popular books to let the public know what we do. An excellent example of "readable social science" is journalist Joe Garreau's (1992) thoughtful and well researched *Edge Cities, Life on the New Frontier.* We also need to think more about "sound bite" definitions of geography. Anyone who has organized a symposium with non-academic participants or who has gone to potential sponsors for subsidies should know the importance of a terse and satisfactory answer to the question "what is geography?" The situation becomes even more critical when dealing with the media. Most people think that they know what an economist does. Few people know that geographers exist. Experiences with the annual French Geography Festival illustrate these assertions. The goal of the Festival is to popularize geography and the staff includes both professional geographers and communications professionals assigned to both print and electronic media. Despite substantial efforts to be media friendly, misunderstandings are common and the effectiveness of media coverage is all-too-often dimished despite the fact that 500 newspapers, and numerous television networks are involved in the coverage.

Many of the social and behavioral sciences have journals aimed at the general public. These journals may not reflect the cutting-edge trends in psychology, sociology, or economics but they heighten awareness of these disciplines and build "brand recognition." Whereas it is difficult to imagine an airport news stand carrying *Economic*

Geography Today, our discipline would benefit from more articles in magazines and trade publications authored by people identified as geographers. Magazines such as *National Geographic* help even though professional geographers are not often contributors. Writing for a lay audience can build identity and it might stimulate demand for geography solutions but equally important, it encourages authors to be market oriented. Many societies, such as the British Royal Geographical Society, have developed market oriented magazines. But when all is said and done, geography has a long way to go in its efforts to reach a broad audience. We need to raise our profile to improve the awareness of the public by creating more magazines, festivals, and geography awareness weeks and as well by contributing to the daily print and electronic media through informal commentary on the "geographic" significance of contemporary events.

CONCLUSIONS

Whereas it looks like geography has, at least to this point, lost out on its bid to be an easily recognized and indispensable academic discipline, it is alive and well especially in some industrialized countries. The usefulness of geography as a research and teaching discipline and as an applied science is much, much greater than its prestige or public acceptance. Its potentials for serving as the disciplinary foundation for professional fields such as GIS (and GISc), urban and regional planning, and economic and social development are enormous. Geography's institutions (Buttimer, 2000) are positioned to bring together scholars from the academy and professionals from the world of practice. Each year geography offers researchers who share an interest in spatial perspectives an attractive menu of meeting venues. Further, geography supports a variety of high-quality scientific journals that are well thought of by geographers and by the non-geographers who read them.

Inasmuch as many geographers are related to an allied field, (e.g., the environmental sciences, regional science, planning, computer science etc.), loyalties are sometimes divided and identities are frequently obscured. This tends to diminish our clout and make even more critical proactive initiatives to aggressively promote our discipline, our institutions, and our publications. This proactive agenda needs to stress the retention of familiar cohorts, i.e., faculty in academic departments and it needs to reach out to new constituencies both within the academy and outside in public agencies and private firms (Golledge, 2000). Given the proper action agenda, the status quo should be easily maintained and the goal of an expanded and even more energetic applied geography community should be within reach.

RECOMMENDATIONS

Geography is a powerful discipline with unrealized potentials for solving critical problems in the natural and social sciences. To assure that geography reaches its full potential we offer the following recommendations:

1. Design instructional programs at all levels to emphasize geography's value as an applied science demonstrating explicitly its relevance to both business and government as well as to NGOs and communities at all levels of scale. Curicula should incorporate developing skills in measurement, modeling, and forecasting tools that enhance the evidence base for decision-making by the "clients" of applied geography.
2. Make the recognition of applied geographers and their work an explicit, not just implicit, part of the instructional package.
3. Develop networks which bring academic geographers into direct and regular contact with professional geographers working in the United Nations, NGOs, government agencies, businesses, and consultancies and the private research sector and the administrators who manage these organizations and agencies.
4. Create an academic culture which values the benefits of external funding and which is supportive of both public-public and public-private partnerships which support geographic applications.
5. Be aggressive in cultivating the media to encourage public recognition of the contributions of geographers to solving problems for business, for governments, and for NGOs.

ACKNOWLEDGEMENT

We would like to thank the colleagues who offered comments on various drafts of this chapter. We are especially grateful to Professor William M. Bowen and to Dr. Lorraine Craig for their detailed and thoughtful suggestions.

ANTOINE BAILLY
University of Geneva, Switzerland

LAY JAMES GIBSON
University of Arizona, U.S.A.

REFERENCES

Association of American Geographers (March 2001): *Bulletin,* AAG, Washington.
Bailly, A.S. and Ferras, R. (2001): *Eléments d'épistémologie de la géographie;* Colin, Paris.
Bailly, A.S. and Coffey, W.J. (1994): Regional science in crisis: a plea for a more open and relevant approach, *Papers in Regional Science,* 73, 1: 3–14.
Bailly, A.S., Coffey, W.J., Gibson, L.J. (1996): Regional science back to the future, *The Annals of Regional Science,* 30: 153–163.
Bowen, W.M. (2001): Personal communication.
Buttimer, A. (2000): The International Geographical Union: Agora for the Twentyfirst Century: *Acceptance speech to the IGU General Assembly,* Seoul, Korea.

Cutter, S., Golledge, R., Graf, W. (2002): The Big Questions in GEography, *The Professional Geographer*, 54, 3: 305–317.

Ehlers, E. and Krafft, T. (1998): *German global change research 1998,* German National Committee on Global Change Research, Bonn, Germany.

Garreau, J. (1992): *Edge cities, life on the new frontier,* Anchor Books, New York.

Geenhutzen, M., Van and Nijkamp, P. (1996): Progress in regional science: a European perspective, *International Regional Science Review*, 19, 3: 223–245.

Gibson, L. (1994): Fixing the fix we are in, *Papers in Reg. Science*, 79, 1: 19–25.

Gibson, L. (1998): Institutionalizing regional science. *Annals of Regional Science*, 32: 459–467.

Gibson, L. (2000): Size matters: why regional science needs to think bigger. *The Review of Regional Studies*, 30: 71–73.

Golledge, R.G. (2000): President's Column, NEVER Be Ashamed of Being a Geographer; *AAG Newsletter*, June: 3, 6.

Gordon, I. and Cheshire, F. (1998): Locational advantage and the lessons of territorial competition in Europe, *Unpublished Paper prepared for Workshop held in Uddevaiia,* June 1998.

Gould, P. and Bailly, A.S. (2000): *Mémoires de géographes,* Anthropos, Paris.

Hägerstrand, T. (1973): What about people in regional science? *Papers of the RSA,* 24: 7–21.

Isard, W. (1999): Regional science: parallels from physics and chemistry; *Papers in Regional Science*, 78: 5–20.

Ives, J.D. and Messerli, B. (1990): Progress in theoretical and applied mountain research, 1973–1989, and major future needs, *Mountain Research and Development,* 10: 101–127.

Jensen, R.C. and West, G.R. (1995): Regional science and regional practice in Australia. A review and comment, *Australasian Journal of Regional Studies,* I, 1: 7–20.

Kates, Robert W. (2001): Queries on the human use of the earth. *Annual Review of Energy and Environment,* 26: 1–26.

Kuhn, T. (1962): *La structure des révolutions scientifiques,* Paris, Flammarion.

Lacoste, Y. (1973): *La géographie, ça sert à faire la guerre,* Maspero, Paris.

Lee, Y. and Gaertner, R. (1994): Technology transfer from university to industry, *Policy Studies Journal,* 22, 2: 384–399.

Rees, J. (1999): Regional science: from crisis to opportunity; *Papers in Regional Science,* 78: 101–110.

Rodwin, L. (1987): On the education of urban and regional specialists: A comparative perspective; *Papers of the RSA,* 62: 1–11.

Royal Geographical Society (2000): *Benchmark statement for Geography,* RGS, London.

Schlanger, J. (1983): *L'invention intellectuelle,* Paris, Fayard.

Sorensen, T. (1997): Interfacing regional development theory and practice, *Presidential Address* to the 21st Australia and N. Zealand RSA Meeting, Wellington, New Zealand.

Thomas, M. (1977): Some explanatory concepts in regional science; *Papers in Regional Science,* 39: 117–123.

Wellar, B. (1998): *Personal communication.*

White, G.F. (1972): Geography and public policy; *The Professional Geographer,* 24: 101–104.

White, G.F. (1958): Introductory graduate work for geographers; *The Professional Geographer,* 9: 6–8.

MICHAEL PACIONE

CHAPTER 2
THE PRINCIPLES AND PRACTICE OF APPLIED GEOGRAPHY

INTRODUCTION

Questions on the usefulness of geographical research and the relationship between theory and practice are central to debate over the place and value of Geography as an academic discipline for the third millennium. Such issues also constitute the core of applied geography.

It is important at the outset to identify the place of applied geography within the discipline as a whole. Rather than being considered as a *sub-area* of Geography, (akin to economic, social or historical geography), applied geography refers to an *approach* that crosscuts artificial disciplinary boundaries to involve problem-oriented research in both human and physical geography. This overarching perspective acknowledges the complexity of real world problems, as evident in the role of human agency in landscape modification, (as in deforestation, desertification and flooding), or conversely in the impact of earthquakes on cities or of coastal erosion on transport routes. The epistemological importance of this perspective is underlined by the regrettable fact that only a minority of human geographers read papers on physical geography, (and vice versa). Indeed, many geographers "have abandoned the possibility of communicating with colleagues working not only in the same titular discipline but also in the same department. The human geographers think their physical colleagues philosophically naïve; the physical geographers think the human geographers lacking in rigour" (Stoddart 1987:320). While it would be absurd to represent applied geography as a Rosetta stone for a divided discipline one of the strengths of the applied geographical approach is that it rejects artificial academic boundaries and highlights linkages between different geographical phenomena.

The relevance and value of applied geographical research has never been more apparent given the plethora of problem situations which confront modern societies – ranging from extreme natural events (such as floods, drought and earthquakes) through environmental concerns (such as deforestation, disease and desertification) to human issues (such as crime, poverty and unemployment). An applied geographical approach has the potential to illuminate the nature and causes of such problems and inform the formulation of appropriate responses.

23

A. Bailly and L. J. Gibson (eds.), Applied Geography, 23–45.
© 2004 *Kluwer Academic Publishers. Printed in the Netherlands.*

This chapter provides an introduction to the principles and practice of applied geography and discusses the definition and development of the approach. Consideration is given to the relationship between "pure" and "applied" research, and the particular concept of "useful knowledge" is introduced. Different approaches to the conduct of applied geography are examined and a general protocol proposed for research in applied geography. Finally, the question of the value of applied geography for contemporary societies is discussed.

THE DEFINITION OF APPLIED GEOGRAPHY

An indication of the nature and content of applied geography may be gained by examining a selection of available definitions of the approach. One of the earliest statements on applied geography was offered by A. J. Herbertson in 1899 in a lecture to the Council of the Manchester Geographical Society. In this he defined applied geography as "a special way of looking at geography, a limitation and a specialisation of the study of it from one point of view. For the business man this point of view is an economic one, for the medical man a climatic and demographic one, for the missionary an ethic and ethical one" (p.1). While the second part of this definition presents a somewhat restricted view of the context of applied geography even at the end of the nineteenth century, the opening sentence has proved to be a prescient statement which, as we shall see, remains relevant today.

More recent attempts to define applied geography are also instructive as far as they reflect a particular view of the subject. In reviewing several definitions of applied geography Hornbeck (1989, p.15) identified two common factors in that applied geography "takes place outside the university, and it deals with real world problems". While the latter observation is apposite the exclusion of academic research in applied geography reveals an excessively narrow perspective which, in part, reflects the situation in North America where many applied geographers employ their skills beyond the walls of academia. The extra-mural focus in applied geographical work is also central to Hart's (1989, p.15) definition which saw applied geography as "the synthesis of existing geographic knowledge and principles to serve the specific needs of a particular client, usually a business or a government agency". The suggestion of uncritical "service to a specific client, whether business or public agency" (p.17) implicit in this definition ignores the volume of critical analysis undertaken by academic applied geographers.

In a more broadly-based statement Sant (1982, p.1) viewed applied geography as the use of geographic knowledge as an aid to reaching decisions over use of the world's resources. More specifically, Frazier (1982, p.17) considered that applied geography "deals with the normative question, the way things should be, a bold but necessary position in dealing with real world problem resolution. In the process, the geographer combines the world of opinion with the world of decision". This latter perspective is closer to the definition of applied geography favoured here.

I propose a definition of applied geography which reflects the central importance of normative goals and which acknowledges the involvement of both academic and non-academic applied geographers in pursuit of these goals (Pacione 1999). Accordingly, applied geography may be defined as *the application of geographic knowledge and skills to the resolution of social, economic and environmental problems.*

The question of how best to attain this goal will be addressed later in the discussion. Here it is appropriate to conclude these introductory comments by examining the academic niche for applied geography, and in particular the question of whether applied geography constitutes a subfield of geography or an approach to the subject. These issues represent more than a simple question of semantics. In essence, a subfield of a discipline is expected to generate its own body of theory and methodology, whereas an approach has its rationale founded on a particular philosophy (such as relevance or social usefulness) and can employ appropriate theory, concepts and methodology from across the discipline and elsewhere. Designating of the area as a subfield of geography invites criticism of applied geography as lacking a coherent structure and characterised by a pragmatic approach. Johnston (1994, p.21), for example, concluded "there is no central theoretical core or corpus of techniques; rather the subfield has been characterised by ad hoc approaches to the problems posed, drawing on the perceived relevant skills and information". This critique, which could be levelled at many subfields of geography, is based on a misunderstanding of the appropriate academic niche for applied geography. Identification of a theoretical core or a unified concept (such as the hydrological cycle in hydrology or the energy budget in climatology) is necessary only for a subject area which seeks to establish itself as a distinct subfield or branch of a discipline. Applied geography does not harbour such parochial ambition and is best viewed as an approach which can bring together researchers from across the range of subfields in geography either in the prosecution of a particular piece of research or in terms of an enduring commitment to the ethos of the approach. For applied geography the unifying concept is not a specific model or theory but the fundamental philosophy of relevance or usefulness to society. This "core", which extends beyond the confines of any single subfield, represents a powerful and clearly articulated rationale. Furthermore, applied geographers would contend that the identification and application of relevant theory, concepts and techniques both from within geography and across disciplinary boundaries is a positive strength, not a weakness, of the applied geography approach. Definitions and critiques that seek to establish applied geography as a branch or subfield of geography are misplaced. As Herbertson indicated a century ago, applied geography is best seen not as a subfield but as an approach which can be applied across all branches of geography.

THE CONCEPT OF USEFUL KNOWLEDGE

The concept of useful knowledge will no doubt upset a number of practising geographers. Those who do not see themselves as applied geographers may interpret the concept as indicating a corollary in the shape of geographical research which is less useful or even useless. This would be a misinterpretation. The concept should be seen as an expression of the fundamental ethos of applied geography rather than a design to alienate "non-applied" geographers. The use of the concept represents a deliberate decision to get off the fence and make explicit the view that some kinds of research are more useful than other kinds. This is not the same as saying that some geographical research is better than other work – all knowledge is useful – but some kinds of research and knowledge are more useful than other kinds in terms of their ability to interpret and offer solutions to problems in contemporary physical and human environments.

We can illustrate this point by comparing the contents of a recent volume on *Applied Geography* (Pacione 1999) with two other geographical agenda, separated by a time span of fifty years. The first of these is the "mission statement" delivered by the eminent historical geographer H. C. Darby in his inaugural lecture in the University of Liverpool. In Darby's (1946) view his goal as a teacher of geography was to help students learn to read their morning newspapers with greater intelligence and understanding, and to take their evening walks or their Sunday drives with greater interest, appreciation, and pleasure. While few modern applied geographers would regard this as an adequate definition of their work there is a degree of overlap between Darby's agenda and the goals of applied geography in that, from a realist standpoint, Darby's activities could be regarded as emancipatory and an example of critical science.

The second example is taken from a more recent "call for papers" issued in May 1997 on behalf of the Social and Cultural and the Population Geography research groups of the Royal Geographical Society/Institute of British Geographers. In preparation for a session at the annual conference offers of papers were requested on the theme of "the body". Additional guidelines for prospective contributors were as follows – "Is the body dead? Has it been 'done'? This joint session seeks to explore current and future critical, geographical perspectives on 'The Body', as a discourse, as a centre for conflict, consensus, rebellion or domination. Participants are encouraged to consider ways in which their own bodies can be used to em-body their presentations. All/any form(s) of (re)'presentation' are welcomed". Clearly the research topics of interest to participants in this conference session would hold little appeal for many applied geographers. Indeed some may even be stimulated to recall Stoddard's (1987) impatience with "so-called geographers ... who promote as topics worthy of research subjects like geographic influence in the Canadian cinema, or the distribution of fast food outlets in Tel Aviv" (p.334).

The distinction between the contents of Pacione's (1999) book on *Applied Geography: Principles and Practice* and the proposed agenda for the 1998 IBG conference session serves as a useful primer for our subsequent discussions. Those who study the kinds of topics identified in the call for conference papers might legitimise their agenda by pointing to the eclectic nature of geography and the value of "pure" research. For these and other geographers the idea of applied geography or useful research is a chaotic concept which does not fit with the cultural turn in social geography or the postmodern theorising of recent years.

We shall return to this question later but, in the meantime, it is useful to make explicit the views which underlie the kind of applied geography favoured here. We can do this most clearly by comparing the applied geographical approach with an alternative postmodern perspective. One of the major achievements of postmodern discourse has been the illumination of the importance of difference in society as part of the theoretical shift from an emphasis on economically-rooted structures of dominance to cultural "otherness" focused on the social construction of group identities. However, there is a danger that the reification of difference may preclude communal efforts in pursuit of goals such as social justice. A failure to address the unavoidable real-life question of "whose is the more important difference among differences" when strategic choices have to be made represents a serious threat to constructing a *practical politics* of difference. Furthermore, if all viewpoints and expressions of identity are equally valid, how do we evaluate social policy or, for that matter, right from wrong? How do we avoid the segregation, discrimination and marginalisation which the postmodern appeal for recognition of difference seeks to counteract. The failure to address real issues would seem to suggest that the advent of postmodernism in radical scholarship has done little to advance the cause of social justice. Discussion of relevant issues is abstracted into consideration of how particular discourses of power are constructed and reproduced. Responsibility for bringing theory to bear on real world circumstances is largely abdicated in favour of the intellectually-sound but morally-bankrupt premise that there is no such thing as reality. As Merrified and Swyngedouw (1996, p.11) express it "intriguing though this stuff may be for critical scholars, it is also intrinsically dangerous in its prospective definition of political action. Decoupling social critique from its political-economic basis is not helpful for dealing with the shifting realities of (urban) life at the threshold of the new millennium". In terms of real world problems postmodern thought would condemn us to inaction while we reflect on the nature of the issue. (As we shall see below, a similar critique may be levelled at the Marxist critique of applied geography which was prevalent during the 1970s and 1980s).

The views expressed in the above discussion do not represent an attempt to be prescriptive of all geographical research but are intended to indicate clearly the principles and areas of concern for applied geography. It is a matter of individual conscience whether geographers study topics such as the iconography of landscapes or the optimum location for health centres, but the principle underlying the kind of useful geography espoused by most applied geographers is a commitment to improving existing social, economic and environmental conditions. *There can be no compromise – no academic fudge – some geographical research is more useful than other work; this is the focus of applied geography.*

Of course there will continue to be divergent views on the content and value of geographical research. This healthy debate raises a number of important questions for the discipline and for applied geography in particular. The concept of "useful research" poses the basic questions of useful for whom?, who decides what is useful?, and based on what criteria? All of these issues formed a central part of the "relevance debate" of the early 1970s which we examine later. The related questions of values in research, the goals of different types of science, and the nature of the relationship between pure and applied research are also issues of central importance for applied geography. These are addressed in the following sections.

THE RELATIONSHIP BETWEEN PURE AND APPLIED RESEARCH

According to Palm and Brazel (1992, p.342) "applied research in any discipline is best understood in contrast with basic, or pure, research. In geography, basic research aims to develop new theory and methods that help explain the processes through which the spatial organisation of physical or human environments evolves. In contrast, applied research uses existing geographic theory or techniques to understand and solve specific empirical problems".

While this distinction is useful at a general level it overplays the notion of a dichotomy between pure and applied geography, which are more correctly seen as two sides of the same coin. There is, in fact, a dialectic relationship between the two. As Frazier (1982, p.17) points out "applied geography uses the principles and methods of pure geography but is different in that it analyses and evaluates real-world action and planning and seeks to implement and manipulate environmental and spatial realities. In the process, it contributes to, as well as utilizes, general geography through the revelation of new relationships". The conjuncture between pure and applied research is illustrated clearly in geomorphology where, for example, attempts to address problems of shoreline management have contributed to theories of beach transport, the difficulties of road construction in the arctic have informed theories of permafrost behaviour, while problems encountered in tunnelling have aided the development of subsidence theory (Brunsden 1985). Applied research provides the opportunity to use theories and methods in the ultimate proving ground of the real world, as well as enabling researchers to contribute to the resolution of real world problems. More generally, Sant (1982) envisaged theory as essential in applied geography at two levels. First, it provides the framework for asking questions about the substantive relationships embodied in a problem (as, for example, where a model of a hydrological catchment illuminates the potential effects of a proposed flood prevention scheme). Secondly, social theory provides a normative standard against which current and future social conditions may be judged in terms of defined moral goals (which may address issues such as whether a minimum wage and basic standard of living should be a legal entitlement in advanced capitalist societies).

There is little merit in pursuing a false dichotomy between pure and applied research. A more useful distinction is that which recognises the different levels of involvement of researchers at each stage of the research and specifically the greater engagement of applied geographers in the "downstream" or post-analysis stages. The applied researcher has a greater interest than the pure researcher in taking the investigation beyond analysis into the realms of application of results and monitoring the effects of proposed strategies. Researcher participation in the implementation stage may range from recommendations in scholarly publications or contracted reports (a route favoured by most academic applied geographers, though not exclusively) to active involvement in implementation (more usually by applied geographers employed outside academia). Between these positions lie a variety of degrees of engagement, including acting as expert witnesses at public enquiries, dissemination of research findings via the media, field involvement in, for example, landscape conservation projects, and monitoring the effects of policies and strategies enacted by governmental and private sector agencies.

The balance between pure and applied research within a discipline varies over time in relation to the prevailing socio-political environment. When external pressures are at their greatest disciplines will tend to emphasise their problem-solving capacity while during periods of national economic expansion "more academic" activity may be pursued in comfort. Taylor (1985) equated these cycles with the long waves of the world economy, and identified three periods in which applied geography was in the ascendancy (in the late nineteenth century, inter-war era, and mid-1980s) separated by two periods of pure geography (in the early twentieth century, and during the post-1945 economic boom). (Table 1).

Our exploration of the link between pure and applied research is not to imply the superiority of one form of knowledge over the other. Rather, it focuses attention on the fundamental question of the use to which the results of geographical research may be put. More specifically, the applied geographer's interest in the application of their research findings is of particular importance given the role of values in the formulation of political decisions. As Harvey (1984, p.7) observed "geography is far too important to be left to generals, politicians, and corporate chiefs. Notions of 'applied' and 'relevant' geography pose questions of objectives and interests served. . . . There is more to geography than the *production* of knowledge" (emphasis added). This conclusion underlines the need for explicit consideration of both the role of values in applied geography and the value of the applied geographical approach. These questions are considered below.

THE VALUE OF APPLIED GEOGRAPHY

A fundamental question for those working within the framework of applied geography concerns the value of a problem-oriented approach. We have examined this issue already in our discussions of useful knowledge and the relationship between pure and applied research but we return to it here to address the specific critique of applied geography

Table 1. Cycles of pure and applied geography

First Applied Period (late nineteenth century)	Geography created as an applied discipline to serve the political, military and commercial interests of the Prussian state.
First Pure Period (early twentieth century)	Based around the holistic philosophy encompassing both physical and human phenomena and focused on the core concept of the region and regional synthesis.
Second Applied Period (inter-war)	A period of war, followed by Depression, and war again demanded geography demonstrate its usefulness in fields such as land use planning.
Second Pure Period (post 1945 boom)	Rejection of ideographic, regionalism replaced by spatial science and the quantitative revolution; demise of holistic approach and emergency of subfields within the discipline
Third Applied Period (mid 1980s)	Extension of the concept of useful research into new areas of concern relating to social, economic and environmental problems; applied geographers working both in academic and in public and private sectors. Applied geography as an approach rather than a subfield crosscuts the artificial boundary between physical and human geography and emphasises the dialectic relationship between pure and applied research. Acknowledgement of the role of human agency and values in research and environmental change, and the need for a pluralist view of science.
Third Pure Period (?)	Characteristics unknown but speculatively – a return to a more holistic philosophy reflecting the growing importance of environmental issues and the combinatory perspective of applied geography

(*Source:* Michael Pacione 1999 Applied Geography: Principles and Practice London; Routledge)

which has emanated from Marxist theorists. While the power of the Marxist critique has been much reduced by its own success in exposing the value-bases of research it still offers a useful perspective on the value of applied research.

The essence of the Marxist critique of applied social research is that it produces ameliorative policies which merely serve to patch up the present system, aid the legitimation of the state, and bolster the forces of capitalism with their inherent tendencies to create inequality. For these radical geographers participation in policy evaluation and formulation is ineffective since it hinders the achievement of the greater goal of revolutionary social change.

In terms of praxis the outcome of this perspective is to do nothing short of a radical re-construction of the dominant political economy (a position which, as we have seen, may also be reached from a different direction by postmodernist theorists). Although the analytical value of the Marxist critique of capitalism is widely acknowledged its political agenda, and in particular opposition to any action not directed at revolutionary social change, finds little favour among applied geographers. To ignore the opportunity to improve the quality of life of some people in the short term in the hope of achieving possibly greater benefit in the longer term is not commensurate with the ethical position implicit in the problem-oriented approach of applied geography.

Neither does the argument that knowledge is power and a public commodity that can be used for good or evil undermine the strength of applied geography. Any knowledge could be employed in an oppressive and discriminating manner to accentuate inequalities of wealth and power but this is no argument for eschewing research. On the contrary, it signals a need for greater engagement by applied geographers in the policy-making and implementation process provided, of course, that those involved are aware of and avoid the danger of co-optation by, for example, funding agencies.

Furthermore, access to the expertise and knowledge produced by applied geographical research is not the sole prerogative of the advantaged in society, but can be equally available to pressure groups or local communities seeking a more equitable share of society's resources. As Frazier (1982, p.16) commented, applied research "involves the formulation of goals and strategies and the testing of existing institutional policies within the context of ethical standards as criteria. This should not imply a simple system maintenance approach to problem solving. Indeed, it is often necessary to take an unpopular anti-establishment position, which can result in a major confrontation". For practical examples of this we need only refer to the pragmatic radicalism practised by the Cleveland City Planning Commission (Kraushaar 1979); the recommendations of the British Community Development Projects which advocated fundamental changes in the distribution of wealth and power and which led to conflict with both central and local government; as well as more recent policy-oriented analyses of poverty and

deprivation in which the identification of socio-spatial patterns is used to advance a
critique of government policy (Pacione 1990).

VALUES IN APPLIED GEOGRAPHY

At each stage of the research process the applied geographer is faced with a number
of methodological and ethical questions. Decisions are required on defining the nature
of the problem, its magnitude, who is affected and in what ways, as well as on the
best means of addressing the problem. All of these require value judgements on, for
example, the acceptability of existing conditions (what is an acceptable level of air
pollution? or of infant malnutrition?). Values are also central to the evaluation and
selection of possible remedial strategies, including comparative analysis of the benefits
and disbenefits of different approaches for different people and places. In some cases
the applied geographer may seek to minimise such value judgements by enhancing the
objectivity of the research methodology (for example, by employing a classification of
agricultural land capability to inform a set-aside policy). In most instances, however, it
is impossible to remove the need for value judgement. As Briggs (1981, p.4) concluded,
"whether objectivity is ever achieved is a moot point. In most cases the subjectivity
is merely transferred from the client (for example the politician or the planner) to the
research designer". The impossibility of objective value-free research is now axiomatic.

One issue of particular concern refers to the values that condition the selection, conduct and implementation of research, a dilemma highlighted by the aphorism "he who
pays the piper calls the tune". J. T. Coppock (1974, p.9), an advocate of public policy research by applied geographers, expressed this in terms of "doubts over whether
government departments will commission necessary research into the effectiveness and
consequences of their own policies and there is a real danger that constraints will be
imposed over publication, especially if this contains criticisms of the sponsors or explores politically sensitive areas". Applied geographers must beware of any restrictions
imposed by research sponsors and aware of the ways in which their research results may
be used. Applied geographers must seek to ensure that their work contributes to human
welfare. In practice this goal may be approached by careful selection of clients and research projects, by ensuring freedom to disseminate results and, where possible, through
engagement in the implementation and monitoring of relevant policy or strategies.

TYPES OF APPLIED RESEARCH

In deciding how to engage in applied geography practitioners have recourse to three
principal kinds of science (Habermas 1974). These are:

(a) the empirical-analytical in which the goal is to predict the empirical world using
     the scientific methods of positivism.

(b) the historical-hermeneutic, with the goal of interpretation of the meaning of the world by examining the thoughts behind the actions that produce the world of experience.

(c) the realist-emancipatory, where the goal is to uncover the real explanations governing society and encourage people to seek a superior social formation.

A key feature of Habermas' approach to knowledge is the recognition that different types of science have different goals. Each of these is of relevance for the practice of applied geography. The empirical-analytical approach using positivist scientific explanation remains the principal route to knowledge in applied physical geography where a primary goal is the understanding, prediction and eventual control of environmental events. Despite the availability of powerful computer algorithms, however, the complexity of many physical environmental processes can confound this prime objective (we need think only of the accuracy of long-range weather forecasts or our primitive attempts at earthquake prediction). In addition, despite a continuing attachment to positivist science applied physical geographers, in particular those working on environmental problems and management issues, recognise the importance of human agency in environmental change and the role of values in decision-making and policy formulation. Slaymaker (1997), for example, argues for a pluralist problem-oriented geomorphology in which the predominant science of positivism is augmented by a realist philosophy which acknowledges the effect of social structures and human geography.

The goal of prediction and control within human geography – often referred to as "social engineering" or the manipulation of society towards certain ends – is even more problematic: (despite the availability of sophisticated macro-economic models few governments can claim to control their own economic destiny). Generally, social engineering, such as that attempted in the neighbourhood planning of the early post-war British new towns, has been discredited as both ineffective and ethically unacceptable. Positivist science, though of continuing value in applied physical geography, has limited relevance for applied research in human geography which draws its methodology from a larger pool.

In applied terms the goal of historical-hermeneutic science is to increase both self-awareness (by assisting people to reflect on their situation) and mutual awareness (by promoting appreciation of the situations of others). The importance (or usefulness) of inculcating mutual understanding through applied research is seen most clearly in situations where it is lacking – for example, within cities where the stereotyping of areas and social groups can lead to social tension, isolation and conflict. The third route to knowledge, via realist science, builds on the foundations of mutual understanding promoted by historical-hermeneutic or humanistic science and seeks to promote real understanding for people of their position within the socio-political structure and of the factors which condition their lifestyles and living environments. For example, by explaining the factors underlying the closure of a local factory realist science can provide redundant workers with knowledge of the causal forces behind the event and thereby empower their response in the political arena.

Habermas' three-fold typology of science may be used to characterise applied geographers as technicians, agents provocateur, and catalysts for social change respectively (Johnston 1986) but this would be an over-simplification. No matter which route to knowledge the applied geographer adopts and irrespective of the methodologies employed all are moving towards the goal of enhancing human well-being, guided by the shared philosophy of the pursuit of useful knowledge for the resolution of contemporary social, economic and environmental problems.

A PROTOCOL FOR APPLIED GEOGRAPHY

Applied geography is an approach that can be pursued via any of the three main types of science. Accordingly, there is no single method of doing applied geographical research. Nevertheless it is useful to examine one possible protocol which, with appropriate methodological modifications to suit the task in hand, can provide a framework for many investigations in applied geography.

The procedure may be summarised as *description, explanation, evaluation* and *prescription* (DEEP) followed by *implementation* and *monitoring* (Figure 1). The "DEEP" procedure represents a useful analytical algorithm. However, the clarity and organisation of the scheme does not imply that simple answers are expected to contemporary social, economic or environmental problems. Normally, in order to understand the nature and causes of real world problems it is necessary to untangle a Gordian knot of causal linkages which underlie the observed difficulty. In some cases, such as the link between ground slippage and building collapse cause and effect are relatively straightforward. But in most instances the cause of a problem may be more apparent than real. Thus while the immediate cause of the problems faced by a poor family on a deprived council estate in Liverpool may be a lack of employment opportunities following the closure of a local factory, the root cause of the social and financial difficulties confronting the family may lie in the decisions of investment managers based in London, New York or Tokyo.

As Figure 1 indicates, as well as describing the nature and explaining the causes of problems the applied geographer also has a role to play in evaluating possible responses and in prescribing appropriate policies and programmes which may be implemented by planners and managers in both the public and private sectors, or by the residents of affected communities. In performing these tasks the applied geographer will be confronted with a variety of potential responses for any problem. The selection of appropriate strategy is rarely straightforward. The decision must be based on not only technical criteria but also on a wide range of conditioning factors including the views and

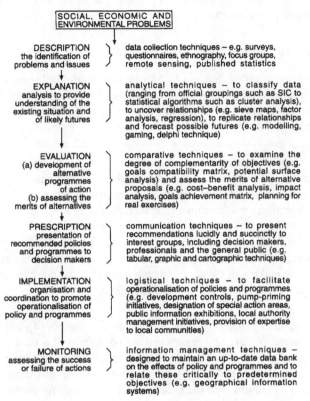

Figure 1. A Protocol for applied geographical analysis
(*Source:* Michael Pacione 1999 Applied Geography: Principles and Practice
London: Routledge)

preferences of those affected by the problem and proposed solution, available finance, and externality considerations or how the strategy to resolve a particular problem (such as construction of flood control levees) may affect other problems (such as increased flooding of downstream communities).

As indicated earlier, applied geographers, in contrast to "pure" geographers, may also be involved in the implementation stage of the research, normally in a supervisory or consultancy capacity to ensure effective application of a strategy. The nature of any engagement is potentially wide ranging, for example, from overseeing the setting-up of a computer-based route planning system for a private transport company or public ambulance service to making one's expertise available to community groups seeking to establish a housing co-operative or local economic development initiative. Finally, as Figure 1 reveals, applied geographers may be involved in monitoring the impacts of policies and programmes implemented to tackle a problem, and in relating these critically to pre-determined normative goals.

THE HISTORICAL DEVELOPMENT OF APPLIED GEOGRAPHY

Applied geography has a long history. As Martin and James (1993) indicated "there has never been a time when the search for knowledge about the earth as the home of man has not been undertaken for practical purposes as well as for the satisfaction of intellectual curiosity".

The earliest geographical research was, of necessity, useful, concerned as it was with describing the nature of the earth as an aid to exploration and human survival. The applied tradition was central to the earth measurement and cartographic research of mathematical geographers working under the direction of Eratosthenes at Alexandria (Bunbury 1879). Strabo in his account of the utility of geography in Greek and Roman society identified a central role for geographic knowledge in politics and warfare. During the first millennium A.D. the same motives stimulated the development of geographical knowledge and map making in China. In Islamic lands this was complemented by the production of travel guides as an aid to pilgrims making the haj.

These stimuli to applied geography were boosted by the voyages of discovery of the fifteenth and sixteenth century at a time when the possession of accurate geographic knowledge bestowed enormous advantage. During the sixteenth century geographical research was undertaken with the principal purpose of enabling European ships to navigate the world and return with the produce of distant lands for commercial profit (Taylor 1930).

Significantly, few of these early practitioners of applied geography would have described themselves as geographers – explorers, adventurers, sailors, traders, astronomers, cartographers, cosmographers, natural scientists, mathematicians, historians, philosophers, surveyors or topographers – but few outright geographers. (What has in more recent times been referred to as the cocktail party syndrome – "Oh, you're a geographer! What do you do?" – also has a long history).

A second point of note is that the early acquisition of geographical knowledge was designed to facilitate domination by merchants or rulers, and the ways in which the knowledge was applied often had negative consequences for those peoples brought within the ambit of the emerging capitalist economic system. While modern applied geographers have been sensitised to the socially-regressive consequences of the mis-use of geographical research (by the work of anarchist geographers such as Kropotkin and by the Marxist critique of positivist science) at the beginning of the seventeenth century such concerns were far from the thoughts of those practising geography.

Varenius, one of the founders of geography as a formal academic discipline, justified the subject on three grounds –

(a) its value being well suited to man as the dominant species of earth,
(b) its being a pleasant and worthy recreation to study the regions of the earth and their properties, and –
(c) "its remarkable utility and necessity, since neither theologians, nor medical men, nor lawyers, nor historians, nor other educated persons can do without knowledge of geography if they wish to advance in their studies without hindrance" (Bowen 1981, p.282).

The commercial and political nature of applied research continued as an important feature of the discipline throughout the eighteenth and nineteenth centuries. In one of the earliest published references to *Applied Geography* Keltie (1890) sought to demonstrate the importance of geographical knowledge for history and especially industry, commerce and colonisation. Similarly, in North America the early efforts of the American Geographical Society at the turn of the century supported exploration and expeditions in the hope of producing "not only new scientific data but facts of practical use to the merchant or missionary" (Wright 1952, p.69).

In the early decades of the twentieth century the development of applied geography was advanced by A. J. Herbertson (1910) who envisaged the role of geographical prospector mapping the economic value and potential of regions and P. Geddes (1915) who was both the founding father of planning and an advocate and exponent of applied geography based on his dictum of "survey before action". The scope of applied geography was broadened by "action-oriented" research undertaken in the 1930s in the U.K. by G. H. Daysh on the problems of distressed areas; by A. E. Smailes on the conurbations, local administrative boundaries, the concept of a city-region, and, most perceptively, on the possible role of regional parliaments; and by L. D. Stamp who employed the methods of survey and analysis in his land use studies of Britain (Stamp 1946).

A similar concern with land use issues characterised applied geography in North America between the wars. The tradition of resource inventory encapsulated in the numerous explorations and surveys of the American west was continued in the work of C. Sauer on land use classification. Sauer's land use survey of Michigan was of the same genre as the First Land Utilisation Survey of Britain organised by Stamp. In the field of water resource management, between 1935–1938, H. Barrows drew up plans for the distribution of user-rights to the waters of the upper Rio Grande among the states of Colorado, New Mexico and Texas. The importance of applied geographical research was also demonstrated in economic development planning by the Tennessee Valley Authority, as well as overseas as in the preparation of a rural land classification and development plan for Puerto Rico.

Geographical research was also applied in the private sector during the inter-war years, notable examples being C. Thornthwaite's use of knowledge of climatology for the benefit of the dairy industry in New Jersey, and the work of W. Applebaum on the location of new retail outlets for the Kroger company. This latter work pioneered the development of marketing geography as an applied field covering issues such as competitive impact analysis and the application of academic models of travel patterns to the business sector (Applebaum 1961).

Applied geographers have also been called into service in times of war and its aftermath. Skills of terrain analysis, air-photograph and satellite imagery interpretation, intelligence gathering, weather forecasting, mapping, route planning and logistics are all activities of vital importance for military planning. Geographic knowledge and skills are of equal value during the ensuing peace in, for example, adjudicating boundary disputes. The American geographer I. Bowman played a major role as Chief Territorial Specialist in the Versailles Peace Conference following the First World War, and was involved in the resolution of territorial disputes both within the USA (notably between Oklahoma and Texas) and between a number of Latin American States during the inter-war period including those between Chile and Peru in 1925, Bolivia and Paraguay (1929) and Columbia and Venezuela (1933).

The growing academic importance of applied geography was recognised by the creation in 1964 of an International Geographical Union Commission on Applied Geography. An indication of the concerns of contemporary applied geography is provided by the programme for the 1972 meeting of the Commission. This included study of –

(a) problems relating to the management of resources in developing countries,
(b) planning for urbanization,
(c) forecasting the impact of technology and development programmes in different countries,
(d) problems of water supply and environmental pollution, and
(e) exploration of new methods of research using computers in all branches of applied geography.

The continued emphasis on land use issues is apparent in the IGU agenda, as well as in the main themes of applied geography in British universities identified by Freeman (1972). These documents provide a snapshot of key contemporary issues (such as regional planning), emerging specialisms (e.g. mathematical modelling), issues of continuing concern (including environmental pollution, and conflict over urban sprawl), as well as the notable absence of themes which have come to the fore subsequently (such as poverty and deprivation, the geography of AIDS, and applications of global positioning systems). Probably unwittingly, as he was talking in the particular context of land use issues, Freeman also gives a hint of the relevance debate which was shortly to impact on geography in his observation that "no geographical study can have validity unless the wishes of people are taken into full consideration" (p.41).

The last quarter of the twentieth century saw the greatest change in the practice of applied geography. Foremost among these developments was the emergence of a welfare-oriented socially responsible applied geography. This was stimulated by theoretical and methodological changes within the discipline, and more generally within wider society. After two decades of relative prosperity the economies of Britain and America began to experience difficulties during the late 1960s as the post-war boom faltered. At the same time, major societal events such as the US involvement in the Vietnam war, racial unrest in American cities, the Civil Rights movement, feminism and consumer rights and environmental groups contributed to a concern over the general issue of "quality of life". In contrast to the optimistic growth-centred outlook of earlier years poverty and inequality were re-discovered in the American city. "By the end of the 1960s urban policy in the United States was in disarray, and by any measure the American central city was in severe distress". (Ley 1983, p.1). For some radical geographers these trends provided clear signs that "the late twentieth century will be a period of continuing and escalating societal crises, the likes of which we have not yet known" (Peat 1977, p.1). Similar tensions were being experienced to varying degrees in Britain and Europe.

These societal influences were reflected in the changing substantive concerns of applied geography. The direction of change is indicated clearly by the content of papers delivered to annual meetings of the American Association of Geographers in the early 1970s. At the 67th annual meeting, held in Boston in 1971, themes included geographical perspectives on poverty and social well-being, ethnic and religious groups, and urban policy, with a general session devoted to discussion of the problems and strategies facing "socially-responsible geographers". This trend was continued at the 68th annual meeting in Kansas where topics included a session on metropolitan spatial injustice, and other action-oriented papers on "place utility, social obsolescence and qualitative housing change", "crime rates as territorial social indicators", and "environmental stress and maladaptive behaviour". The "relevance debate" was also taken up by geographers in the UK (Chisholm 1971, Prince 1971, Smith 1971, Dickenson and Clarke 1972, Berry 1972). One result was that while the pre-war consideration of land use issues including urban sprawl, countryside conservation, land use and resource management, continued to attract attention these were displaced from centre stage by questions relating to geography of poverty and hunger (Morrill and Wohlenberg 1971), crime (Harries 1974), health care (Shannon and Dever 1974), ethnic segregation (Rose 1971), education (Kirby 1979) and the allocation of public goods (Cox 1973).

While most applied (human) geographers were agreed on the important issues, the social relevance "movement" was far from united over the question of the best route towards a solution. For some, action within the existing structure of society was preferred, whereas others advocated a more radical approach aimed at a fundamental restructuring of the social order. The liberal approach essentially represented a continuation of the philosophy which underlay much of the applied geography and land use planning of the inter-war and immediate post-war periods. Work on social issues in the liberal tradition included the mapping of spatial variations in quality of life (Knox 1975) as an input to planning and as a means of monitoring the distributional effects of social policies. Other researchers were more willing to embrace the radical alternatives to liberal formulations. The argument in favour of a Marxist approach was presented by Folke (1972) who considered that geography and the other social sciences are "highly sophisticated, technique-oriented, but largely descriptive disciplines with little relevance for the solution of acute and seemingly chronic social problems ... theory has reflected the values and interests of the ruling class" (p.13). The Marxist critique of capitalism was also a critique of empirical positivist science and, understandably, applied physical geographers found the relevance debate largely irrelevant to the conduct of their research. While we acknowledge the profound and largely beneficial influence of the relevance debate on applied human geography this does not amount to castigation of applied physical geographers or spatial scientists for a failure to adopt the same precepts. As we have seen there is more than a single type of science, more than a single route towards knowledge and enlightenment, and all modes of analysis have the capacity of contributing to the applied geographer's goal of addressing real world problems.

The development of applied geography has been accompanied by debate over the relative merits of pure and applied research. Critics such as Cooper (1966) and more recently Kenzer (1989) warned against the application of geographical methods as a threat to the intellectual development of the discipline. Conversely, Applebaum (1966) took the view that "geography as a discipline has something useful to contribute to man's struggle for a better and more abundant life. Geographers should stand up and be counted among the advocates and doers in this struggle" (p.198). In similar vein, Abler (1993) considered that "too many geographers still preoccupy themselves with what geography is; too few concern themselves with what they can do for the societies that pay their keep" (p.225). There is no reason why an individual researcher cannot maintain a presence in both pure and applied research. The eminent American geographer C. Sauer was both a "scholar" who conducted research on agricultural origins and dispersals and an "applied geographer" who developed a land classification system for the State of Michigan. The terms pure and applied are best seen as the ends of a continuum rather than unrelated polar opposites.

APPLIED GEOGRAPHY: PROSPECTIVE

The practical value of the applied geographical approach has been demonstrated in the foregoing discussion of the principles and practice of applied geography and is illustrated by the wide range of research work referenced in this chapter. Applied geographers are actively engaged in investigating the causes and ameliorating the effects of 'natural' phenonema including acid precipitation, landslides and flooding. Key issues of environmental change and management also represent a focus for applied geographical research with significant contributions being made in relation to a host of problems ranging from the quality and supply of water, deforestation and desertification to a series of land use issues including agricultural de-intensification, derelict and vacant land and wetland and townscape conservation. Applied geographers with a particular interest in the built environment have, in recent decades, directed considerable research attention to the gamut of social, economic and environmental problems which confront the populations of urban and rural areas in both the Developed and Developing realms. Problems of housing, poverty, crime, transport, ill-health, socio-spatial segregation and discrimination have been the subject of intense investigation, while other topics under examination include problems ranging from boundary disputes and political representation to city marketing. The application of techniques in applied geographical analysis is of particular relevance in relation to spatial analyses where the suite of problems addressed by applied geographers ranges from computer mapping of disease incidence to simulation and modelling of the processes of change in human and physical environments.

The list of research undertaken by applied geographers is impressive but there are no grounds for complacency. While applied geographers have made a major contribution to the resolution of real world problems, particularly in the context of the physical environment, in terms of social policy formulation in the post-war era the influence of applied geography has been mixed and arguably less than hoped for by those socially-concerned geographers who engaged in the relevance debate a quarter of century ago.

Several reasons may be proposed to account for this. The first refers to the eclectic and poorly-focused nature of geography and the fact that "geographical" work is being undertaken by "non-geographers" in other disciplines. This undermines the identity of geography as a subject with something particular to offer in public policy debate. The very breadth of the discipline, which for many represents a pedagogic advantage, may blur its image as a point of reference for decision-makers seeking an informed input. Geographers wishing to influence public policy must compete with other more clearly identified "experts" working on similar themes.

A second reason for the relatively limited influence on public policy may be the apparent reluctance of (human) geographers to "get their hands dirty" – an attitude redolent of the eighteenth century distinction between gentlemen, who derived a livelihood from the proceeds of landownership, and those who earned a living through trade. This applies less to research in physical geography where a basis in empirical science and positivist methodology has ensured that applied research has attracted support and acclaim more readily both from within the discipline and from external agencies. Significantly, the growth of environmentalism and accompanying convergence of the philosophy and methodology of physical and human geography has gone some way to bridge the gap between the two major sub-areas of the discipline and may represent an route for applied geographers to increase their policy influence.

The changing content and shifting emphases of human geography during the last quarter of the twentieth century represents a third factor underlying the limited social impact of applied geography. Over the period the replacement of the earlier land use focus in applied human geography by questions relating to the geography of poverty, crime, health-care, ethnic segregation, education and the allocation of public goods brought applied geographers into direct confrontation with those responsible for the production and reproduction of these social problems. Unsurprisingly, since policy-makers are resistant to research which might undermine the legitimacy of the dominant ideology, social policy remained largely impervious to geographical critique, particularly that which emanated from the Marxist analysis of capitalism.

The failure of applied geography to exert a major influence on social policy, however, does not signal the failure of applied geography to promote any significant improvement in human well-being which, as we have seen, may be achieved by means other than via public policy. Any assessment of the contribution of applied geography to the resolution of real world problems must balance the limited success in the specific area of social policy against the major achievements of applied geographers in the large number of other problem areas outlined above. Rather than dwelling on the limited impact to date of applied geographical research in the field of social policy applied geographers can draw encouragement from their unwillingness to compromise a critical stance in return for public research funds or public acceptability of research findings. Furthermore, much of the applied social research undertaken achieves the goal of addressing real world problems via its emancipatory power to expose the structural underpinnings of contemporary socio-spatial problems and by encouraging exploration of alternative social arrangements.

Applied geography is an approach whose rationale is based on the particular philosophy of relevance or social usefulness and which focuses on the application of geographical knowledge and skills to advance the resolution of real world social, economic and environmental problems. As the present discussion demonstrates applied geographers are active across the human-physical geography divide and in most sub-areas of the discipline. The range of applied research presented in the book illustrates not only the contribution that applied geography is currently making towards the resolution of social, economic and environmental problems at a variety of geographic scales, but also the potential of the approach to address the continuing difficulties which confront humankind. Applied geography is a socially-relevant approach to the study of the relationship between people and their environments. The principles, practice and potential of applied geography to engage a wide range of real world problems commends the approach to all those concerned about the quality of present and future living conditions and environments on the planet Earth.

ACKNOWLEDGEMENTS

The author would like to acknowledge the assistance of his son Michael John Pacione in preparing the camera-ready manuscript for publication.

MICHAEL PACIONE
Department of Geography
University of Strathclyde
Glasgow
United Kingdom

REFERENCES

Abler, R. (1993) Desiderata for geography: an institutional view from the U.S., in R. J. Johnston *The Challenge For Geography*. Oxford: Blackwell, 215–38.

Applebaum, W. (1961) Teaching marketing geography by the case method. *Economic Geography* 37, 48–60.

Applebaum, W. (1966) Communications from readers. *Professional Geographer* 18, 198–9.

Berry, B. (1972) More on relevance and policy analysis. *Area* 4, 77–80.

Bowen, M. (1981) *Empiricism and Geographical Thought from Francis Bacon to Alexander von Humboldt*. Cambridge: Cambridge University Press.

Briggs, D. (1981) Editorial: the principles and practice of applied geography. *Applied Geography* 1, 1–8.

Brunsden, D. (1985) Geomorphology in the Service of Society, in R. J. Johnston *The Future of Geography*. London: Macmillan, 225–57.

Bunbury, E. (1879) *A History of Ancient Geography Among the Greeks and Romans from the Earliest Ages Till the Fall of the Roman Empire*. London: John Murray.

Chisholm, M. (1971) Geography and the question of relevance. *Area* 3, 65–68.

Cooper, S. (1966) Theoretical geography, applied geography and planning. *Professional Geographer* 18, 1–2.

Coppock, J. T. (1974) Geography and public policy: challenges, opportunities and implications. *Transactions of the Institute of British Geographers* 63, 1–16.

Cox, K. (1973) *Conflict, Power and Politics in the City*. New York: McGraw Hill.

Darby, H. C. (1946) *The Theory and Practice of Geography*. London: University of Liverpool Press.

Dickenson, J. and Clarke, C. (1972) Relevance and the newest geography. *Area* 4, 25–27.

Folke, S. (1972) Why a radical geography must be Marxist. *Antipode* 4, 13–18.

Frazier, J. W. (1982) *Applied Geography: Selected Perspectives*. Englewood Cliffs NJ: Prentice Hall.

Freeman, T. W. (1972) Applied geography in British universities, in R. Preston *Applied Geography and the Human Environment*. Proceedings of the Fifth International Meeting of the IGU Commission on Applied Geography, University of Waterloo, 369–73.

Geddes, P. (1915) *Cities in Evolution*. London: Williams and Northgate.

Habermas, J. (1974) *Theory and Practice*. London: Heinemann.

Harries, K. (1974) *The Geography of Crime and Justice in the United States*. New York: McGraw Hill.

Hart, J. F. (1989) Why applied geography?, in M. Kenzer *Applied Geography: Issues, Questions and Concerns*. Dordrecht: Kluwer, 15–22.

Harvey, D. (1984) On the historical and present condition of geography: an historical materialist manifesto. *Professional Geographer* 36, 1–11.

Herbertson, A. J. (1899) Report on the Teaching of Applied Geography. *Unpublished Report to the Council of the Manchester Geographical Society*.

Herbertson, A. J. (1910) Geography and some of its present needs. *Journal of the Manchester Geographical Society* 16, 21–38.

Hornbeck, D. (1989) Working both sides of the street: academic and business, in M. Kenzer *Applied Geography: Issues, Questions and Concerns*. Dordrecht: Kluwer, 165–172.

Johnston, R. J. (1986) *On Human Geography*. Oxford: Blackwell.

Johnston, R. J. (1994) Applied Geography, in R. J. Johnston, D. Gregory and D. M. Smith *The Dictionary of Human Geography*. Oxford: Blackwell, 20–25.

Keltie, J. (1890) *Applied Geography: A Preliminary Sketch*. London: G. Philip and Son.

Kenzer, M. (1989) *Applied Geography: Issues, Questions and Concerns*. Dordrecht: Kluwer.

Kirby, A. (1979) *Education, Health and Housing*. Farnborough: Saxon House.

Knox, P. (1975) *Social Well-Being: A Spatial Perspective*. Oxford: Oxford University Press.

Kraushaar, R. (1979) Pragmatic radicalism. *International Journal of Urban and Regional Research* 3, 61–80.

Ley, D. (1983) *A Social Geography of the City*. New York: Harper and Row.

Martin, G. and James, P. (1993) *All Possible Worlds: A History of Geographical Ideas*. Chichester: Wiley.

Merrifield, A. and Swyngedouw, E. (1996) *The Urbanization of Injustice*. London: Lawrence and Wishart.

Morrill, R. and Wohlenberg, E. (1971) *The Geography of Poverty in the United States*. New York: McGraw Hill.

Pacione, M. (1990) What about people? A critical analysis of urban policy in the United Kingdom. *Geography* 75, 193–202.

Pacione, M (1999) *Applied Geography: Principles and Practice* London: Routledge.

Palm, R. and Brazel, A. (1992) Applications of geographic concepts and methods in R. Abler, M. Marcus and J. Olsson *Geographies Inner Worlds*. New Brunswick NJ: Rutgers University Press, 342–62.

Peet, R. (1977) *Radical Geography*. London: Methuen.

Prince, H. (1971) Questions of social relevance. *Area* 3, 150–153.

Rose, H. (1971) *The Black Ghetto*. New York: McGraw Hill.

Sant, M. (1982) *Applied Geography: Practice, Problems and Prospects*. London: Longman.

Shannon, G. and Dever, G. (1974) *Health Care Delivery: Spatial Perspectives*. New York: McGraw Hill.

Slaymaker, O. (1997) A pluralist, problem-focused geomorphology, in D. Stoddart *Process and Form in Geomorphology*. London: Routledge, 328–339.

Smith, D. M. (1971) Radical geography – the next revolution? *Area* 3, 153–7.

Stamp, L. D. (1946) *The Land of Britain and How it is Used*. London: Longman.

Stoddart, D. (1987) To claim the high ground: geography for the end of the century. *Transactions of the Institute of British Geographers* 12, 327–36.

Taylor, E. (1930) *Tudor Geography 1485–1583*. London: Methuen.

Taylor, P. (1985) The value of a geographical perspective, in R. J. Johnston *The Future of Geography*. London: Methuen, 92–110.

Wright, J. (1952) *Geography in the Making*. New York: American Geographical Society.

MICHEL PHLIPPONNEAU

CHAPTER 3

HISTORICAL FOUNDATIONS OF APPLIED GEOGRAPHY

The multiple aspects of applied geography are as old as geography itself, which was for long considered as a scholastic and academic discipline, " revealing Humanity to itself " to quote Max Sorre. From the very beginning, geographical sciences have permitted an enlargement of the known world: an accurate ordnance survey of the already acquired knowledge makes new explorations easier and leads to rub out the white spots from the world map.

It also ensures the organization of human occupancy, the improvement of the living conditions, with the help of a rational development of space, and it answers the tangible problems concerning relations between Man an its environment. It is mainly this aspect which was emphasized by a British geographer, at the end of the 19th century, when he first used the term "*applied Geography*", to have the practical applications of geographical research works at least acknowledged.

These applications were, in the first place, realized by academic researchers in geography, who had themselves understood that their works, with a scientific finality, could also leads to applications. Their experience, research methods, the comparisons they had to multiply could encourage users and decision makers to entrust them to undertake and realize research works under contract; the results of which could then enlighten their actions. An academic could also be called upon as adviser or consultant, on an occasional basis for a precise task, or permanently as a "qualified person".

This first form of contribution of the academic geographer to applied research works must always retain a fundamental importance, for it is his own researches, or those in partnership, that will allow him to improve his knowledge, his techniques, and the quality of his judgements. It is all that he will have to teach his students who will become "*professional geographers*" and not "*academic geographers*" dedicated to teach in secondary education.

As the applications of geography diversify, the number of users multiply, a new job is born and now thriving: the "*professional geographer*" working full time for a user, an administration, a local authority, a firm which find it more interesting to ensure the

47

A. Bailly and L. J. Gibson (eds.), Applied Geography, 47–67.
© 2004 Kluwer Academic Publishers. Printed in the Netherlands.

permanent and exclusive services of a geographer, than to offer expertise and research works under contract to academics.

Evidently the teaching programmes in the universities geography departments are deeply modified. The training of students who are preparing their entrance examinations for secondary teaching can take a lesser place than that of students training to become *"professional geographers"*. Even if there is, for both courses of study, a common basis, which should be broadened with some benefit, experiences, methods, tools are specifically different for each course. It is difficult to draw up the story of these different phases in the evolution of applied geography on a world-wide scale. They showed up earlier in Great Britain and in the USA, than in France, but today the latter is giving a more important place to "professional geography" than most of the European countries. Outside Europe, Brazil had soon taken an early lead among developing countries, but progress are also very fast in the Far East. In the second part of this book, five authors will precise which phase of development applied geography has reached in the great parts of the world they are in charge of.

In the present chapter, after a reminder of the first forms of applications of geography in world discovery and of the oldest forms of spatial development, the first major initiatives, during the interwar period, in the United States and Great Britain will be brought up after the phase of dissociation between geography and action The second world war speeds up and, above all, extends the movement, which, in the sixties, spreads to the, until then, reluctant countries. The creation in 1964 of the Commission on Applied Geography of the IGU is significant of the globalization of a phenomenon, which, under various forms, concerns, as soon as the early eighties, almost every country.

GEOGRAPHY APPLIED TO DISCOVERIES

The use, for practical purposes, of geography is as old as geography itself, the sorting out of knowledge related to the spatial structure encourages an extension of this space. Close links exist between geography and explorations. The latter are bringing in new material to geographical science thus arousing and making further explorations easier to organize (Phlipponneau 1960; Bailly 2001).

In the ancient times, when the Greek and Phoenician navigators are looking for new commercial roads, the first geographical descriptions are revealing, at first, a practical usefulness. But geography also answers an intellectual concern: Eratosthene taking the measures of the Earth, Ptolemy summarizing the geographical knowledge of the ancient world, are mainly spurred by disinterested scientific concerns.

And it is precisely these scientifically minded works which have, despite or perhaps because of their errors, exerted a fundamental influence on the great discoveries of the

Renaissance era. If the discovery of the shorelines of Greenland and North America by Scandinavian sailors did not have a practical use for further explorations, it is because the lack of cartographic representation and written descriptions did not allow to keep a precise memory that would have justified them.

During the great discoveries era, while the map of the world was widening so swiftly, the links between geography and the progress of the explorations, appear to be extraordinarily bound. Cartography, but also descriptive geography benefit from the support of sovereigns who appreciate their practical interest in the extension of their overseas possessions. From his observatory in Sagres, Prince Henry of Portugal is studying the works of ancient geographers but above all the stories of contemporary travellers before launching the crews of his own navigators towards the African coastlines; the acquired results of an expedition helping him to prepare the following one. The sovereigns of the other maritime powers are following his example; the *"King's* (or *Queen's*) *Geographers"* in England, France and Spain are not the only ones to prepare new explorations and draw the lessons from the past ones; an upsurge of cartographers-geographers, at first "freelance" Flemish and German scholars and scientists are publishing the first atlas, the first descriptions of the newly discovered lands. They come up to public expectations and their publications are even able to impose the name of a continent on the basis of a map printed in Saint-Dié, a city that will be recognized, five centuries later, as the privileged location for an International Geographical Festival.

In the 18th century scientific concerns are put forward by the maritime powers, to justify the expeditions launched towards unknown territories, with navigators such as La Pérouse or James Cook. King Louis XVI, who, just before ending up under the guillotine, asks for news from his geographer, is the best evidence of the interest that a sovereign could then take in the advancement of geography. But aren't the admiralties, more than often, led by other more practical considerations? During the first half of the 19th century, the speed races in the Pacific Ocean, between the French and British navies show different concerns: the cartographic survey is the first step of the taking over of a territory.

During the phase of the great continental explorations of the 19th century, the same evolution can be observed. The first explorers, prompted by the taste of adventure, religious or scientific concerns, have only limited means at their disposal. Then the Geographical Societies, backed if not sponsored by the business world and the governments themselves, give them an increasing financial assistance going as far as turning explorations into State Affairs. René Caillé, the sole and lonesome discoverer of Timbuktu, will open the trail to military expeditions pursuing simultaneously the discovery and the pacification of the Sahara to constitute an homogeneous between Maghreb and Black Africa. The explorations led by Livingstone for humanitarian and scientific purposes will be followed by those of Stanley, who is working for the Press capitalizing on the interest of the public opinion, and later on, for the International Congo Association set up by the great tycoon and landowner, King Léopold 2 of Belgium. The economic and political concerns are becoming forefront issues.

The last explorations, especially those concerning the polar regions, are recording similar evolutions. Individual and sporting achievements relying on scientific concerns, will be followed by organized and fully chartered expeditions, and will lead to a permanent occupation of territories, under the responsibility of States, anxious, as for the Antarctic, to avoid an exclusive exploitation.

The evolution of the Geographical Societies, well studied on the occasion of the 175th anniversary of the eldest of them, the Paris Society, founded in 1821, clearly shows the links between geography and discovery of the World (A.L. Sanguin. 1997). The latter fascinates the public, while, on the eve of the first world war, the Colonial Empires are building up. The Societies are not only benefiting from the intervention of enlightened sponsors, but also from the upper crust of the states ruling class. The number of members of the Paris "Societé de Géographie" exceeds the 2 500 mark in 1885; thirty provincial societies are gathering more than 20 000 subscribers. The movement is spreading to the whole world; to Great Britain, where "*all along the Victorian Age, Geography was the Science of British Imperialism*" (Sanguin 1997) as well as to Russia and the USA. Outside France, in 1914, 54 societies in 33 different countries are listed. They play an important role in the development of the teaching of geography, particularly with the creation of University Chairs and Professorships. After a first International Congress in 1871, held in Antwerp, ten others will follow until 1913. After the creation of the International Geographical Union (IGU) in 1922, geography becomes mainly the matter for scholars and academics, and the role of the societies declines together with the number of new explorations. The number of members of the Paris Geographical Society is falling down to 1 300 in 1939 and 379 in 1991, but it is growing up again appreciably, to reach 769 in 2002 (including 65 foreign members). The ties with the Explorer's Club, the attraction of the public for discovery travels, nature and environment, the multiplication of documentary films and dedicated T.V. programmes, of specialized magazines such as "*National Geographic*" or "*Geo*", show that the leading role of the applications of geography is not about to disappear.

THE FIRST FORMS OF APPLICATIONS IN SPATIAL PLANNING

This other utilitarian aspect of geography often supplements the first one. A discovery represents a real interest from the very moment this newly found and explored territory is being organized and planned for various purposes. The knowledge of the characteristics of a territory helps to adjust its development. Geography is not only a basic discipline in the development of new countries, it can also be useful to transform, if necessary, the characteristics of territories which have known the most ancient occupation.

From the earliest times, preliminary geographical surveys appear to be essential for the development of newly occupied territories. Augustus and Caesar are employing thousands of surveyors to measure off the roads innervating the Empire and to make out an inventory of the economic resources of the Roman World, in order to increase them. But the needs of users and decision makers do not show up during the periods of anarchy and political divisions as well as in the contemporary period of economic liberalism.

During the great discoveries age, after contributing to the success of the explorations, geographical research works play a large part in the development of the new territories (Gottmann. 1952; 1954; Phlipponneau. 1959; 1960). It is the case in Venezuela, as well as in Canada. After Jacques Cartier, geographer-discoverer, Samuel de Champlain, geographer-colonizer, working in the first place for a fur trading company, succeeds in persuading the royal government to settle down France in Canada, thanks to his remarkable gifts for geographical observations. Later on, S. de Vauban, Marshal of France, who can be considered as the "*Father of Applied Geography*", relying on the Canadian experience, works out a model of geographical as a preliminary to the development of a new country (Vauban, 1910).

Marshal Vauban, well before the creation of "*Hérodote*" magazine, has discovered the "*Army Staff Geography*". But Vauban, during his "*Oisivetés*", is relying on new cartographic and statistical means, added to his prodigious sense of observation to show how to improve the characteristics of territories of ancient occupation, by making up real "*Pilot Zones*" experimenting new methods. Research and geographical description must be used to propose the means of improving an existing situation, to prepare the decisions of the man of action. In a country such as France, combination of original geographical sectors, it is a must to know their own characteristics, in order to intervene efficiently.

The ideas of Vauban, which did not meet a response from the geographical circles before their "*discovery*" by Jean Gottmann, have nevertheless led the political circles to realize the practical interest of this type of research. At the end of the 18th century, the progress of geographical knowledge was such that it permitted the Constituents to draw up, in a record time, a new administrative geography of France whose permanence is amazing.

Napoléon 1, with his ability to synthesize and his sense of space, was necessarily compelled to realize the interest that geography represented for the strategist as well as for the administrator; drawing up of the "Carte d'Etat-Major" (Ordnance Survey maps), first population census, prefectoral reports for each constituency, monographs of foreign countries written up by diplomats and military attachés. He also intends to initiate, in the Collège de France, a teaching for Geography, with specialized Chairs for each part of the world, so that the administration "*would have to hand the most accurate information's, the most precise notions of the latest discoveries and of the lately occurred changes"* (Vox, 1943).

Unfortunately for geography, as well as for the State, the Emperor didn't have enough time to achieve his project, that neither the geographers, nor the governments, thought to take up under another form, as it was accomplished in the United States, with *"the Geographer"* of the State Department. Throughout the 19th century, besides the Army Geographical Service, actually limited to mapping, the administration develops a statistical apparatus, multiplies sectorial economic surveys, and writes up communal, departmental and regional monographs, without even thinking of entrusting the geographers with their preparation, supervision and exploitation. Geography, as it improves, on a scientific point of view, also furthermore dissociates from action.

ACADEMIC GEOGRAPHY SEVERED FROM ACTION

Throughout the 19th and the first half of the 20th century, geography, while becoming a really scientific discipline, turns from descriptive to explanatory analysis, thus neglecting applications. The known world, which spreads so rapidly, supplies researchers with an extraordinary amount of material; the related sciences are improving and the geographers must, at least, be able to use their methods and results. So they are devoting most of their time to basic research and the training of scholars, who will be in charge of teaching geography to youngsters. Isn't the use of a too young science dangerous because of the incidences induced by basic errors, such as the evidences given by the deviations of *"Geopolitik"*.

Geography as defined by Vidal de la Blache and his disciples does not represent the same dangers. Their conception of a regional geography, studying the *"marvellous variety of combinations"* presented by the phenomena of the Earth surface could have initiated fertile practical applications. But during the first half of the 20th century, the French geographical didn't even think about it.

It is true that the general conditions of the civilization did not lend to a practical use of the *"Vidalian geography"*. When an almost total freedom is granted to individuals to plan their space as they please, in their closest and most immediate interest, what is the use of a discipline describing and explaining spatial combinations? What is the use of research or urban geography, when the builders are entirely free to build whatever they want anywhere? Doesn't the geographer appear to be the trouble maker, who can frustrate lucrative initiatives, when in the same time he can point out that they are contrary to common interest?

It is mainly in Europe, and more especially in France where the Geographical School benefits from a primacy after the first World War, that such a withdrawal phenomenon can be observed, as P. Pinchemel, a fine connoisseur of the Geographical Thought, emphasizes: *"Geography appeared to be professional as well as in phase with the political and economic problems of the time. Geography was then an applied science that was holding the attention of the political and economic leaders. Since, Geography, although it was evolving as a more and more academic science, became progressively-*deprofessionalized- *to the profit of teacher's training"* (Robic, 1966).

The French Geographical School tries to transform the IGU into a *"Scholarly City"* ignoring in some way the *"amateurs"* of the geographical societies. Until the end of the second World War, while the foreign schools had started in the thirties to multiply the applications, to promote the employment of professional geographers, some of the masters of the French School were still admitting, in 1956, without any concern, that *"Cartography is, at the time being, the only branch of geography that offers jobs opportunities to young geographers outside the educational system"* (Cholley, 1957). Max Sorre executes *"this applied geography, to which young colleagues are referring to . . . Some think that there is an opportunity for planning, and that geography is useful for planning. I do not trust planning, for I don't feel a vocation to rebuild the World. I don't think it is the appropriate object of geography. Isn't it its profound sense to reveal Humanity to Itself?"* (Sorre, 1954). Nevertheless this clear condemnation was less dangerous than the secret intrigues aiming to bar those *"young colleagues"* from the Higher Educational System, and from the training of professional geographers. But this period of ostracism could not be perpetuated while the evolution of the world was pushing problems of geographical nature at the foreground.

THE INTERWAR PERIOD

The creation of the IGU in 1922 shows an upsurge of academic geography, if compared to the more pragmatic evolution of the societies of geography; the new organization is based on a States level, so officializing the services that they could expect from geography and geographers (Robic, 1996) Two of the secretaries to be of the IGU, Emmanuel de Martonne and Isaiah Bowman, acted as experts during the 1919 Peace Conference. Forty years later, Georges Chabot, who had himself acted as committee secretary during the conference, admitted that it was a *"remarkable example of applied geography"*(Chabot, 1972).

However, while I. Bowman was combining research, teaching, and expertise for American government leaders, E. de Martonne, who had in mind to which deviations "Geopolitik" could lead, wanted to maintain geography in an only scientific approach, not only by giving priority to physical geography, but also in his regional researches, proving an absolute objectivity as in his volumes of the *"Géographie Universelle"* on Central Europe.

Nevertheless, a return to practical concerns is progressively coming out during the IGU Conferences, in Warsaw, 1934, and Amsterdam, 1938. The specialists in economic geography are trying to emphasize their abilities as experts, especially in the field of transport. But it is mainly researches and applications in town and country planning that are mentioned during the IGU Congresses. The latter will play an essential role in the diffusion of the new types of research and applications, by making them known to the geographers of the participating countries, which number keeps on growing: from 7 members in 1922, to 30 members in 1938. But these applications, realized by expert geographers, researchers, academics and professionals only concern a small number of countries, and will essentially appear with the effects of the 1929 Crisis, which will, while deeply altering the geographical characteristics of whole regions, not to say entire countries, lead to the intervention of geographers.

In the USA, the applications of geography had already come out, well before the crisis, which reached there its pitch, thus explaining their amazing development (Phlipponneau, 1960. American Geography – Inventory and prospects.1954). But, by itself, the American geographical framework with the hugeness of its distances and its incidences on the economy and the living conditions, definitely requires, and this much more than in Europe, the intervention of the geographer. This one is facilitated by the practical mind, the sense of business, but also the solidarity towards the community, of the American citizen. The administration, the local authorities and the business circles have soon understood the interest of geographical researches and are offering either research contracts or more directly plain jobs, while universities are meeting and even anticipating the demand. The famous two months long tour that William Morris Davis organizes in 1912 for some forty European geographers reveals the importance taken by research. If the geographical knowledge of young Americans have been acquired more by the reading of *"National Geographic"* than by the lessons delivered at high school; many universities are then adding geography to their study programmes thus immediately promoting the job hunting for professional geographers. The number of geographers members of the *"Association of American Geographers"* grows up from 48 members in 1902, to 1350 in 1948 and the review *"The professional Geographer"* promotes the creation of jobs, and the recruiting of young geographers.

However it is the psychological shock provoked by the crisis, which, starting from favourable conditions, played a decisive role. America became aware of the dangers represented by an excess of liberalism upon town and country planning and by the lack of long term forecasts in the economic field. A great number of problems to be solved were presenting a geographical aspect. One has to fight against the phenomenal wasting of fertile lands, swallowed up by erosion, to draw up an inventory of natural resources for a more rational exploitation, but also to avoid the building up of unsalable stocks, to promote underdeveloped regions, to redevelop urban areas clogged up by overloaded transport systems and threatened by leprous slums, to establish new industrial plants and new commercial centres taking into account a long term evolution. An ideal experience field for the intervention of the geographer was the complete and all over development of the Tennessee Valley, the most famous example of regional planning, a model for the whole world. As researchers, consultants and managers of planning and resources conservation institutions, geographers, such as Gilbert. White who took part in the deliberations of the IGU Commission on Applied Geography, have opened up to geography more and more fields of intervention.

Three main types of job opportunities already existed on the eve of the second World War. Type One is the geographer civil servant working for the federal government which recruits in all the administrations and at every level. These *"civil servant geographers"* can be found in the states administrations and more often in towns administrations which, in spite of the competition with "planners" employ a fair number of them. But in this special field of rural and urban planning, geographers can also fin a higher job position as free lance experts or consultants, by working in a research department, or keeping a part-time academic position. A last geographers working in business circles remained for a long time a specifically US category. The teaching of geography in the American Business Schools has undoubtedly shown the economic managers, the interest of recruiting geographers on a full time basis in their research departments, for issues such as logistics, location of new plants or commercial centres, development of transport systems. But here again geographers can belong to a research department, keep an academic function, while working under contract or as consultants. *"Applied geography is business geography"* and the latter should provide them with *"interesting jobs, socially useful and well paid"*(Applebaum, 1956).

Naturally these characteristics of American geography, strengthened during the war and its aftermath, but even before they had already given it a marked advance in the field of applications under every form.

In 1938, during the Amsterdam Congress, geographers from the north-western European countries can present to their north American colleagues, who had, by their participation to the TVA project, illustrated their experience in regional planning, ex-periences of the same type. The planning zones of the Netherlands, studied by Dutch geographers, are already covering a fair part of the territory, and the first experimental polders will be realized with the help of geographers. With W.E. Boerman and the School of Rotterdam, geographers begin as soon as 1939 to work for economic plan-ning, as well as for business circles (Boerman, 1952). But the Amsterdam Congress also offers the British geographers the opportunity to introduce their applied researches (Robic, 1996).

The analysis of the origins of applied geography in the United Kingdom was the object of a detailed study by John W. House, presented in 1965 at the first meeting of the Com-mission on Applied Geography in Praha (J.W. House). If the term *"applied geography"* had been used, in the first place, by A.J. Herbertson at the very end of the 19th century and was then related to the links with business circles, it is in the interwar period and essentially with the 1929 crisis, that the applications of geography will multiply with policies of national, regional, urban and industrial planning, and above all with the *"Land Utilization Survey"* carried out by L.D. Stamp (Stamp, 1948). Unlike the USA, land of huge open spaces, Britain has to use for the best the tiniest plot of land, and indeed the LUS enabled the country a swift conversion of its agriculture during the war, but it has also contributed to the protection of the environment and landscapes. With the crisis that affected millions of jobless people in the *"depressed areas"* of the *"Black Country"*, could one leave the destabilization to deteriorate with the London Area, where new industries were then booming?

The first applied works were due to academics and their students, and even, in the case of the LSU, to school teachers, working on research commissioned by local authorities and administrations. It is only with the creation of official institutions for urban, regional and national planning, to challenge the conversion of depressed areas, that non academic geographers, that is to say *"professionals"* are recruited on a full time basis. It is mainly during the war itself that the phenomenon is spreading, with the creation in 1943 of the *"Ministry of Town and Country Planning"*, the *"Ministry of Regional Rural Land Utilization"*, and the *"Ministry of Public Works and Planning"*. Dozens of geographers were then employed by these civil services, even before the Labour government of Clement Attlee passed, in 1949 the *"Town and Country Planning Act"*, which will lead to further recruitment for professional geographers.

On the eve of the second world war, the applications of geography were not limited to North Western Europe and the USA. The IGU counts then 30 member countries; many of them have long established geographical societies, geography teaching at school and university level, and researchers of international fame, but where the applications of geography remain few and little known by the international community. At last the European dominance is still very definite: Asia is only represented by Japan alone, Africa by Egypt, Morocco and South Africa, but Belgian, British and French geographers play an active part in their own colonies. In Latin America, Brazil, which is to play an eminent role in the development of applied geography, becomes a member country only in 1938. USSR, where geographers are already assigned, with the five-years plans and the great development programmes, to work on applications, will join the IGU only after the death of Stalin (1953), willing then to play a major role, together with its satellite countries, in the field of applied geography.

THE DEVELOPMENT OF APPLIED GEOGRAPHY IN THE AFTERMATH OF THE SECOND WORLD WAR

If the British and American geographers had, in the thirties, developed working and intervention methods, the second world war, with its multiple consequences, was to accelerate this evolution, and give it a world dimension, as was the conflict itself.

The task was ahead: one had to rebuild towns and sinister areas, to carry out the transition from a war economy to a peace economy, while nevertheless the possibilities of an opposite movement, to handle and reduce the shortages, to adapt the social and economic structures to new technologies and generalization of free trading, to reduce the imbalances, on a State basis as well as on a world basis, by a true development of the Third World.

The need for a rational planning of the spatial environment, in which men are living, working and moving about, has become universal, from the rural family unit to the megalopolitan areas, from the region to the State and interstate communities on the creation, to international organizations such as FAO and the World Bank which are intervening in favour of developing countries.

The first experiences had shown that geographers could play a useful role to solve the problems which were to deteriorate and multiply in that after war new world. The progress realized in their methods of research, and those of the related sciences allow the success of their interventions achieved at the request of users, economic and political leaders, administrations, local authorities, industrial or commercial firms. If the geographers don't meet their requirements, in order to preserve the purely scientific characteristics of their discipline, other specialists will not have the same scruples; it is precisely during the post war period, and more especially in France, that economists, engineers, architects will take over the key posts of the new administrations concerned by town and country planning. Because of the esprit de corps, even today, young geographers find it difficult to penetrate the reserved domains of other specialists. In the mean time, being unable to demonstrate the usefulness of their discipline and acquire the material resources which would have enabled them to reinforce it, they suffer from the decline of geography, considered as an instrument of culture and youth training, for the benefit of history and economic and social sciences.

Fortunately, in the fifties, in most countries, young geographers are becoming aware of the urgent need to turn to applications, they also succeed in convincing some elders, and some fifteen years later, one can witness a real development of applications in the first concerned countries, those where classical geography was already well represented, as well as in new countries, those of the third world which are creating their own geographical schools or calling for specialists from the former colonial powers. In the meantime geographers from the socialist countries publicize their realizations among the international community, which will, with the help of the IGU, play an essential role in this booming of applied geography.

In 1949, during the first post-war IGU Congress in Lisbon, the *"frenchification of international geography seemed to be acquired"*. This sounded rather reassuring for a young French-Canadian geographer (Hamelin, 1996). Emmanuel de Martonne, General Secretary from 1931 to 1938, is triumphantly appointed as Life President of Honour. However the American influence is growing: the congress of 1952 takes place in Washington, and President George B. Cressey causes a great sensation by revealing to many his country's realizations in the field of applications. Since the Lisbon congress

Jean Gottmann is chairing a commission on regional planning, which will publish one of the first essential books on this basic branch of applications of geography (Gottmann, 1952).

The Rio de Janeiro Congress in 1956, the first to be held in the Tropics, demonstrates that a developing country was able to create its own geographical school, immediately focused on practical applications, thus justifying the place held by the Brazilian example in my book in 1960. The *"National Geographical Council"*, directly referring to the Presidency of the Republic, formed together with the *"National statistical Council"*, the *"Brazilian Institute for Geography and Statistics"* which will play an essential role in the planning and spatial development of the Brazilian territory, for which geographers will contribute in the choice of the location of the new capital. The students, whose training abroad was under completion, the foreign geographers who were in Brazil directing research works or coaching students have managed to form a geographical school, still characterized by the place given to applications.

The Rio de Janeiro Congress is also marked by the internationalization of applied researches. Jean Tricart becomes the Chairman of a commission on applied geomorphology. To prepare the Stockholm Congress of 1960 the IGU also creates a section for applied geography. 1960 is also the year of the simultaneous publication of *"Applied Geography"* by I. Dudley Stamp and *"Geography and Action"* (Phlipponneau, 1960). During the congress 21 papers are presented to the section for applied geography (Phlipponneau, 1996). A small team of geographers, fascinated by the orientation offered to their discipline, is set up, the hard core of the future Commission on Applied Geography. While the *"Section"* of a Congress is only bringing together the participants to deliver their papers, the *"Commission"* ensures the continuation of a collective work, the setting up of questionnaires and surveys, the organization of official meetings and the publication of the results of the primary researches, thus constituting the *"most important organ of the IGU"* (Collignon. in Robic, 1996). The General Assembly invites geographers to answer three questions which would constitute the main objectives of the Commission, finally and officially created in 1964 during the London Congress. It suggests to organize national and international symposiums such as those organized by Canadian geographers in Quebec 1958, or by the British and the Poles in Niewborow in 1959.

THE GREAT START: THE STRASBOURG SYMPOSIUM ON APPLIED GEOGRAPHY (1961)

At the instigation of Jean Tricart and Etienne Juillard, the Centre for Applied Geography, created in 1956, fulfils the invitation of the IGU. It organizes, with the assistance of the CNRS, a symposium rallying sixty French geographers from almost every University Geography Department; some professional geographers and users, together with some fifteen foreign geographers. The originality of this three days symposium is due to the presentation and discussion of the questionnaires sent to the universities, by eleven reporters, relying on the answers to bring out the essential ideas to be submitted to participants, for discussion. And indeed these different reports have helped to take stock of the situation of applied geography in France and paved the way for the initiatives that will multiply in most departments of geography, thus inducing those of the Administration and users, to lead to the present situation. The report on professional geography had hardly listed some 18 geographers coming close to that qualification, thus meaning that every year a very few of them were deciding on such a *"profession"*, for which there was not yet a specific training.

Of course, reticences and even violent oppositions showed up. During the meeting, Pierre George wonders: *"Is there an applied geography?"* (P. George. 1961) The answer is specifically explained in a collective opus, the title of which *"Active Geography"* sounds nevertheless like borrowed from "Geography and action" (George,1964; Phlipponneau, 1960). Such oppositions will mark the return to Scientism of a certain number of geographers, who in their majority interested in counting how many angels can dance on the head of a pin.

But, if the Strasbourg symposium marks the starting point of applied geography in France, it is also because it reveals its significant backwardness in comparison with numerous foreign countries. Its international character was also definite. On the 241 pages of the proceedings of the meeting, 72 of them were devoted to the international report, and on each of the themes, the answers of our foreign colleagues, gathered in the appendix of the publication had been integrated to the discussion.

As I was in charge of the international part in the organization of the symposium, the contacts I had established during the Stockholm Congress enabled me to distribute questionnaires to 80 foreign geographers representing 22 different countries. They sent me back numerous publications about the situation of applied geography in their own countries, and a collective answer was drafted by Belgian, Norwegian, Polish and Czech geographers. The Belgian report alone, published in the proceedings, justified the place that Professor Omer Tulippe and the University of Liège were to take in the Commission on Applied Geography. But the participation of the Polish and Czech geographers also showed the role that the applications of geography were already playing in Eastern European countries. An enlargement towards Scandinavian countries, Austria, Switzerland, Spain, Latin America, Canada, Morocco and Japan was also appearing. The situation was mature enough for the creation of a Commission on Applied Geography of the IGU (Phlipponneau in Robic, 1996).

THE COMMISSION ON APPLIED GEOGRAPHY: WORLD WIDE SCALE AND NEW THEMES

During the London Congress in 1964, the General Secretary of the IGU, Hans Boesch showed his total support for the creation of a Commission *"on Applied Geography"*, an appellation which was preferred by the General Assembly to *"Commission on the applications of geography"*. Omer Tulippe assumes the Presidency of the Commission, being myself in charge of the secretariat. Four other *"Active"* members are designated to represent the main geographical sectors: Peter Nash (USA) for North America, V. Shafi (India) for developing countries, V. Sotchava (USSR) and L. Strascewicz (Poland) for socialist countries. Very soon, eighty corresponding-members, a quite exceptional figure, give a world wide scope to the works of the Commission. As shown by its publications the balance between the two working languages (English and French) was well secured, even if some Anglo-Saxon colleagues had baptized it, with an ironic condescension, the *"French Commission"*.

To fulfil the objectives set up in Stockholm and confirmed in London, how, in the largest number of countries, could the role, the contents and the outlines of applied geography be defined? What were the forms of participation followed by the geographers, their relations with the users? What were the career prospects for young geographers? What complementary, or totally specific, training could help them to become *"professionals"*.

Two complementary methods were to be used. The distribution of questionnaires had the advantage to encourage geographers from the same country to carry out joint studies on the situation and prospects of applied geography, or even to organize national symposiums such as the one of Strasbourg: eight of them were actually organized. However, it is mainly the international symposiums that were organized by members of the Commission which allowed to devise, to follow up the answers to questionnaires and also to bring out, by field visits, a concrete view of the works in applied geography. The publication of the proceedings of these symposiums progressively constitutes first rate material on all the aspects of applied geography in the whole world for the period from 1964 to 1984.

The first symposium, organized in 1965 in Praha, by M. Strida and the Czechoslovakian Academy of Sciences met perfectly the wishes of the Commission: *"Applied Geography in the World"*. One hundred and ten answers to the two questionnaires that were sent were supplying valuable contributions about twenty-eight different countries and were discussed by the sixty-one participants attending the symposium. Even if I had in 1960 granted a fair place to the applications of geography in the USSR and Eastern European countries, the Praha symposium was certainly their best introduction to western geographers. In 1964 already, the translation of *"Geography and Action"* in Russian, with a preface by I.P. Gerassimov, was prompting the geographers from socialist countries to get acquainted with the methods of western geographers. As a matter of form, the Big Boss of Soviet Geography was however emphasizing on *"the contradictions, that are tearing up the capitalist world, thus making a planified and rational reorganization, to which geographers could take part, quite impossible.... while, in the USSR, the grandiose programme of the construction of Communism opens up the widest ranging prospects to soviet geographers"*. The great interest of the Praha symposium and of the subsequent and very active participation of the soviet, Czech, Polish and Hungarian geographers to the works of the Commission, was to encourage comparisons between methods themes and forms of applications, such as the major role played by the Academies of Sciences, in the socialist countries, and by the Universities in the Western World. Two main types of training are also confronting each other: an early specialization in the socialist countries where researchers and professional geographers are trained in the universities and future teachers in teachers' training institutes, while in western countries the universities are mostly training teachers of the secondary school system, a further training, at different levels is organized for the future professional geographers.

In 1966, in Kingston (Rhode Island), Peter Nash organizes a second symposium to which seventy-one geographers are attending, representing eleven countries. Its principal interest is due to the discussions held on the multiple forms of applications in the United States, and mainly to the participation of private and public users, from the State department to chains of supermarket stores. The West European geographers, as well as their colleagues from eastern Europe, were then enabled to learn what the American users were expecting from geographers, researchers, academics, working, or not, under contract, as consultants and professionals.

The third symposium is organized in Liège in 1967, by O. Tulippe. Geographers from twenty-one countries are taking part in a world wide general survey. Since 1964, eight national meetings had been organized on the theme of applied geography. As the first two symposiums had not supplied much information about training, three of the workshops of this one are devoted to this theme. The examples of applications undertaken by the Belgian geographers are analyzed on the spot, in the fields of industry, commercial urbanism, country planning. When in February 1968, O. Tulippe dies, the 832 pages of the proceedings of the three symposiums he has chaired, are testifying of the success of the mission for which the IGU had appointed him London.

From 1968 to 1976, I replace O. Tulippe as president of the commission which pursues its activities at a slower pace: to a period of national surveys succeeds a period of in depth studies on methods, and of research on new themes. In 1968, in New Delhi and Aligarh, the commission, recognized by the IGU as one of the four "*standing commissions*" retains as main theme, the applications in the different methods of soil use, and modes of feeding, in the tropical zones; Mohammed Shafi, the organizer of the symposium, will obtain the creation of a commission specialized in these problems. In 1971, during the Rennes symposium, only one theme is proposed: "*Geography and long term prospects*". Planning was then advancing from short term to middle and long term prospects, that is to say, from four to five years long, to fifteen years long prospects. I had already given examples for Brittany, Canada and Turkey. The role of the geographer, in these types of research, is gaining in importance, as observed in Rennes for new countries such as Algeria and Japan.

The one theme formula is kept in 1972, for the Waterloo (Ontario) Symposium that precedes the Montréal Congress of the IGU. P. Nash, who was then organizing his second symposium, is thus enabled to publish a volume entitled "*Applied Geography and the human environment*", a major contribution, for our discipline, to solve the problems which are becoming more crucial and important. In 1975, at José Sporck's instigation the University of Liège is again organizing a symposium which gathers 95 geographers from 22 different countries around two themes relating to urban renewal and commercial urbanism. The latter will be taken over by a working group, and later on by a commission of the IGU.

Before the Moscow Congress in 1976, V. Sotchava is organizing a symposium in Tbilissi, but the foreign attendance is limited, and the IGU goes back over the notion of "*Standing Commission*". The Commission on Applied Geography, created twelve years earlier is doomed to disappear. An in fact, it is replaced by a working group that I will chair until 1980, when I give way to Hiroshi Tanabe, who had remarkably organized a symposium in Yokohama, which was attended by 42 colleagues from 16 different countries. Its main theme was about the training of professional geographers to be employed by administrations, planning authorities and business circles. Before the Paris Congress, in 1984, the working group is holding meetings in Zurich and Lausanne, but to respect the duration rule it is not renewed

THE LAST TWO DECADES

While in the sixties, the very notion of Applied Geography gave rise to controversy, its existence became so evident that the IGU did not even feel the need to recreate a working group on this theme, before the Congress of Seoul in 2000. Since 1960, the collective effort, undertaken under the auspices of the commission on applied geography, helped to popularize the multiple experiences, thus enabling geographers as well as users, from new countries, to follow them as examples, after an adaptation to their own situations and needs.

When the UNESCO asks me, in 1981, to prepare a summary on "*the development of Applied Geography*", relying on a hundred of publications, it is easy to bring out the common features, but also the differences characterizing the main groups of countries (Phlipponneau, 1981). Everywhere, and proportionately speaking, the applications are due to researchers and academics, but also to consultants and professional geographers; specific training is appearing. All the branches of geography are concerned, with a prevalence of physical geography in the USSR. The research methods are already well advanced, with the use of computers, GIS, and satellite imaging to complete aerial photography. The essential elements are thus appearing in this picture of applied geography in the eighties. We can start from there, to reach the contemporary situation, analyzed sector by sector in the present opus, and in the last chapter of "*Applied Geography*" (Phlipponneau, 1999). All we have to do is to point out the evolutions that re already part of the history of the affected countries.

The most sensible differences are mainly affecting the USSR and its "*socialist satel-lites*" since the "*Collapse of the Berlin Wall*". The applications of geography were related to the importance assigned to planning in every field, and for the USSR by the "*grandiose programmes of transformation of the environment*" thus explaining the im-portance assigned to physical geography (Phlipponneau, 1981). The role of professional geographers matters more than the one of geography teachers educated in the Teachers Training Institutes. The future professionals, after passing a difficult entrance exami-nation, are admitted to the faculties of geography endowed with huge resources, are trained for at least five years; training becoming more and more specialized according to the needs. The greater number of jobs is offered by the multiple Planning Institutes, the technical ministries, and the management of the big industrial complexes. The most original element is the role played by the Institute of Geography of the Academy of Sciences which was then relying on hundreds of researchers. Their works were di-rected towards practical applications and met the needs expressed by the government, the ministries and the leaders of the major planning works programmes. This role of the Academy of Sciences was then identical in the socialist countries of eastern Europe, as well as in Cuba and China.

In spite of the major transformations of the political and economic system, the general organization of geography seems little changed. In Russia, the Academy of Sciences still plays a major part (Kotlyakov, 1996) totalling, in the three main institutes of Moscow, Irkoutsk and Vladivostock, more than a thousand researchers to which those of the great state universities and specialized institutes must be added. The role played by the state, as sponsor of fundamental and applied research has been drastically reduced. The predominance of the themes of physical geography is lessened. Young geographers are, giving up the official public sector for the private sector, undertaking research work for foreign companies in different fields such as marketing, industrial location, distribution and transport systems. In the meantime, the role of the teachers training institutes, once considered as a minor one, is becoming more important when new types of secondary education institutions are springing up. In a logical way, the evolution of geography, applied science and discipline of culture, is complying with that of the whole system, by getting closer to the evolution of geography in the western World.

This evolution was easier in the eastern European countries: the geographers had man-aged to maintain their contacts with western countries. They played an important role within the commission on applied geography, by keeping their freedom of thought if not of speech, before fully recovering the latter. If the system of the Academy of Sci-ences has been kept, the universities are still training researchers as well as professional geographers, and the differences with the western countries are fading away.

In 1981, in my description of the World situation, I was emphasizing on the pre-eminence of the US, as I had depicted it twenty earlier in all the fields of applica-tion of geography. During these twenty years, Canada had joined its American neigh-bour, thanks to the influence of the French geographers in Québec and of American colleagues in the english speaking Provinces. The Commission on Applied Geography had held two symposiums, in Kingston (1966) and Waterloo (1972), and numerous american geographers took part to its works. Nevertheless, in 1999, in the last chapter of "*Applied Geography*", I can ask the question: "*Can North America keep its supremacy?*" Is the answer so obvious? Thus the *National Geographic* magazine holds a readership of 40 millions, the Association of American Geographers counts 7 000 members, 5 000 university diplomas in geography are granted every year by some 400 colleges and universities, of which 76 are preparing for doctorates. However, american geographers have noticed by themselves a state of crisis marked by the closure of prestigious geography departments, replaced by research centres which are soon abandoning geography as a global science, to favour specialities and tools, descending sequels of the "*Almighty Computer*". Boosted by a decade long advance, persuaded that the most complex problems can be solved with the help of GIS and the quantitative methods of the "*New geography*", many american geographers have mistaken tools and the material to treat. Because of an excessive specialization, they end up joining related disciplines. At the same time american geography shuts itself from the rest of the world, and ignores haughtily all the non-all-english publications and those not worthy of a translation. Its members no longer attend so assiduously international meetings and conferences. However, this crisis, marked, from 1985 to 1990, by a fall of the membership of the AAG and of the number of graduates in geography, seems to be coming down and by my own observations as a participant to an AAG Assembly in 2001. In Canada, it is the applications that seem to "*wake up a geography*" sent to sleep with theoretical abstractions, by encouraging geographers to work on the field.

As soon as 1980, in all the European countries, the characteristics of applied geography had already been fixed, and the applications simply multiplied with an increasingly important role played by professional geographers, especially in France that made up for lost time, and where the training of professionals is well placed in the geography departments of most universities. The geographers of almost all European countries took part indeed in the works of the commission on applied geography, organized international symposiums as those of Liège, Wolfsburg, Roma and conferences in Great Britain, Switzerland, Italy and Portugal.

Due to this generalization, Great Britain could keep as much advance as it has acquired for a long time. The number of students in geography and of professional geographers is continuously increasing until the seventies when it is dramatically interrupted in 1979, with the advent of "*thatcherism*" and its combined effects of neo-liberalism authoritarianism of the central power together with harsh budgetary restrictions. After 1990, the British educational system knows a radical transformation, and the situation recovers. The applications still make up a larger job opportunity than secondary degree teachers training, but the geographers usually follow a further specific training in "*planning, local government, business and tourism*", before becoming professionals.

Differently from the United Kingdom, applied geography does not seem to have encountered such crisis in the countries where it had already largely developed about 1960 (see chapter Applied Geography in Western and Southern Europe): Belgium, the Netherlands, as well as Scandinavian countries, Germany, Austria, Switzerland. In the countries of southern Europe, a remarkable growth is to be noticed, especially in Italy, where earlier on the geographers had necessarily to deal with tragic spatial imbalances. In Spain and Portugal, where the political transformations have favoured a parallel development of academic geography and professional geography, it is also the entry in the European Community that has promoted the multiplication of works in applied geography, and the creation of jobs for professional geographers. In 1964, "*Geography and action*" had been translated into Portuguese, in 2001 "*Applied Geography*" was translated into Spanish; this can be regarded as an evidence of the interest that the geographers from the Iberian peninsula, as well as those from Latin America, are expressing for French methods.

During the eighties, outside Europe and North America, one can distinguish the countries which have already created their own autonomous school of geography, directed towards applications; this orientation became generally more pronounced as observed in the second part of this book. These countries have sent their students abroad to complete their studies, and have also called for external co-operation. It is the case of Brazil which has maintained its advance, of Mexico and Venezuela. The New Delhi Congress has shown in 1968, how the leading British universities had achieved the creation of an Indian geographical school, and how its applied researches could interest other tropical countries. Similarly the Sydney Congress in 1988, the regional conferences of Palmerston (1974) and Durban (2002), have shown the large place taken by applications, in Australia, New Zealand, and South Africa.

The case of Japan is slightly different since it was already part of the seven founding members of the IGU in 1922. The balance between the French and Anglo-Saxon influences maintained for long and from the creation of the Commission on Applied Geography, the Japanese geographers have expressed their interest for the applications, before Hiroshi Tanabe became Chairman of the working group. Radical transformations of the academic world linked to the consequences of the demographic evolution and of the economic problems, will lead to a development standing the comparison with that of the United States. The Seoul Congress, in 2000, has shown the extremely fast development of geography, since the end of the Korean War. As soon as 1973, with the growth of the applications, the "*Korean Association of Professional Geographers*" publishes a quarterly review. The american influence is not exclusive, some French geographers have found in Korea great subjects for their thesis.

In 1980, while visiting several Chinese universities with a group of French colleagues, I was struck by the similarities between the organization of Chinese geography and that of Soviet geography: prevailing role of physical geography, of the Academy of

Sciences, of the recruiting of geographers in the Planning institutions. Ten years later, during the regional conference of the IGU in Beijing (1990), the themes of research and the methods had diversified, leaving the general organization unaffected. One has to wait for the return of the Chinese students, sent abroad to complete their studies, to check if the evolution is really appreciable.

On the African continent, the characteristics of applied geography have not changed since the eighties. Yet, the regional conference of Lagos in 1978, the Presidency of the IGU, by A.L. Mabogunje, from 1980 to 1984, the enrolment of 25 countries to the IGU in 1988, all suggest a successful take-off. But the political, economic, and sanitary difficulties have broken down this development; only South Africa recovers and takes an eminent place, materialized by the participation of its geographers to different meetings and by the organization of the regional conference of Durban in 2002. In the early eighties, two forms of application of geography were coexisting. That of geographers, coming from the former colonial powers, who had been for long working on concrete problems, often under contract with institutions such as the Applied Geography Laboratory of the University of Strasbourg, and more often, under the auspices of the ORSTOM. This formula tends to be reduced to the advantage of applications undertaken by French and mainly African geographers belonging to autonomous geographical schools, shoots of the newly created universities in Black Africa and Maghreb. In France, the most remarkable fact is the multiplication of the number of thesis, prepared by African students who are there to complete their studies. Very often, they choose, for subject of thesis, some practical problems of applied geography in relation with their countries. As all of them can't become higher education academics, they will definitely have the opportunity to become professional geographers.

A lesson is outcoming from this flying over the last twenty years evolution: the absence of this incomparable instrument of observation, information, comparison and of initiatives that a commission of the IGU can represent. Of course, the commission on applied geography had not achieved all of its objectives. Since its first meetings, it had recommended the creation of a bilingual review. "*Applied Geography*" was indeed created in 1981, with quite a success, by British colleagues, but it only publishes in English. The commission also tried, in 1972, to interest the major international institutions in recruiting professional geographers and to set up international research teams working under contract, on subjects related to neighbouring countries. The idea should be taken up again.

Since 1992, I have made proposals to the IGU to recreate the commission on Applied Geography, in order to give an up to date report on the evolution and prospects. The working group, created in Seoul in 2000, has become commission again, with Antoine Bailly as chairman, and will be able to present in Glasgow, in 2004, together with this book and new proposals, the prospects of development of applied geography for the 21st century.

REFERENCES

Actes des réunions de la commission de géographie appliquée de l'UGI. 9 vol. Références in Phlipponneau. 1999. Bibliography.

American Geography. Inventory and prospects. Syracuse. 1954.

Applebaum W. (1956) What are geographers doing in business? *The Professional Geographer.* (01/1956).

Bailly A. & et al. (2001) *Les concepts de la geographie.* Paris, Colin.

Boerman W. E. (1952) La géographie économique et son importance pour la solution des problèmes économiques contemporains. In *50 ème Anniversaire du Laboratoire de géographie de l' Université de Rennes.*

Chabot G. (1972) La géographie appliquée à la Conférence de la Paix en 1919. In *La Pensée géographique contemporaine. Mélanges Meynier.* Saint-Brieuc. P.U.B.

Cholley A. (1957) Tendances et organisation de la géographie en France. In *La géographie Française au milieu du 19ème siècle.* Paris Baillière.

Collignon B. (1996) Les commissions. in *Robic (1996).*

Colloque National de Géographie Appliquée de Strasbourg (1961). Paris. C.N.R.S. 1962.

George P. (1961) Existe-t-il une géographie appliquée? *Annales de Géographie* p. 337–344. Paris.

George P. & et al. (1964) *La géographie active.* Paris. P.U.F.

Gottmann J. (1952) *La politique des Etats et leur géographie.* Paris. A. Colin.

Gottmann J. (1954) *L'aménagement de l'espace. Planification régionale et géographie.* Paris. Colin.

Hamelin L. E. (1996) L'internationalité de l'UGI au milieu du 20ème siècle. In *Robic (1996).*

House J. W. (1967) Applied Geography in Britain. *Acts of Praha meeting, 1965.*

Kolyakov V. M. (1996) *Geography in Russia 1992–1995.* Moscou. Academy of Science.

Phlipponneau M. (1959) Vauban et la géographie appliquée au Canada. In *Mélanges Raoul Blanchard.* Québec. Université Laval.

Phlipponneau M. (1960) *Géographie et Action. Introduction à la géographie appliquée.* Paris. A. Colin.

Phlipponneau M. (1981) L'essor de la géographie appliquée. *Revue internationale des sciences sociales.* p. 148–176.

Phlipponneau M. (1996) La commission de géographie appliquée. In *Robic (1996).*

Phlipponneau M. (1999) *La géographie appliquée. Du géographe universitaire au géographe professionnel.* Paris. Colin.

Pinchemel P. (1996) Présentation. In *Robic (1996).*

Robic M. C. & et al. (1996) *Géographes face au monde.* Paris. L'Harmattan.

Sanguin A. L. (1997) Les Sociétés de Géographie. *Acta Geographica.* N° 111.

Sorre M. (1954) L'orientation actuelle de la géographie humaine. *Norois.* 1954.

Stamp L. Dudley (1948) *The land of Britain. Its use and misuse.* London, Penguin.

Stamp L. Dudley (1960) *Applied Geography.* London. Penguin.

Vauban S. de (1910) *Oisivetés et correspondances.* Paris. Berger-Levrault. (réédition)

Vox M. (1943) *Correspondance de Napoléon.* Paris. NRF.

KINGSLEY E. HAYNES, QINGSHU XIE AND LEI DING

CHAPTER 4
POLITICAL GEOGRAPHY, PUBLIC POLICY AND THE RISE OF POLICY ANALYSIS

ABSTRACT

This paper discusses the importance of considering the geographic context in policy-making and highlights the necessity of incorporating policy analysis in the study of political geography. We begin with a brief review of literature on political geography and a categorization of scale issues. This is contrasted with views of the policy process typically used by policy analysts. The result is a shift from theory to operational considerations which often leads to a loss of the important spatial dimension. Two simple applied cases studies are presented on how geographic factors affect public policy analysis and how applied public policy studies can be carried out with sensitivity to geographic considerations. The two case studies are regional income convergence and the rise of regional transportation management institutions. The first is analytic in character while the second has a policy evolution perspective but both indicate how the application of a geographic perspective invigorates policy analysis.

INTRODUCTION

Political geography is a broad term with many subtle dimensions, which examines the geographical factors involved in political systems. The link between political geography and public policy is direct and substantive. Political geography spans the study of territory and political economy, territory and the state, politics of identity, politics of borders and sovereignty, electoral geography, and environment policy. In fact, geography as a policy factor has an important impact on policy-making and public policies often have their own spatial dimensions. Further the effects of a public policy vary greatly across space. A thorough understanding of such a mutual determinative relationship between geography and public policy is a key element for the efficient and effective execution of public policy. Although political geography has become increasingly important in the last two decades, with few exceptions policy analysis as an element of political

The authors wish to express their appreciation to NSF/EPA grant #SES-9976483, NSF grant #ECS-0085981 and US DOT Cooperative Agreement #DTRS98-6-0013. Any errors are the responsibility of the authors.

A. Bailly and L. J. Gibson (eds.), Applied Geography, 69–93.
© 2004 *Kluwer Academic Publishers. Printed in the Netherlands.*

geography has been missing either in case study terms or in comparative research. Furthermore, geographers in the last two decades have often ignored public policy as a central component of their research (Martin, 2001).

This paper discusses the importance of considering the geographic context in policy-making processes. We begin with a brief review of literature on political geography and a categorization of scale issues in political geography studies. This is contrasted with views of the policy process typically used by policy analysts. The result is a shift from theory to operational considerations often with a loss of the important spatial dimension. This paper then presents two simple applied cases studies on how geographic factors affect public policy analysis and how applied public policy studies can be carried out with sensitivity to geographic considerations. The two case studies are regional income convergence and the rise of regional transportation management institutions.

RECENT THEORIES IN POLITICAL GEOGRAPHY

Marxist political economy theory

Theories of the state were among the topics of central interest in political geography in the early to middle 1980s. Concern was (1) with the extent to which the state could be viewed as an autonomous social entity, (2) with understanding the concrete institutional forms the state takes under capitalism, and (3) with how these structures change over time. Marxist political theory provides one important perspective on state formation. For the Marxists, the state is seen as a mechanism of oppression of the exploited class by the ruling class, and as a part of the societal superstructure in contrast to the economic infrastructure (Johnston, 1984). Consequently, the emergence of the state is the outcome of the formation of class struggle. The development of technology, which is viewed as the increased sophistication of the forces of production, is seen as a driving force in social evolution. The state is thus seen as a superstructural institution, called into existence by the basic "imperative" of the social mode of production. Under capitalism, Johnston (1984) maintains, the state must promote and legitimize capital accumulation. The form and character of the capitalist state is derived from these basic principles. Individual capitalist states are simply examples of the capitalist state-in-general at different points in an evolutionary progression – a classic time-space substitution.

The Marxist political economy perspective is but one example of so-called a "society-centered" analytical framework where the state is the decision unit. Within this frame-work the state embodies, expresses or reflects the will, interest, or need of individuals, classes or an entire social structure. In his review of political geography and state formation, Driver (1991) indicates that the "society-centered" approaches to state theory are defective in several respects. First, they abstract the state-in-general from its specific territorial form and almost totally ignore space and location factors. They also interpret state behaviors only in terms of a number of preconceived functions such as the mode of production and they tend to over-abstract state policy as a response only to social pressures and economic forces on the class system.

Even Fukuyama's contrasting view in "The End of History" (1992) with its radical alternative perspective is built on an interactive society with state-centered concepts where its dialectics are similar to Marxist theory in terms of its change dynamics. Of course his work is not based on class conflict and Marxist ideology but it serves as an alternative bridge from the social-centered framework to the state-centered one we discuss below.

State-centered framework

Within the state-centered framework, the state is viewed as more than a reflection of or response to underlying socioeconomic structures. This perspective is character-ized by its focus on institutions, its emphasis on questions of territorial forms, and its recognition of context for processes of state formation (Driver, 1991). The state is con-ceived as a set of centralized institutions which exercise power over a specific territory (Mann, 1984). So the state is viewed primarily in institutional rather than functional terms. Organization and communication issues, historical patterns of development and geographical variations become important. Driver (1991) indicates that this approach adopts an explicitly historical perspective on the various conditions and effects of state structures in different contexts.

However, state-centered theory is not without its critics. Gilbert and Howe (1991) challenge state-centered theories and argue for the convergence of state institutional capacity and class capacity in their historical analysis of the development of the U.S.'s new deal agriculture policy as they focus on region-specific class conflict as the policy dynamic.

Postmodernism

Postmodernism is a complicated set of ideas, which has only emerged as an area of academic study since the mid-1980s. According to Lyotard (1984), the postmodern would be that which in the modern invokes the unpresentable in presentation itself, that which refuses the specification of correct forms, refuses the consensus of per-mitting a common experience, and inquires into new presentations – not to use them as a new basis structure, but to better produce the feeling that there is something unpresentable. And the "postmodern condition" for Lyotard was one where human emancipation and liberation, and big science (progress and freedom) were coming un-done and being replaced by a proliferation of local discourses and pragmatic languages (Lyotard, 1984; Toal and Shelley, 2002). According to Sarup (1993), modernity implies the progressive economic and administrative rationalization and differentiation of the social world while postmodernity refers to the incipient or actual dissolution of those social forms associated with modernity. In this view postmodernism is the culture of postmodernity.

Although the issues in "postmodernism" disrupted academic discourse, it opened up a range of new political-geographic areas. These included studies of the spatial dimensions of the politics of identity, which ranged from considerations of the politics of cultural formations, national identity, sexual identity, and etc. (Toal and Shelley, 2002). "Postmodernism" was also a sign used to describe specific methodologies with qualitative and individually interpreted meaning like semiotics, deconstruction and discourse analysis.

The politics of place

In attempting to suggest how space matters in social life, human geographers have advanced three general perspectives (Buck, 1990; Reynolds, 1994): (1) they suggest that people have ideas about place and that places have significant and different social consequences; (2) they claim that social processes are constituted differently in different places, and (3) they assert that the costs of overcoming distance ensure that social life takes place in relatively circumscribed spatial contexts. The third contention forms the basis for the first two claims. The present authors' view of these three perspectives is (1) maybe (2) sometimes but generally no and (3) yes. However, Harvey's (1985) analysis of regional class alliances provides a basis for appreciating the application of these ideas in place-based politics.

The geographic behavioralists viewpoint put place in a social utility context making experience and localized knowledge more immediately valuable to the individual than distant and unexperienced places. It also indicates a high price in time and effort to overcome the disutility of new places but with the expectation that this disutility and its associated costs decreased with time for local participants (for example as in-migrants becoming residents).

Paasi's (1991) made progress in supplying a framework for a place-based analysis of politics. He sees previous approaches divided into those in which regions are social creations due to a specific mode of production while other regions are layered products of individual and collective consciousness without any separate organizing structure. He proposes a more specific categorization of a historical and cultural approach that connects the material conditions of social life to its experiential dimensions by linking individual biographies with the histories of social practices occurring at various spatial scales.

Following on from her earlier work with Fincher (Chouinard and Fincher, 1987), Chouinard (1990) also made an important effort to link theory and application by suggesting how attempts to extend and deepen the dominant state forms of social control and regulation can influence the politics of place by shaping peoples subjective experiences with the state and with political life in general. She does this in an empirical study of struggles over state regulation of community legal aid clinics in Ontario, Canada by focusing on how these struggles have helped shape possibilities for oppositional legal relations in Ontario localities (Chouinard, 1990).

In another attempt to apply theory to a place perspective but in contrast to Chouinard's, Cox and Mair (1988) adopt an essentially "functionalist" view of local politics. They have three "classes" of actors in their model: businesses, individuals, and local units of government. According to Cox and Mair (1988), the key to understanding politics in a locality is the degree to which the reproduction of businesses, individuals, and local governments are tied to the locality. Since local governments are completely "locally dependent", businesses are least locally dependent and individuals (primarily labor) fall somewhere in between. As a result of these variations in local dependence, business coalitions which are dominated by those units which are more locally dependent than others supposedly form and chart the future economic and social course of the locality. Cox and Mair (1988) argue that it is precisely because of these particular economic circumstances in which they view local politics to be place-bound that their concepts of local dependence describes that element.

While the above has been a cursory review of relevant theories from political geography, it has moved from the general to the specific, from the state to the local perspective and from theory to application. This was purposeful because we wish to show that the former "international and state" perspective is important and linked to the "local and specific" perspective by theory, methodology and history. However in an applied geography sense it is the latter (local and specific) that links political geography to policy analysis for better or worse. Policy analysis operates in the "localized" world of the *politics of place* (all politics are local) and the *politics of issues* as specific and narrow (place specific) as they may be. For this reason grand theories of political geography, although real and valuable, are often at an intellectual distance from the worlds of the policy process and policy analysis.

SCALE ISSUES IN POLITICAL GEOGRAPHY

Studies on the topics of political geography are conducted at different levels. If forced to classify these categories, three scales can be summarized: micro-scale (local level), meso-scale (state level), and macro-scale (global level).

Micro-scale (local level)

Perhaps the best example at this scale is electoral geography which is micro at the participatory level even if its outcomes can be aggregated to macro consequences. This cluster of studies of electoral geography includes the outcomes of elections and other democratic and quasi-democratic processes in domestic countries, as well as the implementation of democratic processes in previously non-democratic countries. Because of the practical importance of and the general public interest in elections, political geographers have been active in identifying and interpreting electoral patterns. With the growth of trade and globalization, with enhanced telecommunications, and with the global diffusion of democracy, there has been a revitalization of the realm of electoral geography.

Before the end of Cold War, studies of voting patterns in the USA and the UK had dominated the electoral geography literature. After the cold war with the reintroduction of elements of electoral democracy in the former Socialist countries a new supply of interesting election returns for geographic analysis has been generated. Studies of elections at the national scale have extended beyond the USA and the UK to include South Africa, Russia, Ukraine, Slovakia, Hungary, Poland, Moldava, Turkey and Mexico (Reynolds, 1993; Flint, 2002; Toal and Shelly, 2002). Of course, issues of ethnicity and national identity are central concerns in political mobilization and the geography of electoral participation.

Earlier electoral geography was criticized as excessively mechanistic and overly reliant on rational choice and economic interpretations of voter behavior while ignoring social and cultural factors that influence voter decisions (Painter, 1995). Recent electoral geography involves much greater recognition of the role of local context in electoral outcomes (Eagles, 1995) as well as an explicit treatment of cultural influences on local and regional voting outcomes. Gender, cultural factors, and nationalism have to varying degrees influenced electoral geography in the post-Cold War United States (Shelley and Archer, 1994; Shelley, Archer, Davidson, and Brunn, 1996; Archer, Lavin, and Martis, 2001) and Europe (Davidson, 1996; Agnew, 1995). Today's electoral geography has successfully added social and cultural elements which complement the long-standing tradition of focusing on economic considerations (Flint, 2002).

Electoral geography is also making advancements in its knowledge of how voters perceive the locale (Johnston et al., 2000; Pattie and Johnston, 2000), the role of party organization, the importance of financing (Johnston and Forrest, 2000), and how elections are linked to other political processes and structures, such as patriarchy (Webster, 2000) and civil society (O'Loughlin and Bell, 1999). Further, the interactions of the local and global scales are being addressed, for example, as a reaction to global migration flows local voters' support for the extreme right has changed (De Vos and Deurloo, 1999). Geographers have also investigated the effects of boundary delineation at the local level on public policy outcomes by studying elections of representatives to local and national legislative bodies in territorially defined districts in the U.S. (Grofman, Handley and Niemi, 1992). However given the variety and quality of national and international polling data its use by geographers is still quite limited.

Meso-scale (regional level)

Uneven regional development and regional competition are two examples of these meso-scale concerns. Previously, analysis of the political economy of regions tended to focus on relative factor endowments (natural or man-made), privileged relations, or regional dominance within a national system, nowadays the emphasis is on uneven development in a multi-scalar context and on regional competition.

Uneven regional development

Differential regional growth has been revived to become one of the major topics in political geography and regional economics following the publication of two influential papers on endogenous economic growth in the 1980s (Romer, 1986; Lucas, 1988). The new wave of regional growth studies has two interrelated elements: the development of endogenous growth theories and their application (Johansson, Karlsson, and Stough, 2001) and empirical testing of the regional convergence hypothesis. Apart from a large number of new models, there have also been numerous empirical studies on economic convergence at the international and national levels. These studies have two purposes: to test regional growth theory; and to provide information to policy makers with regard to regional income gaps.

There are at least two perspectives on the convergence of regional development. The first sees regional differences in performance and affluence as temporary or the result of government interference in market allocation of resources. In the UK and the US, a number of scholars rehabilitated neoclassical growth models for closed economies (such as those of Solow, 1956; Koopmans, 1965) and applied them to the problem of regional differences. These models suggests that the rate of growth in per capita incomes is inversely related to the per capita starting income. If poor regions have access to similar technology, they will grow faster than richer ones, promoting convergence (i.e., poor regions will catch up with rich ones). The other perspective focuses on globalization, where they saw differences in regions and localities in the developed world as fading away in social importance as social and cultural practices standardize around the globe. Commentators on globalization claimed that in the developed world technological and economic changes mandated an increased priority of time over space. The outcome is a "deterritorialized" world in which a space of flows increasingly substitutes for a space of territories or regions (e.g., O'Brien, 1992; Castells, 1996). Only at a global level, split between those with and those without access to the new technologies, does geographical difference continue to operate.

As the performance of businesses seemed increasingly sensitive to their regional context, and as political movements arose based on claims to regional discrimination or uneven development, the uneven aspects of regional development was increasingly appreciated. The primary arguments of continued uneven development are summarized as being due to: the role of different physical geographies in economic growth (e.g., Gallup and Saches, 1998); the role of the accumulation of "social capital" in regional economic growth (e.g., Putnam, 1993; Storper, 1997); the renewed prominence of regionalism within government structures (e.g., Keating and Loughlin, 1997; Keating, 1997); and the importance of cooperation or lack of it among regional economic and political elites (e.g., Piattoni, 1997). These studies point in many directions. They serve as positive responses to issues of uneven regional development. However, although they respond in one way or another to lessons learned since the 1970s, curiously they have a dearth of appreciation of the role of spatial interaction, regional interdependency or spatial dynamics on patterns of regional economic development.

Regional competition
Regional competition reflects the emerging importance of regions as centers of global competition. Studies indicate that regional development requires a set of social conditions relating to the nature of both local and state involvement in regional economies (e.g. Agnew, 2000). There is evidence of a bond between the character of regional social structure and economic growth (Florida and Lee, 2001). Entrepreneurship usually has precursors in terms of social stimulation, validation, and reinforcement as well as in terms of government expenditures. Subnational governments, from US states to regions and localities all over Western Europe, are becoming active in global affairs particularly through economic competition. This occurs often independently of their respective national governments and it is a reflection of decentralization of national decision-making and increased demands for local control. Of course this happens at a time when other sovereign powers are moving from nation states to multi-national bodies as nations become part of multi-regional trade blocs – EU, NAFTA, APEC etc. As a consequence, firms are developing a locational perspective about making investment decisions at the geographical scales of regions and localities rather than the national territories that once dominated their geographical mindset (e.g., Jones, 1997; Wright, 1998; Pantulu and Poon, 2003).

This is a difficult interpretation since historically firms have always invested locally but this was a consequence of the place specific aspect of any investment and the related capital that historically was locally derived. As capital pools became national and investments were not as locally tied, regions with different potentials for returns to investment competed with each other. In the U.S. this was reflected in the debates about the gold vs. silver standard, the role of the interstate commerce clause, the fight over interstate banking, and the issue of public investment in transportation to access agriculture and natural resource regions. In the U.S., this was typical of the late 1800s' robber baron period, where national pools of capital looked for different returns across regions and peripheral regions fought back to preserve their capital bases.

It is speculated that as capital has globalized the search for returns moved from the regional to the national or multi-national scale (such as EU, NAFTA, and APEC).

However, with the rise of technology, leading regions may have driven capital returns to become more targeted again, shifting backdown from the international to the national and even the regional scale (e.g., venture capital is investing in Bangalore but not necessarily elsewhere in India or in Austin, TX not necessarily elsewhere in the U.S.) in search of differential rate of return opportunities.

Regional competition also concerns the increase in coordinated actions within regions to improve competitive positions relative to other regions. In this sense regional competition policy should be distinguished from traditional national top-down regional reallocation policy. It is different in terms of its regional origins and its competitive verses reallocation goals. Indeed, it is prosperous city-regions that have been the most active participants in regional competition. In Europe, regional development policies of this type have expanded enormously (Dicken and Tickell, 1992; Cheshire and Gordon, 1998). In the USA such policies have been in operation for much longer. For example almost all states have tax subsidy, infrastructure and labor training policies designed to attract external investment since the 1920s. In recent years subsidies have been particularly strong in some southern states such as South Carolina, Alabama and Arkansas.

Results have been mixed as to the stimulative effects of such policies, although recent research suggests a number of conclusions. There appear to be clear winners and losers for regions that have pursued such policies. Those with local assets (economic rents) linked to areas of innovation benefit the most (Cheshire and Gordon, 1995). Certain types of policy seem to be more successful than others. Attracting mobile external capital with subsidies seems wasteful both from a regional and global perspective but getting regions organized to compete does seem valuable. Those policies to enhance growth in one region without detracting from growth elsewhere such as training, fostering entrepreneurship, business advice, infrastructure investment, uncertainty reduction and coordination seem to do best (e.g., Wins, 1995; Cheshire and Gordon, 1998). It is precisely the activity of taking stock of regional assets and organizing for effective interregional competition that appears to aid regions stimulating entrepreneurship and local public investment strategies thereby making a region better off. This might be interpreted as the basis for endogenous growth activity.

Macro-scale (global level)

Discussion of the processes of globalization dominates contemporary political geography writing at the international scale (Flint, 2002). This paper interprets globalization as referring to capitalism's latest spatial division of labor with its international gateways, its front office and back office divisions, its subcontracting web in support of flexible manufacturing, its export processing and free trade zones, and its "just-in-time" production and distribution systems (Cox, 1997; Daniels and Lever, 1996). Equally important is the area of global finance and its headquarters in gateway cities wired to major world markets often with crucial offshore centers beyond national financial regulations (Leyshon and Thrift, 1997).

Globalization also refers to the regions of the globalized gateway city. Regions associated with these large gateway centers show signs of emerging as the key nodes in the new global economy (Stough, 2000). It is through these gateways that the world economy is increasingly organized since major service industries and corporate headquarters are located there. However, the regions of economic development associated with these gateway cities are rarely regions in a political or institutional sense. This introduces an important question: although economies of North American, Europe, and Asia are structured around city-regions, politics and policy-making are still locked in a framework of territory organized as provinces or states and large cities often have limited political autonomy. What does this situation mean for the future coordination of political and economic power?

Even with these changes in regional economic structure theoretical frameworks that already exist are still useful. The centripetal and centrifugal forces of globalization may also be interpreted using the Marxist notions of equalization and differentiation (Harvey, 2000; Smith, 2000). The world-systems perspective maintains that the world-economy is structured territorially into core and periphery by interclass and intraclass conflict between capitalists and propertyless workers (Chase-Dunn, 1989). Such a competition is seen as producing a division of labor in which capital-intensive production is increasingly focused in economically dominate states. Literature on the cultural politics of globalization also focuses attention on alternative perspectives in the modern world. New and innovative theories are still needed to interpret the trend of globalization and its spatial consequences.

POLITICAL GEOGRAPHY AND PUBLIC POLICY

As indicated previously, the link between political geography and public policy is direct and substantive but not well developed in the field of geography. The lack of a full understanding of the mutual determinative relationship undermines the efficiency and effectiveness of a public policy. However, with few exceptions policy analysis as an element of political geography has been missing both in case studies and in comparative research. As mentioned earlier the few exceptions to the former is Chouinard and Fincher (1987), Chouinard (1990), and Cox and Mair (1988) among others. But even the Chouinard and Fincher's work on the service delivery program in southern Ontario has as a focus only as a critique and demonstration of theoretical issues rather than as a policy analysis with an alterative design for effective delivery. This is fine from an academic perspective but inadequate as a policy analysis. Most policy related work in geography falls into two categories: a theoretical based academic analysis or a highly focused place specific "local" study with a limited spatial dimension. Both are often extremely competent but usually neither is able to respond to the policy maker's needs often because they are too general or too specific. And often the specific studies which are closest to decision maker if not policy maker needs do not focus on spatial issues where geographers have an intellectual advantage. The loss of geography in policy analysis is the dimension that needs to be retrieved and made more central to political geography if it is to play an important role in applied geography.

Blow two simple applied cases studies are presented: regional income convergence and the rise of regional transportation governance institutions. They are used to demonstrate how geographic factors affect public policy analysis and how applied public policy studies can be carried out with sensitivity to geographic considerations. The first is analytic in character while the second has a policy evolution perspective but both indicate how the application of a geographic perspective invigorates policy analysis.

GEOGRAPHY APPLIED TO REGIONAL INCOME CONVERGENCE: SPATIAL ANALYSIS CONSIDERATIONS

It seems obvious that geography should have a role in the analysis of economic convergence because economies are not really dimensionless points but occupy territory. Unfortunately, geography is often not considered as a factor in many if not any economic studies of convergence. This is reflected in the neglect of spatial scale effects and spatial dependency aspects of aggregation analysis. It has been demonstrated that the results of spatial data analysis are impacted by geographical scale (Gehlke and Biehl, 1934; Yule and Kendall, 1950) and the aggregation level chosen for analysis. This phenomenon is often referred to as MAUP (modifiable areal unit problem) in recent geography literature (Openshaw and Taylor, 1981). MAUP has not been widely recognized as a problem in economic convergence analysis although it is well known among geographers (Wong and Amrhein, 1996). The lack of awareness of the impact of MAUP on economic convergence analysis may lead to misunderstanding of spatial economic problems. This section discusses the impact of geography, i.e. geographical scale and aggregation, on the analysis of regional income convergence.

At the international level, it seems to be curiously acceptable to omit geography as a factor in across-country convergence analysis because countries are regarded as independent political units. Nevertheless, this argument is actually facing great challenge under the influence of increasing economic globalization and with the formation of regional free trade blocks such as NAFTA, and even super-state regions such as the European Union. National policy makers often have to meet the requirements of a super-state entity during certain policy making processes. Despite the existence of super-state entities, the nations within such super-state entities are still geographical entities of great sovereignty and can thus be accepted as a relatively reasonable units of convergence analysis. At the regional level, people and other factors of production flow across regions more freely than across countries; regions within a nation do not have such power as national sovereignty; and regions are not really independent from each other. Furthermore, a nation can be disaggregated into different levels of regions.

Which level of regions should be used as the spatial unit for analysis is not well defined and requires the judgment of researchers based on the context and research purpose of the study. The neglect of geography, both scale and aggregation, may cause serious problems such as conflicting results and implausible policy recommendations. As early as the mid-1970s, the role of geographical scale and aggregation in the study of regional income inequality was noted by Gilbert and Goodman (1976) who criticized the arbitrary use of regional divisions in the well-known study by Williamson (1965). Xie and Stough (2001) indicate that geographical scale is one of the factors that lead to a debate over the trend of regional income convergence in the reform period in China. Rey (2002) demonstrates that a change in geographical aggregation affects the measurement of interregional and intraregional inequality components in the case of United States and related inferences.

To illustrate the impact of geography on the analysis of regional income convergence, the cases of the European Union, China, and the United States are briefly reviewed. Regional income convergence in the European Union has been analyzed in many convergence studies. The time periods and regional units vary with different studies, resulting in different and even conflicting conclusions on the trend of convergence. Barro and Sala-i-Martin (1991) show evidence of regional income convergence across the regions of seven European countries over the period from 1950 to 1985. Button and Pentecost (1995) find no convincing evidence of regional income convergence over the entire 1975–1988 period, but especially over the latest 1981–1988 period they examine. Besides the differences in time periods and model specifications, geographical scales are very different in the two studies. In the study of Barro and Sala-i-Martin (1991), seven countries are portioned into 73 regions and regions in France and Italy are much smaller than in the UK and Germany. Due to the constraint of data availability, the partitioning of regions in Barro and Sala-i-Martin (1991) is not made on a comparable basis of area, population, density, economic units or income levels but as a simple assembling of regional datasets for each of the countries. The overrepresentation of French and Italian regions undoubtedly has an impact on the measurement of regional income convergence. In contrast, Button and Pentecost (1995) use 51 EU level 1 statistical regions of nine countries as their spatial units for analysis over 1975–1988 and 54 EU level 1 regions (including three Greek regions) over the 1980s. The equal status of regions makes the results of this study more convincing. Furthermore, there are some studies that employ EU level 2 regions (NUTS level 2)[1] as analysis units (Baumont, Ertur and Le Gallo, 2001; Le Gallo and Ertur, 2003). Among 138 regions (127 NUTS 2 and 11 UK NUTS 1 regions), there is only a very weak convergence of growth rates (Baumont,

---

1. NUTS is the abbreviation of "Nomenclature des Unités Territoriales Statistiques (Nomenclature of Territorial Units for Statistics)", which is a system for organizing regional statistics in European Union. See European Commission, "European Regional Statistics Reference Guide" at the EU's official website http://europa.eu.int/ for details.

et al, 2001). Using spatial analysis techniques, North and South spatial regimes are identified and no convergence is found within the North regime and only a weak one within the South. Hence, the mild general convergence is only a reflection of the weak convergence in the South. Regrettably, no data at NUTS 2 level for the UK were used in this study resulting in a mixed use of regions at different scales, which is similar to the Barro and Sala-i-Martin (1991) problem.

There have been a large number of studies on China's regional income convergence (inequality) in the reform period. After a review of over 20 empirical studies, Xie and Stough (2001) find that the choice of geographical scale has a great impact. The studies with different geographical scales often seem to provide conflicting results. This problem is further complicated by the use of multiple income indicators, variations in study periods, different inequality measures, and alternative data deflation procedures. When the studies are sorted out by income indicators and other factors, it is found that the results of many studies are not really conflicting but complementary because they provide information on regional income convergence at different levels of geographical scales. Take an example of the studies with GDP or national income as regional income indicators. Among the regional units for analysis are three alternative regional economic clusters, i.e. the East, the Central and the West; the Coastal and the Inland regions; and the 30 provinces. Reviewing the results of comparable studies with respect to other factors like study period and data deflation, Xie and Stough (2001) summarize the trends of regional income convergence in China's reform period as follows: (1) interregional inequality (across the three regions or the coastal and inland regions) has increased in the reform period; (2) interprovincial inequality decreased in the 1980s; (3) interprovincial inequality started to rise in the 1990s; (4) significant intra-provincial inequalities continue to exist.

The above findings suggest that intraprovincial inequalities should not be neglected while the focus of China's regional policy is switched to the Great West region. What China needs is not only a regional policy at the three economic region level but also regional policies at multiple levels of geographical scale.

In the United States, the spatial units used in regional income convergence include states, Census regions, eight regions defined by the Bureau of Economic Analysis (BEA), metropolitan statistical areas (MSA), and counties. States are usually the primary spatial units while there are some studies based on other spatial partitioning. Xie (2002) compares the trends of interstate and interregional income convergence and finds that there is great similarity between the trends at the two levels over the 1929–1998 period. But the level of income inequality at the state level is greater than at the BEA region level. This is not an unexpected result because the variance decreases with the increase in aggregation level. One question is what spatial unit is most suitable for the analysis of US regional income convergence. Fan and Casetti (1994) argue that the states are the most appropriate geographical units for income inequality analysis in the United States. From the view of policy analysis, their argument can be justified since the state is a political entity that has the power to develop important economic guidance. By contrast, the MSA is not endowed with political power and is thus less important in policy making process but it is a unit that is economically integrated, viable and consistent. To understand the sources of regional economic growth the MSA may be a better candidate. However an MSA is often a region that involves more than one state and is entangled in interjurisdictional fragmentation if not conflicts.

Rey (2002) demonstrates how a change of geographical aggregation may affect the magnitude and relative importance of interregional and intraregional inequality components using Theil's decomposable inequality index. Three regional divisions are used: four Census regions, nine Census divisions, and eight BEA regions. Not unexpectedly the interregional inequality component is smaller than the intraregional (interstate) inequality with the four Census regions but is greater with nine Census divisions and eight BEA regions. The aggregation by BEA region generates the highest interregional inequality component while the aggregation by four Census regions produces the lowest. With the county as the basic spatial unit, the ranking of the magnitude of interregional inequality component across geographical aggregation generally is states, BEA regions, nine Census divisions, and four Census regions with an exception that the interregional inequality components with eight BEA regions and nine Census divisions are almost the same after 1985.

In addition to MAUP, spatial effects (spatial autocorrelation and spatial heterogeneity) should also be taken into account in regional convergence analysis. Rey and Montouri (1999) find that regional income inequality and spatial autocorrelation are closely correlated. Baumont, et al (2001) identify the presence of spatial regimes in the EU regions. Using a neighborhood disparity index as a spatial inequality measure, Xie (2003) shows that the influence of spatial autocorrelation on regional income inequality can be greatly reduced. More advanced spatial analysis techniques are described for use in future studies (Fotheringham, Brunsdon, and Chartlton, 2000).

The above discussion suggests that geography is an important factor in regional income convergence analysis. Lack of knowledge on this will only lead to partial understanding at best and even misunderstanding of regional income convergence. To better inform policy makers, multiple scale and alternative spatial aggregation approaches are preferable in regional income convergence in the future. Furthermore, advanced spatial statistical approach will also be helpful in analyzing the impact of geography in regional income policy.

GEOGRAPHY APPLIED TO GOVERNANCE: Evolution of Regional Transportation Authorities

Below we look at the institutional framework of transportation management with a particular focus on localized regionalization issues in US metropolitan centers. This section begins with a discussion of the spatial evolution of metropolitan transportation planning institutions and initiation of regional operating organizations (ROOs), followed by a discussion of the increased need and opportunity to operate transportation facilities at a regional scale if they are to be sustainable. Then the emergence of new regional transportation organizations and the benefits these offer to regions and some of the institutional and political problems associated with the creation of these spatially focused organizations are reviewed.

Traditionally, US transportation institutions have been primarily at the local and state levels with national decision systems becoming important only as a part of national integration and defense strategies. This was true in both the railroad era (mid to late 1800's) and the interstate highway era of the mid to late twentieth century (1957–1987). Structurally most transportation investment responsibility in the United States has been at the state and local level with sharply increasing professionalization over time and with state and local political systems being the basis of institutional legitimacy. With the establishment of the Interstate Highway System and its application in urban areas, a special purpose organization was often established specifically to carry out studies in areas where many jurisdictions and departments were involved. In Creighton's (1970) words, these organizations were created to accommodate the "bewildering variety" of institutional arrangements necessary to accommodate the "prevailing balance of power between local organizations". States were important but the Interstate Highway System could not rely on them alone. The personalities and institutions of urban jurisdictions were politically powerful within states, and sometimes even nationally and therefore had to be significantly included in their own right in the process. More practically, most state highway departments lacked the requisite expertise in urban design, transit system and land use planning, and economic impact analysis to carry out urban transportation plans. The various jurisdictions and public and private operators in urban areas already had many of these capabilities. Therefore, regional study groups were brought together both to gain political legitimacy and to have the requisite skills to complete the task.

The 1962 highway act was pivotal in that it required "continuous, comprehensive, and cooperative" (3C) planning. All 204 of the nation's cities with population over 50,000 were required to develop and maintain a plan conforming to these specifications. Though this requirement was intended to induce regional efforts to develop transportation plans, FHWA's (Federal Highway Administration) regulations for the 3C process allowed states to work directly with localities and did not require or provide dedicated funding for regional planning organizations. In 1973, efforts to formalize metropolitan transportation planning received a further boost when Congress set aside 0.5% of all federal-aid highway funds for urban planning, apportioned to the states on the basis of urban populations, to be made available to authorities responsible for urban transportation planning. In 1974 Congress required transit and highway planning to adhere to the same long-range planning requirement. Joint highway and transit planning regulations followed the next year, along with a requirement formalizing the requirements for regional coordinating authorities (Weiner, 1997).

Since the late 1960s some form of regional cooperation of local government – particularly in urban areas – has been an element of the US transportation planning process. From the late 1960s, regional management and investment coordination concerns were reflected in the Federal Office of Management and the Budget (OMB), whose A95 Circular required the regional cooperation of local political entities. This has usually been through regional Councils of Governments (COGs) that were required to review and agree to local infrastructure requests in order to receive federal investment, usually as, matching funds. Although this requirement disappeared in the early 1980s, it was replaced by a stronger set of Federal requirements in the Intermodel Surface Transportation Efficiency Act (ISTEA) of 1991, which for metropolitan areas created Metropolitan Planning Organizations (MPOs). This required cooperative strategic planning specifically for transportation across metropolitan areas. These requirements are continued in the follow on Transportation Equity Act for the Twenty-first Century (TEA 21).

New interest in the regional operation of metropolitan transportation facilities and the provision of transportation services in metropolitan areas has arisen in the past decade for several reasons. The national highway system is highly developed and the Interstate Highway System is largely complete. Increasingly ubiquitous and well-organized community opposition and stricter environmental regulations limit new highway planning and construction even when funding might be available. Thus today transportation preservation and operation requires significantly more attention than planning and construction. They also take up a much greater proportion of the available funds. Increased interest in operations has been fueled not only because as roads are built, attention naturally shifts to operations and management along those roads, but also because the growing demand for these roads can no longer be addressed in a timely way by new highway capacity. This situation encourages transportation policy makers to look for increased efficiencies from the better operation and incremental improvement of the existing road network.

At the very time that road construction has slowed, the emergence of just-in-time production processes and complex supply chain management has meant that the reliability and consistency of transportation services are becoming increasingly important. Transportation, and the time a product remains in transit, is an important variable in most production today as firms try to shrink inventory costs. Though the importance of this variable is an additional burden for those responsible for highway operations, the very information and communications technologies associated with the transformation in industry have expanded the technological opportunities and made transportation operations more professionally and commercially interesting. As part of Intelligent Transportation Systems (ITS), these technologies greatly expand the capacity for data collection and monitoring and coordinated or centralized decision-making.

The question of how to implement operations and management systems has raised issues of coordination among jurisdictions. Transportation users would presumably prefer a seamless set of services that did not disconnect at jurisdictional boundaries. State and certainly nationwide implementation, however, is politically unlikely and in most cases technologically and fiscally impractical. Another issue is the role that federal money plays. Many existing regional operating organizations (ROOs) are funded with federal ITS money, and proposals are already being made to increase the share for operations in the upcoming 2003 transportation reauthorization. FHWA itself recently created an operations division in its organizational structure. With the practical benefits and federal involvement, attention has turned to a regional approach for fiscal operations (Haynes, Arieira, Burhans, and Pandit, 1995).

Although as mentioned above the role of MPOs in operation is projected to increase in the future, this does not mean that ROOs are not being created. Instead, in many places operations are being overseen by regional organizations other than MPOs, so-called "new ROOs" (Briggs, 1999). MPOs are typically participants, but not the lead agency, in these new organizations. Here, the focus is on the characteristics of these new ROOs and some of the benefits they might provide over MPOs. In simple terms, the impetus to create a new regional organization appears to come from a specific event, either the need to respond to a crisis or to meet the deadlines and requirements for federal funding. In the case of Transcom in the New York City area (New York, New Jersey, and Connecticut), it was the coincidence of a number of disruptive construction projects that led regional jurisdictions to seek coordination of information on construction, incidents, and congestion. In short, there has to be a need or a problem that is interjurisdictional in scope and there have to be either locally available funds or an opportunity to get outside (federal) funds. Most important, according to Briggs (1999), there has to be leadership in the form of an agency, a politician, or some kind of private advocate that brings the regional actors and funding together to address the problem.

Therefore, the mission of these next layers of spatially focused organizations is initially dictated by a well-defined problem and a set of technologies that are to be implemented to address the problem. Most often the problem and the solution are found in a well-defined area, such as a single corridor or a small number of interconnected highways. What is important here is that although the missions of these organizations are much more narrowly defined (both functionally and geographically) than is the case for most MPOs or regional governments, they are much more flexible than that of a traditional regional infrastructure authority, in which the mission is legally constrained (Gifford and Stalebrink, 2002).

Given the different natures of their mission, the new regional organizations are structured in a different way than are either MPOs or other regional bodies. These organizations tend to be either "virtual organizations" or in some cases private corporations rather than governmental institutions. A virtual organization is not a legal entity.

Instead, it is an alliance of independent agencies. It is the agencies themselves that provide staff, project oversight, procurement, and other needs. Alternatively, a private corporation can be formed to give the organization legal status. Forming a private corporation binds the partners together and entails an enforceable commitment from each. There is another form for new ROOs, the public-private partnership (Lawther, 2001), which is differentiated from the private corporation in that there is private capital participation and a formal partnership with firms in the private sector. In most cases a private organization collects data and information from public entities in the region and provides transportation users with the information and analysis on a regionwide basis.

As to the financing of new ROOs, typically states with Federal government support usually as matching funds are the source on the assumption that spill-over benefits are positive and that investments will contribute to the overall functioning of the network and general levels of economic development (Haynes and Button, 2001). Boarnet (1998) shows that in urban regions in particular, where-highway expansion is just as likely to generate negative spillovers, it is important to be guided by the policy that local benefits should be supported by local funds, and region-wide consideration is needed to manage external costs from those investments.

The new transportation organizations are quite different in their structure and mission from traditional regional authorities. What are the benefits of these new ROOs? A number of the benefits come from making decisions and implementing programs on a regionwide basis. This regionality allows for economies of scale in purchasing and the pooling of resources and expertise, which reduces redundancy, avoids conflicts, and allows the system, which was designed regionally, to be operated more effectively. There are also likely to be cost savings from specialization and streamlined processes. These, however, are all benefits from regionalization in general, and they do not help to clarify the emergence of new regional transportation organizations.

There are other reasons that creating new organizations may be preferable to using existing ones. First, the focused mission of the new organizations allows them to provide a clear service to customers. The agency knows what it does and whom it serves, and the customer is provided with "one-stop shopping" for a well-defined set of services. Thus the new organizations may be more focused and flexible than their public counterparts. Second, there may be political reasons as well for the creation of new organizations independent from planning authorities with a different organization structure. COGs and MPOs have been in existence in some cases for as long as 50 years. Their external relationships with state departments of transportation and local agencies and governments have long been formed and defined. Internally they may have developed bureaucratic norms and subsequently a significant amount of organizational inertia. When COGs/MPOs are presided over by elected officials, they may be considered too political to be trusted with daily operations tasks by many of the operating agencies. There is also often the perception that over the years COGs have become beholden to or infiltrated by political interests that might impede what the operating agencies perceive as operational effectiveness. Besides, these new organizations have the additional benefit that they were not mandated by the federal government. These organizations may be seen as more convenient to start by state and local operating agencies since they avoid the known political difficulties associated with COGs, and at the same time state and local officials may generally feel more ownership of the organization thus contributing to their sustainability.

Clearly the political and institutional benefits are considerably more controversial than the technical and economic arguments. Applying a new organization to the problem of regional transportation management and operation may also create some problems and challenges. The primary problem to be foreseen with the emerging regional transportation environment is that the very problems that both MPOs and new regional transportation organizations were created to address may emerge in a different guise. There is the problem that the geographical jurisdiction of the new organizations does not correspond to that of the MPO. As operations continue to gain in importance for regional transportation, this element of noncorrespondence can be addressed in three ways. (1) "New" new organizations can be created to fill other niches; (2) The MPO can take on other operations tasks as they move to the regional level, or (3) The new organization can grow to take on other operations tasks.

Such an evolution of new forms of spatially organized structures for governance and real world problem solving has interesting applications for political geographers with a sensitivity to policy analysis.

SUMMARY

This paper discussed the importance of considering the geographic context in the policy-making process and highlights the necessity of incorporating policy analysis in the studies of political geography. We began with a brief review of literature on political geography and a categorization of scale issues in political geography studies. This was contrasted with views of the policy process typically used by policy analysts. The result was a shift from theory to operational considerations often with a loss of the important spatial dimension. Two simple applied cases studies are reviewed on how geographic factors affect public policy analysis. The two case studies are regional income convergence, and the rise of regional transportation management institutions. The first is analytic in character while the second has a policy evolution perspective.

Discussion of the role of geography in regional income convergence studies suggests that geography is an important factor in regional income convergence analysis. Lack of knowledge on this will only lead to partial understanding at best and even misunderstanding of regional income convergence. To better inform policy makers, multiple scale and alternative spatial aggregation approaches are preferable in regional income convergence in the future. Furthermore, advanced spatial statistical approach will also be helpful in analyzing the impact of geography in regional income policy.

In the context of the US metropolitan transportation, the interests in the regional operation of metropolitan transportation facilities and the provision of transportation services in metropolitan areas have raised issues of coordination among jurisdictions to facilitate, oversee and monitor the implementation of new technologies in transportation area. The sustainability of new ROOs depends on how they navigate the transition from being new organizations whose focus and jurisdiction do not adequately correspond to regionwide operations issues to being full-fledged regional organizations in an environment populated by existing MPOs and other regional authorities. The particular challenge is how these organizations will be able to grow to meet region-specific needs and challenges while providing sufficient accountability and consistency to facilitate federal participation. The evolution of the US transportation institutions suggests that universal and far-reaching federal mandates may not be the best strategy to achieve these goals and may undermine long-term institutional sustainability.

Applying geography to the analysis of policy inputs such as assessment of regional convergence and policy outputs such as the creation of new entities of regional transportation governance is critical to appreciating the usefulness of geography in solving real world problems. Considering geographic factors in policy analysis is essential to good public policy.

KINGSLEY E. HAYNES*, QINGSHU XIE AND LEI DING
School of Public Policy
George Mason University
MS 3C6, 4400 University Drive
Fairfax, VA 22030-4444, USA
*Email: khaynes@gmu.edu

REFERENCES

Agnew, J. (1995). The rhetoric of regionalism: the Northern League in Italian politics, 1983–1994. *Transactions of Institute of British Geographers*, 20: 156–172.

Agnew, J. (2000). From the political economy of regions to regional political economy. *Progress in Human Geography* 24, 1: 101–110.

Archer, J. C., Lavin, S. J., Martis, K. C., and Shelley, F. M. (2001). Atlas of American Politics, 1960–2000. Washington, D. C.: CQ Press.

Barro, R. and Sala-i-Martin, X. (1991). Convergence across States and Regions. *Brookings Papers on Economic Activity* (1): 107–182.

Baumont, C., Ertur, C., and Le Gallo, J. (2001). The European convergence process, 1980–1995: do spatial regimes and spatial dependence matter? A paper presented at *The 48th North*

American Meetings of the Regional Science Association International, Charleston, South Carolina, November: 15–17.

Boarnet, M. G. (1998). Spillovers and the Location Effects of Public Infrastructure, *Journal of Regional Science* 36(3): 381–400.

Briggs, V. (1999). New regional transportation organizations. *ITS Quarterly*, 7 (3): 35–46.

Buck, N. (1990). Review of Agnew, J. A. and Duncan, J. S., editors, The power of place and Wolch, J. and Dear, M., editors, The power of geography. *Sociology* 24: 555–556.

Button, K. J. and Pentecost, E. J. (1995). Testing for convergence of the EU regional economies. *Economic Inquiry*, 33 (4): 664–671.

Castells, M. (1996). The rise of the network society. Oxford: Blackwell.

Chase-Dunn, C. (1989). Global formation: structures of the world-economy. Oxford: Basil Black-well.

Cheshire, P. C. and Gordon, I. R. editors. (1995). Territorial competition in an integrating Europe. Aldershot: Avebury.

Cheshire, P. C. and Gordon, I. R. editors. (1998). Territorial competition: some lessons for policy. *Annals of Regional Science* 32: 321–346.

Chouinard, V. (1990). State Formation and the Politics of Place: the Case of Community Legal Aid Clinics. *Political Geography Quarterly*, vol. 9 (1): 23–38.

Chouinard, V. and Fincher, R. (1987). State Formation in Capitalism: A Conjunctural Approach to Analysis. Antipode, Dec.: 329–353.

Cox, K., ed. (1997). Space of Globalization: Reasserting the Power of the Local. New York: Guilford.

Cox, K. and Mair, A. (1988). Locality and community in the politics of local economic develop-ment. *Annals of Association of American Geographers* 78: 307–325.

Creighton, R. L. (1970). Urban Transportation Planning. Urbana: University of Illinois Press.

Daniels, P. W. and Lever, W. F. ed. (1996). The Global Economy in Transition. Harlow: Longman.

Davidson, F. M. (1996). The Fall and Rise of the SNP since 1983: Analysis of a Regional Party. *Scottish Geographical Magazine*, 112: 11–19.

De Vos, S. and Deurloo, R. (1999). Right extremist votes and the presence of foreigners: an anal-ysis of the 1994 elections in Amsterdam. *Tijdschrift voor economissche en sociale geografie* 90: 129–141.

Dicken, P. and Tickell, A. (1992). Competitors or collaborators? The structure of inward invest-ment promotion in northern England. *Regional Studies* 26: 99–106.

Driver, F. (1991). Political geography and state formation: disputed territory. *Progress in Human Geography* 15, 3: 268–280.

Eagles, M. ed. (1995). Spatial and Contextual Models in Political Research. London: Taylor and Francis.

Fan, C. C. and Casetti, E. (1994). The spatial and temporal dynamics of US regional income inequality, 1950–1989. *Annals of Regional Science* 28, 177–196.

Flint, C. (2002). Political geography: globalization, metapolitical geographies and everyday life. *Progress in Human Geography* 26, 3: 391–400.

Florida, R. and Lee, S. Y. (2001). Innovation, Human Capital, and Diversity. Presented at the APPAM 2001 Research Conference, November 1, 2001, Washington DC. Available at: http://www.heinz.cmu.edu/~florida//pages/pub/working_papers/APPAM_paper_final.pdf

Fotheringham, A. S., Brunsdon, C. and Chartlton, M. (2000). Quantitative Geography: Perspectives on Spatial Data Analysis. London: Sage Publications.

Fukuyama, F. (1992). The end of history and the last man. New York: Free Press; c1992.

Gallup, J. L. and Sachs, J. D. (1998). Geography and economic development. Cambridge, MA: Harvard University Institute for International Development.

Gehlke, C. E. and Biehl, K. (1934). Certain effects of grouping upon the size of the correlation coefficient in census tract material. *Journal of the American Statistical Association* Supplement, 29 (185), 169–170.

Gifford, J. L. and Stalebrink, O. J. (2002). Remaking transportation organizations for the 21st century: consortia and the value of learning organizations. *Transportation Research.* 36A 7: 645–657.

Gilbert, A. and Goodman, D. E. (1976). Regional income disparities and economic development: a critique, in A. Gilbert ed.: *Development, Planning and Spatial Structure*. New York: John Wiley and Sons. 113–141.

Gilbert, J. and Howe, C. (1991). Beyond 'State *vs.* Society': Theories of the State and New Deal Agricultural Policies. *American Sociological Review* 56 (April): 204–220.

Grofman, B., Handley, L. and Niemi, R. (1992). Minority Representation and the Quest for Voting Equality. New York: Cambridge University Press.

Harvey, D. (1985). The urbanization of capital. Baltimore: Johns Hopkins University Press.

Harvey, D. (2000). Space of hope. Berkeley, CA: University of California Press.

Haynes, K. E., Avieira, C., Burhans, S., and Pandit, N. (1995). Fundamentals of infrastructure financing with respect to ITS. *Built Environment* 21(4): 246–254.

Haynes, K. E. and Button, K. (2001). Transportation systems and economic development. *Handbook in Transport Systems and Traffic Control* (ed. K. Button and D. Hansher), Oxford, UK: Pergamon: 174–192.

Haynes, K. E., Maas, G. C., Riggle, J. D., and Stough, R. R. (1997). Regional Governance and Economic Development: Lessons from Federal States in Danson, Mike, Stephan Hill and Gregg Lloyd (Eds). Regional Governance and Economic Development (European Research in Regional Science Bd. 7), London, UK: Pion Ltd.

Johansson, B., Karlsson, C., and Stough, R. ed. (2001). Theories of Endogenous Regional Growth: Lessons for Regional Policies. Springer-Verlag, Berlin.

Johnston, R. J. (1984). Marxist political economy, the state and political geography. *Progress in Human Geography* 8: 473–492.

Johnston, R. J., Dorling, D., Tunstall, H., Rossiter, D., MacAllister, I., and Pattie, C. (2000). Locating the altruistic voter: context, egocentric voting, and support for the Conservative Party at the 1997 General Election in England and Wales. *Environment and Planning A* 32: 673–694.

Johnston, R. J. and Forrest, J. (2000). Constituency election campaigning under the alternative vote: the New South Wales Legislative Assembly election, 1995. *Area* 32: 107–117.

Jones, B. (1997). Wales: a developing political economy. In Keating, M. and Loughlin, J., editors, The political economy of regionalism, London: Frank Cass, 388–405.

Keating, M. (1997). The invention of regions: political restructuring and territorial government in western Europe. *Environment and Planning C: Government and Policy* 15: 383–398.

Keating, M. and Loughlin, J. ed. (1997). The political economy of regionalism. London: Frank Cass.

Koopmans, T. (1965). On the concept of optimal economic growth. In Koopmans, T., The econometric approach to development planning, Amsterdam: North Holland, 5–18.

Lawther, W. (2001). Effective Public-private Partnership Models in the Deployment of Metropolitan ITS. Presented at the 11th Annual Meeting of ITS America. FL: Miami Beach.

Le gallo, J. and Ertur, C. (2003). Exploratory spatial data analysis of the distribution of regional per capita GDP in Europe, 1980–1995. *Papers reg. Sci.* 82: 175–201.

Leyshon, A. and Thrift, N. (1997). Money/Space: Geographies of Monetary Transformation. London: Routledge.

Lucas, R. E. Jr. (1988). On the mechanics of economic development. *Journal of Monetary Economics* 22 (1): 3–42.

Lyotard, J. F. (1984). The Postmodern Condition: A Report on Knowledge. Minneapolis: University of Minnesota.

Mann, M. (1984). The autonomous power of the state: its origins, mechanisms and results. *Archives Europeenes de Sociologie* 25: 185–213.

Martin, R. (2001). Geography and public policy: the case of the missing agenda. *Progress in Human Geography* 25 (2): 189–210.

O'Brien, R. (1992). Global financial integration: the end of geography. London: Printer.

O'Loughlin, J. and Bell, J. E. (1999). The political geography of civic engagement in Ukraine. *Post-Soviet Geography and Economics* 40: 233–266.

Openshaw, S. and Taylor, P. J. (1981). The modifiable areal unit problem. In N. Wrigley and R. J. Bennett (eds.): *Quantitative Geography, A British View*, 60–70. London: Routledge and Kegan Paul.

Paasi, A. (1991). Deconstructing regions: notes on the scales of spatial life. *Environmental and Planning A*. 23: 239–256.

Painter, J. (1995). Politics, Geography and 'Political Geography'. London: Edward Arnold.

Pantulu, J. and Poon, J. P. H. (2003). Foreign direct investment and international trade: evidence from the US and Japan. *Economic Geography* 3(3): 241–259.

Pattie, C. and Johnston, R. (2000). People who talk together vote together: an exploration of contextual effects in Great Britain. *Annals of the Association of American Geographers* 90: 41–66.

Piattoni, S. (1997). Local political classes and economic development. The cases of Abruzzo and Puglia in the 1970s and 1980s. In Keating, M. and Loughlin, J., editors, The Political Economy of Regionalism, London: Frank Cass, 306–346.

Putnam, R. D. (1993). Making democracy work: civic traditions in modern Italy. Princeton, NJ: Princeton University Press.

Rey, S. J. (2002). Spatial analysis of regional income inequality, presented at the 41st Annual Meeting of Western Regional Science Association at Monterey, California in February 17–20, 2002.

Rey, S. J. and Montouri, B. D. (1999). US Regional Income Convergence: A Spatial Econometric Perspective. *Regional Studies*. 33: 145–156.

Reynolds, D. R. (1993). Political geography: closer encounters with the state, contemporary political economy, and social theory. *Progress in Human Geography* 17, 3: 389–403.

Reynolds, D. R. (1994). Political geography: the power of place and the spatiality of politics. *Progress in Human Geography* 18, 2: 234–247.

Romer, P. (1986). Increasing returns and long-run growth. *Journal of Political Economy,* 94(5): 1002–1037.

Sarup, M. (1993). An introductory guide to post-structuralism and postmodernism. Atlanta: University of Georgia Press.

Shelley, F. M., and Archer, J. C. (1994). Some Geographical Aspects of the 1992 American Presidential Election, *Political Geography*, 13: 137–159.

Shelley, F. M., Archer, J. C., Davidson, F. M., and Brunn, S. D. (1996). The Political Geography of the United States. New York: Guilford.

Smith, N. (2000). Author's response. *Progress in Human Geography* 24: 271–274.

Sollow, R. (1956). A contribution to the theory of economic growth. *Quarterly Journal of Economics* 70: 65–94.

Storper, M. (1997). The regional world: territotial development in a global economy. New York: Guiford Press.

Stough R. (2000). The Greater Washington Region: A Global Gateway Region. In *Gateway to the Global Economy*, edited by Andersson, A. E. and Andersson, D. E. Northampton, MA: Edward Elgar Pub.

Toal, G. and Shelley, F. M. (2002). Political geography: from the "Long 1989" to the end of the post-cold war peace. In *Geography in America*, edited by Gale, G. and Willmott, J. Columbus, Ohio: Merrill.

Webster, G. R. (2000). Women, politics, elections, and citizenship. *Journal of Geography* 99: 1–10.

Weiner, E. (1997). Urban Transportation Planning in the United States: An History Overview, 5th ed. US Department of Transportation. Available at: www.bts.gov/tmip/papers/history/utp/toc.htm.

Williamson, J. G. (1965). Regional Inequality and the Process of National Development: A Description of the Pattern, *Economic Development and Cultural Change* 13 (4) July: 3–81.

Wins, P. (1995). The location of firms: an analysis of choice process. In Cheshire, P. C. and Gordon, I. R., editors, Territorial Competition in an Integrating Europe, Aldershort: Avebury, 244–266.

Wong, D., and Amrhein, C. (editors), (1996). The Modifiable Areal Unit Problem. Special issue of *Geographical Systems* 3: 2–3.

Wright, V. (1998). Intergovernmental relations and regional government in Europe: a skeptical view. In Le Gales, P. and Lequesne, C., editors, Regions in Europe, London: Routledge, 39–49.

Xie, Q. (2002). Convergence or Divergence? A Revisit to the Debate on U.S. Regional Divergence in the 1980s, An invited paper presented at the 41st Annual Meeting of Western Regional Science Association at Monterey, California, February 17–20, 2002.

Xie, Q. (2003). Regional Income Convergence in the United States: 1970–2000. Ph.D. Dissertation. George Mason University.

Xie, Q. and Stough, R. (2001). China's regional income inequality in the reform period. *Policy and Management Review*, 1 (1): 116–163.

Yule, G. U. and Kendall, M. G. (1950). *An Introduction to the Theory of Statistics*. New York: Hafner.

ARTHUR GETIS

CHAPTER 5
THE ROLE OF GEOGRAPHIC INFORMATION SCIENCE
IN APPLIED GEOGRAPHY

ABSTRACT

Applied geography has undergone remarkable changes in the last 20 years. Powerful new technologies have emerged that greatly improve the ability to collect, store, manage, view, analyze, and utilize information regarding the critical issues of our time. These technologies include geographic information systems (GIS), global positioning systems (GPS), satellite-base remote sensing, and a great variety of remarkable software that allows for the analysis of the compelling problems. The issues include globalization, global warming, pollution, security, crime, public health, transportation, energy supplies, and population growth. Geographic Information Science (GIScience) has given rise to an essentially multidisciplinary approach to applied problems. No single person is expert in all of these areas. It is necessary to emphasize coordination and collaboration and to find the bridges that reduce the barriers between disciplines. In this chapter we briefly discuss the new technologies and the way in which they are being used to solve the critical issues. We then make suggestions for an applied geography future vis-à-vis the geographic information sciences.

INTRODUCTION

In the early 1980s, the world of applied geography began to undergo a radical change. The desktop workstation made information storage, retrieval, manipulation, and processing faster and offered greater capacity than geographic problem solvers had ever had before. By the mid-1980s, geographic information systems (GIS) software added the important mapping dimension to data display and data organization. One could use software such as SURFER in the late 1960s and 70s, but the great power and ease of the personal computer in the 1990s changed the ground rules. One no longer had to gather a huge packet of Hollerith cards punched on an IBM 029 and troop across the campus to the computer center for processing. Now, in the mid-2000s, not only is it routine activity to use GIS software, but the nature of GIS itself is changing dramatically. If productivity is any measure of technological gain, it would be fair to say that spatial data analysis is perhaps 10,000 times more powerful in 2003 than in 1983. Moore's Law holds here, that is, about every year or so the speed of computer processors doubles. Data storage for inexpensive personal computers has gone from 64 kilobytes in the

A. Bailly and L. J. Gibson (eds.), Applied Geography, 95–111.

mid-80s to 60 megabytes in the mid-90s to 60 gigabytes today. Soon, terabytes will be the common way to measure computer capacity.

Other chapters in this volume cover traditional areas of concern of applied geography. These subjects have changes only modestly during this period of great technological change. Interest continues in commercial and industrial location, transportation, crime, public facility location, disease distributions, and environmental concerns. But instead of processing data and drawing maps by hand, now we depend nearly entirely on computers to help us in our research endeavors. The traditional research process – problem to hypothesis to model to data gathering to test to decision – has been altered. Because of the power and flexibility of computers, the process has changed appreciably to include: problems, large scale data gathering, data manipulation including data visualization, model building, hypotheses generation, simulation, model validation and testing. The prescribed order of research is much more idiosyncratic now. While much can be accomplished with the new technology, more data are needed in order to satisfy the appetite of the more demanding research process.

National and international geographic information science

In the mid-1990s, a new organization was established in the United States, the University Consortium for Geographic Information Sciences (UCGIS). This was the first formal recognition that GIS is but a subset of the broader field of *GIScience*. Similar organizations have been established in Europe (AGILE) and Japan (GIS Association of Japan). Originally, most GIS practitioners were geographers, but with the burgeoning of the field and the need for more and more sophisticated technology, engineers, computer scientists, and practitioners in fields that have recognized the value of GIS have attached their interests to the GIScience bandwagon. These related interests cover such areas as spatial data acquisition and integration, distributed computing, cognition of geographic information, interoperability of geographic information among computer platforms, and remote sensing. As a recognition of this broadening field, the term GIScience is now being used to express many of these interests, not to the exclusion of GIS but in conjunction with GIS. This has stimulated a multidisciplinary approach to the solution of applied geographic problems. In fact, it is becoming more and more difficult for individual researchers to engage in a research project without some sort of interdisciplinary collaboration. For example, the solution of a facility location problem might require the cooperation of a modeler, GIS specialists in data base management and graphics design, a remote sensing expert, and a spatial statistician.

Several National Science Foundation supported centers have been developed over the last several years. The purpose of these institutes is to inform interested researchers, teachers, and practitioners of the work being done in the spatial sciences and to instruct those new to the field in the techniques of spatial science. These organizations include the Center for Spatially Integrated Social Science (CSISS) and Spatial Perspectives on Analysis for Curriculum Enhancement (SPACE). In this chapter, we outline what GIScience means, how it is used in applied geography, and give examples of its use for a variety of problems often addressed in applied geography.

THE CONTENT OF GISCIENCE

In the last several years the UCGIS has begun to create model curricula that when completed will represent the subject matter of GIScience. The model curricula are designed to satisfy the needs of educators who want to prepare students for the use of geographic information and technology in their post-graduate work. The purpose of having curricula rather than a single curriculum is to afford students from a wide variety of disciplines and emphases the opportunity to follow relevant paths toward a useful knowledge of GIScience. For those working in applied geography, nearly all of the subject areas covered in the model curricula will be of interest. One can convincingly argue that a working knowledge of GIScience is mandatory in order to do research in applied geography. The major elements of the model curricula (as of mid-2003) are called knowledge areas (see www.UCGIS.org). These represent the content of GIScience. The following brief sketches define the various aspects of the knowledge areas.

Conceptualization of space. The nature of space and time is the context for earth-related phenomena. This includes different notions of space and time for differing applications and different disciplines. It includes the means to understand scale, pattern, location, and region, and forms the basis for dealing with the entirety of GIScience.

Formalizing spatial conceptions. Here we cast our conceptual view of space and time into a specific, logical organizational structure from a given application perspective. Formalization incorporates a set of specifications to measure, reference, and locate spatial and spatial-temporal conceptualizations. Examples include modes of measurement, coordinate systems, map projections, spatial relationships, topology, object and object-type categories, diffusion, and network flows.

Spatial data models and data structures. This is the representation of formalized spatial reality through data models, and the translation of these data models into data structures that are capable of being implemented within a computational environment. Examples of spatial data model types are discrete (object-based), continuous (location-based), dynamic, and probabilistic. Data structures represent the operational implementation of data models within a computational environment.

Design aspects of GIScience and technology. *Analytic model design* incorporates methods for developing effective mathematical models of spatial and spatial-temporal situations and processes. *System design* addresses the manner in which existing GIScience concepts and technology are matched with the requirements to create solutions to operational spatial problems. *Spatial Database Design* concerns the optimal organization of spatial data.

Spatial data acquisition, sources, and standards. Data acquisition is required for the development of fundamental data layers within a GIS. These data may have spatial, temporal, and attribute (descriptive) components. Examples of primary data sources include surveying, remote sensing (air photography, satellite imaging), the global positioning system (GPS), work logs (e.g., police accident reports), and surveys. Secondary spatial or spatial–temporal data can be acquired from digitized and scanned analog maps as well as from sources such as governmental agencies. Data standards for spatial data, images, and metadata exist to document data quality, lineage, and appropriate use.

Spatial data manipulation. This includes the ways in which one transforms spatial data into formats that facilitate subsequent analysis. Examples of data manipulation include vector-to-raster conversion, line generalization, attribute aggregation, projection transformation, and transaction management.

Exploratory spatial data analysis. Such analysis includes operations whose objectives are to derive summary descriptions of data, evoke insights about characteristics of data, contribute to the development of research hypotheses, and lead to the derivation of analytical results. This is also called *data driven* analysis.

Confirmatory spatial data analysis. This includes the techniques used to create and test spatial and spatial-temporal process models. It may also be called *model-driven analysis.* This area is tied directly to specialized problems studied in the social, behavioral, and physical sciences. For example, an applied environmental geographer would want to learn the confirmatory analytic procedures that are particularly well suited for modeling a spatial environmental process such as air or water pollution.

Computational geography. This is the application of computationally intensive approaches to the study of the geosciences. The focus of geocomputation is a variety of methods designed to model and analyze a range of highly complex, often non—deterministic, non—linear problems. These include such areas as neurocomputing, fuzzy sets, and genetic algorithms.

Cartography and visualization. This is the creation and effective interpretation of graphic representations of spatial data sets and of the results of spatial analysis activities.

Professional, social, legal aspects of GIScience and technology. This covers the nature and extent of (a) property rights in the spatial information itself, (b) the use of spatial information for land management and other decisions by both public and private actors, and (c) the distribution of the spatial information. Administrative practices are concerned with the nature and extent of how organizations and agencies are organized and made into effective operations that manage the information and its application.

METHODS OF SPATIAL ANALYSIS FROM A GISCIENCE PERSPECTIVE

Because of technological advances, researchers in the social and physical sciences seeking empirical verification of their models are faced with larger data sets made up of more detailed spatial information than has been the case in the recent past. Among the types of expanding data sets are: detailed consumer surveys (household data), the point location of crime and other social variables, detailed land cover information (remotely sensed data), environmental indicators (air and water quality data), the geo-referenced selling price of houses (real estate data), traffic flow (origination and destination data), consumer demand (individual choice data), and migration (individual household movement data). As the new technology evolves, larger and larger data sets will contain information on smaller and smaller spatial units. Small unit data, such as census block data, are currently available in easy to use graphic form. Data pinpointing 911 calls are an example of available detailed spatial information. Of course, in due time detailed historical data of this nature will be available for study. Currently, *The National Historical Geographic Information System* project is preparing for analysis socioeconomic data from United States censuses dating back to 1790. Maps are being prepared that will allow one to follow changes in geographic areas over time.

Coming to grips with the new technology is not a straightforward undertaking for applied geographers who are used to modest sample sizes, aggregate data, and models that require assumptions about data. Nonetheless, the new technology should be welcome for a number of reasons.

- The need to test theory at the scale that the theory requires tends to favor larger scales of analysis over smaller scales. Most social, economic, and environmental theory is based on individual rather than group behavior.
- Applied geographers usually opt for larger rather than smaller data sets, not only so that they may obtain more reliable estimates of model parameters, but to reserve data for sensitivity analyses.
- Increasingly, applied geographers favor analytical schemes in which the strict rules of statistical theory can be relaxed. Fast computers capable of handling large sets of data are ideal for devising simulation schemes that are less dependent on the complications of statistical rigor.
- Exploratory data analysis can become a highly productive prelude to confirmatory data analysis in applied geography.

Data preparation and GIScience

Of great concern to applied geographers is the preparation of data for use in model building and eventually in the confirmatory analysis that model building implies. Small spatial unit data suffer from a variety of inherent problems that make them generally unfit for confirmatory analysis. These problems revolve around the assumptions made about the distribution of data that are to be used to test models. The main issues are:

- Finding the appropriate scale of analysis. A GIScience approach, if the data allow, can be used for the study of a problem in multiple scales.
- Recognizing the degree of stationarity of spatially distributed variables. A GIScience approach allows for the kernel estimation of the density of phenomena over the study surface.
- Identifying the degree and characteristics of any spatial association in the data. Spatial statistics now embedded in GIS software provides measures of the degree of spatial autocorrelation in data.

Although exploratory spatial data analysis (ESDA) techniques have been in existence for a long time, GIScience technology allows for the manipulation of large amounts of data quickly. Models can be developed by means of identifying particularly constructive aspects discovered in an ESDA. In the process of ESDA, models become clearer, data are prepared for analysis, and problem solving can be carried out effectively. This sequence does not obviate the need for looping backwards to ESDA to stimulate further thought on the nature of the proposed model.

Data analysis

The types of statistical methods used in applied geography are a function of both the nature of the problems to be solved and the availability of the new faster technologies. We list below four general areas that are particularly useful for applied geography. Each is described in terms of the kinds of problems being solved, their general formulation, and their usefulness within the GIS community of analysts. Many new techniques have not yet been tied to hypothesis guided inquiry. Such areas of inquiry as spatial neural nets, spatial fuzzy sets, and simulated annealing are just now being developed; they are not discussed here.

Pattern analysis
Popular in the 1960s was point pattern analysis based on the spatially homogeneous Poisson process. It was common to find a researcher working at a light table making measures from pencil-numbered points to the first nearest neighbor of each point. Now, with the use of digitized georeferenced data, we are easily able to take measurements from all points to all other points. In addition, measurements of line segments, distances between line intersections, areas, and characteristics of areas such as perimeter length, neighboring areas, and so on, are basic measuring rods within many GIS.

Pattern analysis in the spatial sciences grew out of a hypothesis testing tradition, not from the extensive pattern recognition literature. Nearest neighbor work continues today, but the work of Clark and Evans (1954) has now been modified for the sake of unbiasedness to take into account the length of the perimeter of study areas and the distance to study area boundaries (refined nearest neighbor analysis) (Boots and Getis 1988). Today, nearest neighbor measurements are a fundamental function of many GIS.

Perhaps the most important developments in recent years are the application of K-function analysis to the study of point patterns and the use of Voronoi polygons for the study of spatial tessellations (Okabe et al. 1992). In addition, fractals study is a promising area for pattern analysis (Batty and Longley 1994). A brief outline of the idea behind the use of K-function analysis and some of the successful applications to spatial phenomena follow.

The K-function is the ratio of the sum of all pairs of points within a pre-specified distance d of all points to the sum of all pairs of points regardless of distance (Ripley 1981). The function is adjusted to identify distances that are closer to the boundary of the study area than to d. The K-function takes into account the need to stabilize variance, and Getis generalized the formula to include the weighting of points, such that the sum of pairs of points became the sum of the multiples of the weights associated with each member of a pair of points (1984). Diggle (1983) has successfully exploited this formulation to show many new features of patterns. For example, not only can one easily show the difference between an existing pattern and a random pattern, but also one can develop theoretical expectations for other than random patterns. In addition, patterns divided into different point types (marked patterns) can be studied easily. For testing purposes, an envelope of possible outcomes under the hypothesis of say, randomness, is usually constructed by means of a Monte Carlo simulation.

Although studies of the spatial distribution of vegetation dominate the empirical literature of K-function analysis (Diggle 1983), the method has been used for the study of population distribution and disease distribution. Bailey and Gatrell (1995) showed that the K-function can be used as an indicator of time-space clustering. That is, one simultaneously finds pairs of points separated by designated units of time and distances in space. This approach is particularly useful for identifying disease clustering over time.

Spatial association
Finding the degree of spatial association (autocorrelation) among data representing related locations is fundamental to the statistical analysis of dependence and heterogeneity in spatial patterns. A fundamental feature of most GIS packages is the Moran and Geary measures of spatial autocorrelation. Moran's statistic, very much like Pearson's product moment correlation coefficient, is based on the covariance among designated associated locations, while Geary's takes into account numerical differences between associated locations. The tests are particularly useful on the mapped residuals of an ordinary least

square regression analysis. Statistically significant spatial autocorrelation implies that the regression model is not properly specified and that one or more new variables should be entered into the regression model. Research in this area has emphasized the distribution characteristics of the statistics under varying spatial resolutions and spatial weights matrices (Anselin and Rey).

Mantel (1967) and Hubert (1979) have shown that statistics of this nature are special cases of a general formulation, gamma, that is defined by a matrix representing possible location associations (the spatial weights matrix) among all points multiplied by a matrix representing some specified non-spatial association among the points. The non-spatial association may be an economic, social, or other relationship. When the elements of these matrices are similar, high positive autocorrelation obtains. Gamma describes spatial association based on co-variances (Moran's statistic, I), or subtraction (Geary's statistic, c), or addition (Getis and Ord's statistic, G). These statistics are global insofar as all measurements between locations are taken into account simultaneously.

When the spatial weights matrix is a column vector, gamma becomes local; that is, association is sought between a single point and all other points (I_i, c_i, G_i). Research on local statistics has been especially active recently since these types of statistics lend themselves to kernel-type analyses in a GIS where data sets are large (Anselin 1994, Getis and Ord 1992, Ord and Getis 1995). Local statistics have been used to classify remotely sensed data, and show associations between neighborhood crime rates and the conflict propensities of countries.

Geostatistics
The variogram, discussed earlier in this book, plays a useful role as the function that describes spatial dependence for a regional (georeferenced) variable. The term intrinsic stationarity is used to describe the natural increase in variance between observations of a regional variable as distance increases from each observation. The semivariance, a measure of the variance as distance increases from all points or areas (blocks), eventually reaches a value equal to the variance for the entire array of data locations, regardless of distance. Clearly, at zero distance from a point, the semivariance is also zero, but the semivariance increases until, at a distance called the range and a semivariance value called the sill, the semivariance is equal to the variance. The function describing the semivariance is usually spherical, exponential, or Gaussian. The variogram is essential for kriging, which is a technique for estimating the value of a regional variable from adjacent values while considering the dependence expressed in the variogram. There are many kinds of kriging, each designed to give the highest possible confidence to the estimation of a variable at non-data locations. If there is no bias in the variogram, and all required assumptions are met, the kriged values, as opposed to trend surface, TIN, or other estimation devices, will be optimal. A large literature has developed in geostatistics. The definitive text by Cressie (1991) details many instances where the geostatistical approach has proved helpful. These include studies of soil-water tension,

wheat yields, acid deposition, and sudden infant death syndrome. The variogram has now been introduced into several GIS, and several programs that can be interfaced with GIS are available to help construct variograms and apply the kriging process. The geographic literature in this area is building rapidly.

Spatial econometrics
The fundamental work in this area can be traced to Paelinck in 1966 (Paelinck and Klaassen 1979). Anselin has made spatial econometrics accessible to a wide audience with his text (1988) and now his SpaceStat software. In addition texts by Haining (1990) and Griffith (1988) have helped to widen the appeal of these methods in geography. The approach is as Anselin says, 'model driven;' that is, the focus is on regression parameter estimation, model specification, and testing when spatial effects are present. Regression models constitute the leading approach for the study of economic and social phenomena. The assumptions required for the basic linear regression model, however, do not satisfy the needs of spatial regression models; they must take into account spatial dependence and/or spatial heterogeneity. Spatial dependence occurs when there is a relationship between observations on one or more variables at one point in space with those at another point in space, while spatial heterogeneity results from data that are not homogeneous, for example, population by areas which vary considerably by size and shape.

A variety of spatial autoregressive models have been developed that include one or more spatial weight matrices that describe the many spatial associations in the data. The models include either a single general stochastic autocorrelation parameter, a series of autocorrelation parameters; one for each independent variable conditioned by spatial effects (dependency or heterogeneity), an error term autocorrelation parameter, or some combination of these. Parameter estimation procedures can be complex. The usual approach is to use diagnostic statistics to test for dependence and/or heteroscedasticity among these spatially weighted variables or error term. Fortunately, SpaceStat, designed for the exploration and testing of spatial autoregressive models, is sufficiently user friendly to allow for the development of final autoregressive models.

Several other approaches have been taken to specify the influence of spatial effects in a regression model environment. Casetti's expansion method (Jones and Casetti 1992) is designed to increase the number of variables in a regression model to take into account secondary, but influential spatial variables, such as the x, y coordinates of georeferenced variables. This approach uses the parameters of the expansion variables as the indicators of the spatial effects.

In another development, Getis (1995), Griffith (1996), and (Getis and Griffith 2002) suggests transforming the spatially autocorrelated model into one without spatial autocorrelation embedded within it. By filtering out the spatial autocorrelation, the ordinary least squares model can be estimated and evaluated using R^2. The filtered spatial components are reentered into the regression equation as separate spatial variables.

Of the thousands of examples of the use of GIS in applied research, I have selected a few from each of a number of fields in order to illustrate the rich and diverse nature of GIScience. Most important, however, from an applied geography perspective is the problem that motivates the research. Our point below is that GIScience represents a superb pathway for the solution of applied geographic problems. Thus, in each of the sketches we identify the problem, then point out the role of GIScience for its solution. Every issue of *ArcNews*, a monthly publication of the ESRI company, describes numerous examples of 'GIS in Action' from a variety of fields. Many of the discussions give a website for readers who want further information about a particular subject. In addition, Goodchild and Janelle (2003), present detailed examples of research examples in GIScience and Longley, et al. (1999) provide extensive surveys of practical applications of GIS.

GIScience and disaster response
One of the strongest examples of the use of GIScience took place in the days immediately following the disaster of September 11, 2001. The problem was to help design an evacuation (for debris) and rescue equipment transportation network. With smoke billowing from the site of the attack, and the mayor's planning office destroyed, infrared images were gathered so that one could 'see' through the smoke. A planning system was devised using detailed maps that allowed large equipment to get into and out of the World Trade Center site.

Because natural and human-induced disasters such as tornadoes, hurricanes, earthquakes, wildfires, and floods usually occur suddenly, threatening people and structures, they create chaos and panic (Greene 2002). Increasingly, GIS technology is being used to help communities prepare for disasters, create response plans, track and assess damage resulting from a disaster, and coordinate workers in the field (Briggs et al. 2002). Whether the disaster is a wildfire in Arizona or a tornado in Oklahoma, maps produced using GIS have been invaluable in tasks such as locating houses, identifying property ownership, helping responders decide where to send field crews or rescue workers, generating evacuation routes, and siting field hospitals or paramedic bases. Using GIS, the turnaround time is so short that maps can be updated quickly and given to crews in the field. In the wake of the disaster, maps have been used to locate damaged structures, make property damage assessments, and prepare for debris removal.

GIScience and epidemiology
In what way can infectious disease diffusion be controlled? Here, as in all GIScience solutions, a significant resource is the spatial data management system (SDMS).

Accurately mapped detailed information is needed for the location of the incidence of disease over time, entomological risk factors, and healthcare facilities. These data are linked together in a relational database. Such diseases as malaria, West Nile virus, SARS, AIDS, dengue, and Ebola result from the spread or diffusion of one or more viruses. For dengue, the mosquito species *Aedes aegypti* is the vector that carries dengue viruses from infected people to susceptibles, usually children. The data are statistically analyzed to determine the degree of spatial congruence between infecteds and susceptibles. Recommendations for disease mitigation result.

GIScience and public health
How can the delivery of healthcare service be improved? In a GIScience environment, the solution to this problem requires the detailed management of a large amount of georeferenced socioeconomic data (Cromley and McLafferty 2002). The data are manipulated into healthcare regions of varying bases dependent on such items as physicians services, clinics for a variety of healthcare needs, and transportation networks. The regions are compared and contrasted and evaluated with regard to available financial and human resources.

GIScience and telecommunications
To what extent and in what ways does electronic mail substitute for international telephone flows? In this case, the SDMS includes data on international telephone flows, their origins and destinations, and routes. These data in conjunction with variables such as Internet access, language, tourism, and income are manipulated to identify associations, spatial and non-spatial. Such geographic variables as time zones and nearness are studied using spatial software.

GIScience and demographic research
More and more demographers are finding that the spatial component of their work has much to offer in their explanations of age, sex, birth and death rate differences. For example, the question has arisen about the influence of nearby areas, for example, the neighborhood, that may transcend more theoretically plausible explanations for birth rate decline. This kind of work usually requires extensive georeferenced databases.

In a particular instance, that is, for a study of rapid changes in patterns of fertility in the developing world, it was necessary to identify regional differences among demographic variables. For national planning purposes, the research is geared to finding correlates of fertility changes. Data from census units provide information on age, number of children, socio-economic and neighborhood characteristics. In addition, remotely sensed data may be used as a surrogate to identify the location of regions of urbanization. These are entered into a relational database keyed to the centroid of each census unit. A regression model is used to explain the level of fertility vis-à-vis patterns of marriage and female education.

GIScience and economic research

For many years economists paid scant attention to the location of economic activity. With the arrival of the field of regional science in the 1950s and its enhancement by GIS in more recent years, considerable emphasis is now being given to locational phenomena Bateman et al. 2003). As an example, economists have given new meaning to the convergence hypothesis. That is, economic theory would predict that in a free choice setting regionally disparate incomes would tend to converge to the same level. The study of such a phenomenon requires an understanding of the nuances and influences of the structure of the underlying data. For example, to what extent does aggregated data (say data aggregated from census blocks into census tracts) distort the convergence process? Also, how does spatially correlated data influence results of such studies. This type of analysis is representative of what is called a spatially autoregressive process. Software packages make it possible to model the spatial effects of many types of economic phenomena.

Applied geographers consider the study of the location of economic activity as the fundamental building block of their area of study. A large literature has helped in the modeling of efficient locations for such things as retail outlets and industrial plants. In the past, maps were used mainly to display the location of inputs, outputs, and resulting locations. Now, through the visualization mechanisms of GIScience, one can map, chart, graph, and otherwise "see" the interconnections between production and consumption. Better solutions are available mainly because of the new ability to manipulate data in a myriad number of ways (Birkin 1996).

GIScience and urban and transportation research

Fundamental to urban research is the determination of what may be called optimal land use systems. Optimality is usually defined in terms of the immediate objectives of a study; for example, an optimal transportation system may be one that minimizes delays due to congestion. Again, any success in land use modeling is usually traceable to a detailed, extensive, and well-structured data set. For transportation research this usually means that origin-destination data are available as well as data on the use of various transport modes at different times of day (Lang 1999). A significant challenge is to design such a data system that can be used to study various transportation scenarios. With projections of increasing population, for example, plans can be developed that will shed light on urban congestion.

GIScience and sociological research

The degree of segregation or social exclusion has been a recurrent research theme in sociology and geography. After acceptable and rigorous definitions of what is meant by segregation and exclusion, the sociologist will depend on detailed, usually urban, data to identify clusters of exclusion and inclusion. Spatial statistics and geostatistics can be used to show changing structures of segregation over time.

GIScience and criminal justice
Given limited budgets, police departments must constantly find ways to allocate resources in the most effective ways. The local citizenry is ever alert in making recommendations as to how crime might be reduced, but it is the sophisticated SDMS that
forms the foundation of effective analyses and mitigation. There are many examples of
the use of GIS to identify the location of 'hot spots' of criminal activity so that police
resources can be redirected to them at a moment's notice. The hot spots are of two
varieties: the first are locations of elevated levels of criminal activity. More often than
not, these are expected centers of crime based on population density, economic, and age
data. The second type of hot spot analysis identifies crime levels statistically significantly greater than expected. The hot spots become the locations of out-of-the-ordinary
criminal activity, thus allowing for redeployment of resources. Spatially based software
packages are now readily available to crime solvers for these types of analyses.

GIScience and political science
Software exists that makes it possible to redraw legislative districts under a variety
of criteria and scenarios although, in the final analysis, politicians usually determine
where district boundaries are drawn. Sometimes, of course, courts overturn a particular
districting "solution." GIS speeds up a very slow, mandatory process.

One type of approach to redistricting combines apportionment equality subject to compactness and contiguity constraints. More usual, however, is the approach that maximizes
compactness while satisfying apportionament conditions such as a maximum permitted
deviation from a quota. Most of this work is based on data from the smallest possible
census zones (blocks and tracts). A particularly useful aspect of this is the opportunity to evaluate already existing legislative zones or zones being proposed by various
legislative or lay groups.

GIScience and the environment
No subject area is better suited for study by means of GIScience than the environment.
Numerous environmental problems have been and are currently being studied in a
GIS setting. These subjects include protection of endangered species, reduction of air
and water pollution, the evaluation of fire hazards, wildlife management, optimum
habitat locations, natural resource use, landscape conservation, and the monitoring of
environmental variables. These subject areas require extensive data collection and data
organization. Mapping in an exploratory setting aids in the development of hypotheses.
Models based on environmental relationships can be evaluated using a series of map-
based statistical tests (Skidmore and Prins 2002).

GIScience, government, and public policy
Local and regional governments are constantly faced with the problem of finding efficient locations for services rendered. One recurrent problem is that of school locations

and school service areas. Such techniques as the location allocation procedure of p-median have now been incorporated into GIS software so that alternate "solutions" may be evaluated using socio-economic criteria.

Given the power of GIScience for the exploration of data, it has become common to create systems that efficiently lay out alternative scenarios so that informed judgments can be made. This field has come to be called Spatial Decision Support Systems (Malczewski 1999). Public policy can be enhanced if various alternatives are clear not only to the policy makers but also to their constituents. These systems are designed to take complex spatially based problems and lay them out so that intelligent decisions can be taken.

GIScience and anthropology/archaeological research
To what extent did ancient roads facilitate exchange among prehistoric societies? The first task is to recreate the road network of the ancient society being studies. This might require the use of current remotely sensed data where road networks may still be observed but only from high altitude. Cost-path analysis, available in GIS, could be used to identify the possible links between settlements. It may be the result of such a study that trade was not a significant factor in early settlement patterns, but other factors such as religious rituals and political control were more important (Wescott et al. 2000).

Recent, ongoing studies of archaeological sites in the Mediterranean Sea and the Andes Mountains illustrate how GIS can be used to explore the legacy of the past at remote sites. The University of Haifa's Maritime Civilization Department and the Geography Department's GIS and Remote Sensing Lab are using a marine GIS to answer questions about maritime trade at Tel Shiqmona, located on the southern tip of Haifa Bay in Israel, during the period 538 B.C.–332 B.C. The marine GIS incorporates and integrates data on a number of variables, including coastal geomorphology, tides, currents, sedimentation, and sediment transport.

Since 1999, researchers from the University of California at Santa Barbara have been using GIS-based methods to record excavations at an archaeological site called Jiskairumoko in the Andes Mountains of Peru. Excavations and the recording of site data are typically extremely time consuming tasks, but the UCSB researchers are demonstrating that utilizing GIS for collecting and organizing data directly in the field permits the speedier excavation of a large area of the site and a highly accurate recording of buried archaeological deposits.

THE FUTURE OF GISCIENCE AND APPLIED GEOGRAPHY

One of the side effects of the introduction of GIS into mainstream geographic analyses and data exploration is that societies exposed to GIS are becoming more geographically literate. Many people now have a better sense of space, distance, direction, landforms, climate, political entities, and so on. This salubrious effect encourages more people to turn to geographic solutions to problems. The epidemiologist, economist, health care specialist, or political scientist will naturally think of spatial relations when dealing with georeferenced data, and, georeferenced data will more often be collected, manipulated, and presented. Specialists from several fields will be needed in order to fulfill the needs of the research. This will result in better communication among scientists and standardization in GIS notation and manipulation processes. Better techniques of data gathering, such as the use of computers in the field, more refined satellite images, and a host of new data manipulation technologies will only draw more attention to geographic solutions. In addition, more systems are becoming interoperable with regard to those produced in other parts of the world.

Perhaps the greatest fear in this otherwise positive view is that, as in so many of the new technologies, GIScience may be abused. Scenarios in an SDSS, for example, may be manipulated in such a way as to favor special interest outcomes. The issue of privacy becomes critical, especially when more detailed data are desired. Misinterpretation of results is always a problem. With GIScience, the issue of uncertainty in the data gathering instruments, data manipulation procedures and inappropriate tests on hypotheses are of considerable concern (Heuvelink et al. 1998, Hunsaker et al. 2001). For example, with finer data, the effect on results of spatial autocorrelation becomes magnified. Researchers need to be well informed of this phenomenon, otherwise biased or scientifically faulty results will obtain. There is always the danger that data gathered for one purpose may used unquestioningly for some other purpose. The problem of cost is also an issue. Most of the new technology requires large start up costs for equipment, software, and labor. This may result in having the rich and powerful control data gathering and data analysis techniques. Clearly, for the sake of good science, open systems and freely exchanged information are highly desirable goals.

ARTHUR GETIS
San Diego State University

REFERENCES

Anselin, L. (1988) *Spatial econometrics: methods and models*. Dordrecht: Kluwer.
Anselin, L. (1994) Local indicators of spatial association – LISA. *Geographical Analysis* 27: 93–115.
Anselin, L. and Rey, S. (1991) Properties of tests for spatial dependence in linear regression models. *Geographical Analysis* 23: 112–31.
ArcNews ESRI, 380 New York Street, Redlands CA 92373–8100.
Bailey, T. C. and Gatrell, A. C. (1995) *Interactive Spatial Data Analysis*. Essex: Longman Scientific and Technical.

Bateman, I. J., Lovett, A. A., and Brainard, J. S. (2003) *Applied Environmental Economics: A GIS Approach to Cost-benefit Analysis*. Cambridge: Cambridge University Press.

Batty, M. and Longley, P. (1994) *Fractal Cities*. London: Academic Press.

Birkin, M. et al. (1996) *Intelligent GIS: Location Decisions and Strategic Planning*. New York: John Wiley & Sons.

Boots, B. and Getis, A. (1988) *Point Pattern Analysis*. Newbury Park, CA: Sage.

Briggs, D. J., Forer, P., Jarup, L., and Stern, R., eds. (2002*) GIS for Emergency Preparedness and Health Risk Reduction*. Dordrecht: Kluwer.

Clark, P. J. and Evans, F. C. (1954) Distances to nearest neighbor as a measure of spatial relationships in populations. *Science* 121: 397–8.

Cressie, N. (1991) *Statistics for Spatial Data*. Chichester: John Wiley.

Cromley, E. and McLafferty, S. (2002) *GIS and Public Health*. Guilford Press.

Diggle, P. J. (1983) *Statistical Analysis of Spatial Point Patterns*. London: Academic Press.

Getis, A. (1984) Interaction modeling using second-order analysis. *Environment and Planning A* 16, 173–183.

Getis, A. (1995) Spatial filtering in a regression framework: experiments on regional inequality, government expenditures, and urban crime. In Anselin, L. and Florax, R. J. G. M. (eds.) *New Directions in Spatial Econometrics*. Berlin: Springer: 172–88.

Getis, A. and Ord, J. K. (1992) The analysis of spatial association by use of distance statistics. *Geographical Analysis* 24: 189–206.

Getis, A. and Griffith, D. A. (2002) Comparative spatial filtering in regression analysis, *Geographical Analysis*, 34, 2, 130–140.

Goodchild, M. F. and Janelle, D. G. (2003) *Spatially Integrated Social Science: Examples in Best Practice*, Oxford: Oxford University Press.

Greene, R. W. (2002) *Confronting Catastrophe: A GIS Handbook*. Redlands: ESRI Press.

Griffith, D. A. (1988) *Advanced Spatial Statistics: Special Topics in the Exploration of Quantitative Spatial Data Series*. Dordrecht: Kluwer.

Griffith, D. A. (1996) Spatial autocorrelation and eigenfunctions of the geographic weights matrix accompanying geo-referenced data. *The Canadian Geographer* 40: 351–67.

Haining, R. (1990) *Spatial Dta Analysis in the Social and Environmental Sciences*. Cambridge: Cambridge University Press.

Heuvelink, G., Goodchild, M. F. and Heuvelink, B. M. (1998) *Error Propagation In Environmental Modelling With GIS*. London: Taylor & Francis.

Hubert, L. J. (1979) Matching models in the analysis of cross-classifications. *Psychometrika* 44: 21–41.

Hunsaker, C. T., Goodchild, M. F., Friedl, M. A., and Case, T. J. (eds.) (2001) *Spatial Uncertainty in Ecology: Implications for Remote Sensing and Gis Applications*. Berlin: Springer-Verlag.

Jones, III, J. P. and Casetti, E. (1992) *Applications of the Expansion Method*. New York: Routledge.

Lang, L. (1999) *Transportation GIS*. Redlands: ESRI Press.

Longley, P., Goodchild, M. F., Maguire, D. J., and Rhind, D. W. (eds.) (1999) *Geographical Information Systems* (two volumes). New York: John Wiley.

Malczewski, J. (1999) *GIS and Multicriteria Decision Analysis*. New York. John Wiley.

Mantel, N. (1967) The detection of disease clustering and a generalized regression approach. *Cancer Research* 27: 209–20.

Okabe, A., Boots, B., and Sugihara, K. (1992) *Spatial Tesselations: Concepts and Applications of Voronoi Diagrams*. New York, John Wiley.

Ord, J. K. and Getis, A. (1993) Local spatial autocorrelation statistics: distributional issues and an application. *Geographical Analysis* 27: 286–306.

Paelinck, J. and Klaassen, L. (1979) *Spatial Econometrics*. Farnborough: Saxon House.

Ripley, B. D. (1981) *Spatial statistics*. New York, John Wiley.

Skidmore, A. and Prins, H. (2002) *Environmental Modelling With GIS and Remote Sensing*. 2nd ed., London: Taylor & Francis.

Wescott, K., and Brandon, R. J. (eds.) (2000) Practical Applications of GIS for Archaeologists: A Predictive Modeling Toolkit. London: Taylor & Francis.

LAY JAMES GIBSON

CHAPTER 6
ECONOMIC BASE THEORY AND APPLIED GEOGRAPHY

ABSTRACT

Economic base theory belongs to both economics and geography. The theory per se belongs to economics but it is geographers who have brought it to life by tieing it to real places and their economic landscapes. This paper identifies seven economic development problems commonly faced by development practitioners and illustrates how "best practice" solutions can be drawn from economic base studies. A number of studies are used to illustrate problems and approaches but two are featured; one is a regional economic base study that looks at both a large rural region and at five individual communities within the region. The other is a study of a single community that was initially completed in 1974 and replicated three times between 1974 and 1995.

INTRODUCTION

Geographers and other scientists have long debated the roles of basic research and applied research. Whereas this debate is not likely to be resolved anytime soon, a new debate has emerged to parallel it the debate about who, ultimately, is responsible for the actual implementation of so-called applied research. To a significant extent, this issue is being brought to a head by university administrators who are increasingly concerned about industrial support of university research, commercialization of university research, and the "entrepreneurial university" (see, for example, Blumenthal, Gluck, Seashore Louis, and Wise 1986; Blumenthal, Epstein, and Maxwell 1986; and Smilor, Dietrich, and Gibson 1993). And to some extent, the issue is being resolved by a growing acceptance of papers such as this one which feature a "best practice" approach.

The practitioner is sometimes poorly informed about the nuances of research and the researcher is often unaware of the real needs of the practitioner. My intent in this paper is to provide the geography-trained practitioner with a series of implementable approaches to frequently encountered economic development problems. Further, I will offer the academic applied geographer perspectives on the sort of research questions asked by those engaged in economic development practice.

Economic base analysis has been with us for decades. Early work clearly established the value of economic base approaches to understanding community economic systems

113

A. Bailly and L. J. Gibson (eds.), Applied Geography, 113–131.
© *2004 Kluwer Academic Publishers. Printed in the Netherlands.*

and it has raised a variety of methodological questions and considerations. In Europe, economic base theory can be traced back to Sombart (1907); more recently work by Claval (1968) and Bailly (1975) and even Bailly's chapter on Medicometry in this book provide an European perspective. In the United States Andrews (1953) provided a historical overview of the economic base concept and Alexander (1953) in his study of Madison, Wisconsin, provided a detailed community case study using the economic base approach. Blumenfeld (1955) offered a thoughtful challenge to the conventional wisdom regarding economic base approaches and a suggestion to economic develop-ment planners that enhancement of non-basic activities and leakage reduction can be an effective regional development strategy. His essay was not so much an indictment of economic base theory as it was a reminder that it is easier to understand the economic base concept than it is to apply it, and that it is even more difficult to draw lessons from economic base studies that are useful for informing policy and supporting economic development practice. Finally, the monograph by Tiebout (1962) was a sort of "inventory and prospect" that established a new baseline for economic base studies and it helped reposition economic base analysis as a tool to support economic development practice.

Subsequent work has clarified many fundamental issues about alternative strategies for evaluating the community economic base and it has explored ways that the technique can be extended to answer increasingly sophisticated questions about multipliers and sectoral employment structures. Typical of studies of these sorts are those by Mulligan, 1987, 1988, 1994; Mulligan and Fik, 1994; Mulligan and Gibson, 1984a and 1984b; Mulligan and Kim, 1991; Reeves and Gibson, 1974; Richardson, 1985; Vias, 1995; Vias and Mulligan, 1996; Gibson, Barr, and O'Keefe, 1975; Gibson and Reeves, 1974; and Gibson and Worden, 1981a. Other studies have focused on economic base analysis as a technique which helps sharpen understanding of how communities change when they experience fundamental changes in their economic structure (Gibson, 1987; Gibson and Worden, 1984; Roderique, 1986; Rusden, 1988; and Vierck, 1983).

This paper offers a different perspective than those referenced above. Whereas it deals with a series of questions familiar to those involved with the practice of economic development they are typically not seen as questions with an "economic base answer". There is general agreement among applied researchers that significant goals are to a) inform policy and b) enhance practice. This paper explores ways that economic base analysis can be used to do exactly that.

ECONOMIC BASE STUDIES

The best practices which are explored in this paper are based on economic base studies that have several distinctive attributes. All of the studies except one (Bisbee 1995) are part of the Arizona Community Data Set which was initially introduced by Gibson and Worden (1981) and more recently examined in great detail by Vias (1996). The communities are relatively small (populations are less than 10,000) and data were collected for all private and public sector employers by field survey. The two study areas are the White Mountains in east-central Arizona and Bisbee in southeastern Arizona.

The White Mountain study is unique in the ACDS inasmuch as it is a regional study. The questionnaire was structured in a way that allowed each of the region's communities to be treated separately or as part of a single, larger regional economic base system.

The White Mountain region is composed of seven communities with a total of 13,442 head-count employees in 1,226 private or public sector establishments. 1) Springerville and Eagar and nearby Greer are on the eastern margin of the region near the New Mexico state line. Ranching, logging, tourism, and a large coal-fired electric generating plant are activities which support the contiguous service centers of Springerville and Eagar. (249 firms; 2,735 head-count employees). 2) Pinetop-Lakeside (including Hon Dah and McNary) is a community conspicuously dependent on tourism and retirement. (375 firms; 2,831 head-count employees). 3) Show Low is on a north-south U.S. highway; it is the region's largest and most complex service center. (378 firms; 3,535 head-count employees). 4) Whiteriver is the economic, political, and administrative center of the White Mountain Apache Reservation. (55 firms; 2,408 head-count employees). 5) Snowflake and Taylor are adjacent to one another and incorporated. They are regional service centers with a conspicuous forest products sector. (161 firms; 1,544 head-count employees). 6) Hawley Lake-Sunrise are places of work but not places of residence. They are both recreation-oriented; the Sunrise Ski Area is clearly the major employer and the driving force behind worker in-commuting. (8 firms; 389 head-count employees). 7) "Other", i.e. the interstitial space that separates the six communities from one another. The "other" territory holds a handful of residences including a few ranches but no significant employers.

The other study area which is examined in this paper is Bisbee, Arizona. Bisbee has been the subject of four economic base studies in 1974, 1981, 1986, and 1995. This allows us to take a longitudinal look at what has happened in a community that has experienced substantial change since 1974. Bisbee is a good place to look at change. For decades Bisbee was known as a copper mining town. In late 1974, the Phelps Dodge Corporation initiated a series of layoffs that continued into 1975 (Gibson 1987). The economic base studies that have been done in Bisbee give a detailed picture of how a regional center changes over time once a major employer has left.

THE BEST PRACTICE PERSPECTIVE

This paper is not about economic base approaches per se. It is about using the output of economic base studies in an unconventional way to answer questions that are asked by those who develop economic development policy and by those engaged in development practice. Economic base analysis is a familiar element in many economic development programs, but typically its contribution is a multiplier that is used to estimate project benefits and impacts. This paper explores the notion that a survey-based economic base study produces a number of outputs that have value for the development practitioner. Specifically, the economic base survey leads to the creation of a data set that is not usually otherwise available. Some examples from the studies drawn upon in this paper are as follows:

- All public and private employers can be included in the data set, not just those in industries covered by unemployment insurance or which meet other criteria such as minimum employment levels or sales volumes.
- "Employment" can be defined to include proprietors, those on salary, and family members. In other words, it can describe the economically active population.
- Distinctions can easily be made between full-time, part-time, and seasonal workers and male and female workers.
- In-commuters can be separated on a firm-by-firm basis from employees who both live and work in the community.
- Regional base studies can identify cross-shopping and cross-commuting patterns.
- Basic and non-basic bifurcations can be made on an individual firm basis.
- The significance of transfer payments and out-commuters can be accounted for and they can be described as "synthetic sectors" that are directly compared to conventional employment sectors.

Practitioners are generally receptive to an economic base approach partly because it seems to lead to the sort of baseline document that every community needs and partly because of the promise of income or employment multipliers which are understood to have useful diagnostic value, especially for evaluating the full consequences of regional growth or decline. More often than not, the outputs of the economic base study are underutilized – in part because the practitioner/client does not know what to ask for and in part because the researcher does not really appreciate what he or she might serve up to benefit the client.

The list that follows identifies six fairly typical practitioner needs and it briefly notes what the applied researcher in possession of outputs from an economic base study might bring to the table to meet those needs.

1. REGIONAL APPROACH

It has been popular in recent years to promote a regional approach for economic development initiatives. State development agencies, for example, often favor a regional approach because it is easier to allocate limited program dollars to a group of communities than to a number of individual entities. Further, it is argued that the aggregated resources of the region give it more clout when managing change or in attracting industry than any of its individual communities would have. Community leaders are sometimes

Practitioner needs/research solutions

What the practitioner needs	What the applied researcher can offer
1. Evidence that a region is made up of linked and interrelated communities, not just a series of free-standing towns that happen to be in close proximity.	The regional economic base study provides detailed data on cross- shopping and cross-commuting.
2. A list of target industries, i.e. business types that can be expanded or introduced into the region to meet presently unmet demands for goods and services.	Cross-shopping data from a regional economic base study can serve as the basis for identifying local-serving businesses that are potential targets.
3. Evidence that export activity involves more than just jobs in the resource industries or in manufacturing.	The economic base survey allows the bifurcation of data for individual firms. When aggregated to a standard 10-sector model these data provide a convincing picture of what is export and what is local-serving.
4. Evidence that out-commuters and transfer payments in general, and retirement incomes in particular, may make a significant contribution to the community's economic base.	Data on out-commuters and transfer payments are essential ingredients in economic base multiplier calculations. By presenting these data separately, an out-commuter sector and a "transfer sector" can be created.
5. Data that document the nature of the tourism sector and confirm that it is indeed an export sector.	Data from an economic base survey allow for a distinction between local-serving and export activity on a firm-by-firm basis; the importance of export income can be evaluated in both absolute and relative terms.
6. An objective way of identifying export oriented "engine industries" that might define a "cluster" or industrial agglomeration that is the core of a complex of like and linked industries.	Like firms can be aggregate in terms of both total and export employment to determine which specific activities can legitimately be considered to be core engine industries.
7. A picture of what happens to a community when it loses a large export employer.	A series of community economic base studies conducted at different points in time can provide a detailed picture of changes in the nature of employment, changes in the relationship between local-serving and export employment, and evidence regarding the extent to which lost jobs are literally or effectively replaced.

skeptical – perhaps because they do not want to lose control over the management of development initiatives, but also because the economic relationships between their community and their neighbors are not well understood.

A regional economic base study will not assure that a community does not lose control over development program management, but it can help clarify questions about the relationships between one community and another. Two key outputs of the regional base study are data on cross-commuting and data on cross-shopping. Cross-commuting data distinguish between where the jobs are and where the employees live and, presumably, where their paychecks are spent. Cross-shopping data describe retail and other expenditure leakage. Some out-shopping is expected in any community, but an evaluation of detailed data on out-shopping makes it clear that some communities do a much better job than others at the business of internalizing local demands for goods and services.

Example. The White Mountain study produced detailed data on cross- commuting (Table 1) and cross-shopping (Table 2). These data make it clear that there is a great deal of give-and-take within the regional system. They also clearly show that creating jobs and building the population base are two different things, i.e. it is one thing to create a job and another to supply goods and services to the worker and his or her family. The "winners" in the White Mountain region in terms of resident population are Pinetop-Lakeside which offers 2,140 jobs and accommodates 2,643 workers and Snowflake and Taylor which offer 1,289 jobs and accommodate 1,693 workers. Show Low and Springerville, Eagar, and Greer, on the other hand, come up short using this measure. Show Low offers 2,899 jobs yet accommodates only 2,483 workers and Springerville, Eagar, and Greer offers 2,326 jobs and accommodates only 2,022 workers. Work and residence are more or less in balance in Whiteriver and, as noted elsewhere, Hawley Lake and Sunrise are a special case.

Table 1. Cross-commuting within the white mountain region to and from selected work places in and out of the region (FTE employment)

	Work			
Residence	Pinetop-Lakeside, Hon Dah, McNary	Total living in region	Work outside/ live in region	Total
Pinetop-Lakeside, Hon Dah, McNary	1674.5	2579.4	64.0	2643.4
Total Working in Region	2056.2	10293.1	702.0	10995.1
Work in Region/Live Outside Region	83.9	623.7	0	623.7
Total	2140.1	10916.8	702.0	11618.8

Source: Survey by Economic Development Research Program/University of Arizona. From Gibson.

Table 2. Estimated total fte employment in each community supported by sales to its own residents, residents of other communities in the region, and to the outside world

	Place of business	
Sales: Customer's residence	*Pinetop-Lakeside, Hon Dah, McNary*	*Total*
Pinetop- Lakeside, Hon Dah, McNary	817.4	1369.1
	38%	13%
Total Sales in Region	1231.5	5437.6
	57%	50%
Out of Region Sales	908.6	5479.2
	43%	50%
Total	2140.1	10916.8
	100%	100%

Source: Survey by Economic Development Research Program/University of Arizona. From Gibson.

The data on cross shopping are hardly surprising, but they are useful inasmuch as they show in specific terms sales at three scales the local community, other communities in the region, and out-of-region sales. Further, they show the extent to which local employment is supported by local demand versus demand from other communities both within and outside the region. Pinetop-Lakeside and Show Low are especially effective in capturing demand from neighboring communities.

Interestingly, these data cut two ways. On the one hand, they clearly demonstrate that there are substantial economic ties between communities in addition to the political and social ties that are more intuitively understood. Communities that are effective in attracting business from neighboring communities or effective in housing residents who work in other communities are likely to gain a new appreciation for the regional approach to economic development. On the other hand, communities which see demand for goods and services drained off by neighboring communities or those which see large numbers of local jobs filled by in-commuters might find themselves taking more of an isolationist approach. In either case, however, the conclusions reached will be better informed than might otherwise be the case.

2. TARGET INDUSTRY IDENTIFICATION

One of the most common requests by local development officials is for recommendations for target industries, i.e., business types that should be encouraged to expand in, or enter, the local market. Targeting studies usually focus on export industries and ignore local-serving or non-basic business types. As an alternative to threshold-type studies, the regional economic base study can provide information for identifying local-serving businesses which can be introduced or expanded. Further, the economic base approach can provide insight into the magnitude of unserved or underserved markets.

The regional economic base study includes the standard economic base bifurcation question, but it can also pose a follow-up question which asks the respondent to allocate within-region sales to the other communities found in the region. The percentage values estimated by each firm are then applied to that firm's employment to yield weighted values. Because the question is asked of each firm, it is possible to determine the aggregate cross-shopping for any industry type described at the four-digit SIC level, e.g. grocery stores, simply by summing the firm data. Better yet, it is possible to trace customers back to their community of residence. When the weighted employment values for each sending community are summed, the development planner is given a specific estimate of the local employment being exported by out-shopping.

Example. The case of Whiteriver, Arizona can be used to illustrate this approach. Whiteriver is the largest population center on the White Mountain Apache Reservation. Residents of Whiteriver regularly go to the region's other five communities, especially Show Low and Pinetop/Lakeside, for goods and services. Survey data suggest that out-shopping by Whiteriver residents in the five White Mountain Region communities supports a total of thirty jobs in restaurants, eleven jobs in grocery stores, nine jobs in variety stores, five jobs in automotive repair shops, and smaller numbers of jobs in automotive parts stores, self-service laundromats and various other establishments.

Some cross-shopping is normal in any region given combination trips and the need to access higher threshold goods than those available locally. Still, the approach described here allows the local development official to objectively see which business types seem to be exporting demand and the economic developer is given a quantitative estimate of the magnitude of the exported demand. When evaluating the effectiveness of existing firms and the potentials and prospects for the creation of new local-serving businesses tribal planners would do well to put restaurants, grocery stores, variety stores, and automotive repair shops at the top of their list.

3. EXPORT OR BASIC JOBS AND INCOME

Whereas there is an impressive and growing literature on the service sector as a source of export jobs, this fact is often lost on those involved with policy and practice. There is a tendency to assume that only manufacturing and perhaps resource extraction-type activities such as mining and forestry are truly basic or export-oriented. Further, the problem is often exacerbated by the failure to appreciate that the propensity to export varies from firm-to-firm in a given sector and even within firms.

Bifurcation of employment or sales is a signature feature of any economic base study. In census survey type studies such as those discussed in this paper, the bifurcations are done at the firm level. The data produced allow the applied researcher to give the policy maker and practitioner very specific information on the nature and extent of export activity.

The most recent study of Bisbee's economic base can be used to illustrate the point that export or basic jobs are found in all ten sectors (Table 3).

Table 3. Estimated non-basic and basic employment by sector bisbee and vicinity, 1995 (adjusted full-time equivalent employment)

Sector	Percent non-basic	Percent basic	Total employment
Agriculture	15	85	2.0
Mining	0	100	34.0
Construction	42	58	20.57
Manufacturing	13	87	120.0
Transportation, Communication & Public Utilities	46	54	38.75
Wholesale Trade	21	79	14.0
Retail Trade	63	37	403.52
Finance, Insurance, & Real Estate	67	33	58.18
Services	40	60	503.29
Public Administration	13	87	491.43
Subtotal	36	64	1685.74
Out-Commuters	0	100	368.69
Employee Equivalents From Transfer Payments	0	100	865.10
Total	–	–	2919.53

Mining and manufacturing clearly have a significant export component, but so too do services and public administration in absolute terms these two sectors are easily the major contributors to export employment.

Just as service sector contributions to export employment may be underappreciated, so too are the contributions of out-commuters and transfer income.

4. TRANSFER PAYMENTS, RETIREMENT INCOME AND OUT-COMMUTERS

Economic developers and elected officials typically undervalue the role of unearned income in the regional economy. There are two main types of unearned income, property income (including rents, royalties, interest, and dividends) and transfer payments (including Social Security payments, private pension payments, Aid for Dependent Children, unemployment insurance, and a number of other, often minor programs). The general problem facing the development researcher is that of positioning the unearned income "sector" in the regional economy. A more specialized problem in many areas is that of sorting-out the economic contribution of retirement.

An economic base approach to this problem allows the unearned income "sector" to be compared to other more conventional employment sectors and it specifically allows unearned income's contribution to the export base to be evaluated.

Example. Transfer payments are estimated for the White Mountain Region and its communities using data from the Arizona Department of Economic Security, the Social Security Administration, and the Bureau of Economic Analysis. Transfer payments, rather than all unearned income, are considered here to assure comparability of the White Mountain data to data describing other communities in the Arizona Community Data Set (ACDS). These dollar figures are then converted to employee equivalents by dividing them by the average wage in the region. The original economic base model based on survey results had nine employment sectors. Transfer payments becomes a synthetic tenth employment sector. The transfer income sector has two distinctive parts welfare and retirement. Using Social Security payments as a proxy for retirement income, the importance of this item alone is obvious. Whereas most policy makers and development planners will not make attraction of a welfare population a goal, building the retirement population could be an important element in the community economic development strategy.

The "transfers sector" (welfare and retirement) is enormous; when it is added into the basic employment for the region and for each community its true magnitude can be appreciated. The same is true of the "retirement sector" alone. The roles of transfers in general and retirement in particular are described as follows:

Community	Key "sectors" (ranked by employment)
Pinetop-Lakeside	1. Transfer Payments 2. Services 3. Retirement
Show Low	1. Transfer Payments 2. Services 3. Retail Trade 4. Retirement
Snowflake and Taylor	1. Manufacturing 2. Transfer Payments 3. Retirement
Springerville, Eagar, and Greer	1. Transportation, Communication, and Public Utilities 2. Services 3. Transfer Payments 4. Construction 5. Public Administration 6. Retirement
Whiteriver	1. Public Administration 2. Services 3. Manufacturing 4. Transfer Payments 5. Retirement

Source: Survey by Economic Development Research Program/University of Arizona and secondary sources. From: Gibson.

Data from the 1995 economic base study in Bisbee (Table 3) reinforce the significance of transfers and they underscore the significance of another key element out-commuting. Out-commuters are estimated by contacting major employers outside the base area to determine the number of their employees who live within the base area.

Example. The Bisbee data (Table 3) provide a straightforward illustration of why it is essential to account for both out-commuters and transfer payments. These two variables become Bisbee's first and third largest sources of export income. Transfers, especially in terms of retirement income, are enormous. But also important is the fact that Bisbee is an attractive place of residence for those who work throughout the region. The Army's Fort Huachuca in Sierra Vista, 48 kilometers to the west and the State Prison in Douglas, 34 kilometers to the south and east are just two large establishments which employ Bisbee residents.

5. THE "TOURISM SECTOR"

Historically, economic development programs have tended to emphasize the manu- facturing sector wages are relatively high, it is often export oriented, and many types of manufacturing are footloose enough to consider alternative locations. For a variety of reasons, including the general decline in manufacturing employment in the United States and the realization by many communities that they have no comparative advan- tage for manufacturing, many economic development programs have embraced tourism as a focus. In most cases, it is probably fair to say that the local development commu- nity has been more convincing in its proclamation of interest than in its description of what is included in "tourism". As pointed out previously (Gibson, 1993), "The tourism sector is easy to define in an initiative way during conversation, but much more difficult to define in a formal sense. The reason is obvious: tourists, and especially cottagers or seasonal residents, are likely to consume at least small quantities of all goods and services consumed by year-round residents" (page 154).

The economic base survey with its firm-by-firm bifurcation of employment permits easy identification of those firms determined to be in tourist-serving industry categories. Further, both the relative and absolute importance of tourism-supported employment can be measured.

Example. Once again the White Mountain Regional Study can be used to illustrate the value of the economic base approach. Research surveys identified thirty-eight tourist serving industrial categories in the White Mountain Region. The touristic share of each industry's employment is assumed to be export-oriented. Seven industries (Table 4) are judged to be key tourist serving industries and in absolute terms it is clear that just four drive the White Mountain tourism sector grocery stores, eating places, real estate agents, and motels and hotels. Over three-fourths of the region's tourism employment is concentrated in these four industries.

Table 4. Employment [1] in tourist-serving SICs

Tourist-serving SIC	Total	Export	Tourist-serving SIC	Total	Export
5411 Grocery stores	415.5	101.4	6531 Real estate agents	204.4	120.8
5812 Eating places	692.5	247.6	7011 Motels & hotels	217.0	200.1
5941 Sporting goods	45.0	24.5	7033 RV parks	23.6	21.7
Gift & Souvenir Stores	9.5	4.5	Total 38 Tourist-Serving SICs	2282.4	871.4

[1] Employment expressed in terms of FTE or full-time equivalent employment. From: Gibson, L.J. (1993).

Table 5. Employment by sector and by community for those who live and work in the community [2]

Sector	Pinetop-Lakeside, Hon-Dah, McNary	Show low	Snowflake, Taylor	Springerville, Eagar, Greer	Whiteriver
Retail trade	586.5	723.1	234.5	414.5	161.5
(Tourist portion)	(191.0)	(149.0)	(14.0)	(81.1)	(12.6)
FIRE[3]	172.9	155.8	34.4	61.4	2.4
(Tourist portion)	(66.1)	(39.8)	(4.6)	(10.3)	(–)
Services	536.3	610.0	331.2	486.9	443.7
(Tourist portion)	(123.6)	(96.2)	(6.3)	(62.7)	(4.0)
Total (FTE)	1674.5	1930.7	1116.0	1871.9	1802.6
Total touristic (FTE)	386.8	286.7	25.2	156.1	16.6
Percent touristic	23.1	14.8	2.3	8.3	0.9

[2] Employment expressed in terms of FTE or full-time employment. From: Gibson, L.J. (1993).
[3] Finance, insurance, and real estate. From: Gibson, L.J. (1993).

Tourism employment is compared to total employment in Table 5. Just three sectors typically account for most tourist serving employees in the study communities.

Despite some limitations (see Gibson, 1993, pages 156–159), the data confirm that tourism can be a conspicuous contributor to the community economic base, that touristic activities are scattered among several of the nine "standard" employment sectors, and that by itself, tourism would be a major "sector" in at least two of the five major communities where White Mountain residents live and work.

6. IDENTIFYING "ENGINE INDUSTRIES" AND CLUSTERS

In recent years the work of Porter (1990, 1998a and b, 2000) has had a profound impact on the practice of economic development. Most economic development agendas are organized around key clusters and the identification of the engine industries that drive them. Economic base analysis can be a valuable technique for both identifying the export oriented engines and for identifying the primary indirect firms that effectively add to the magnitude of the engine industries.

Example. The communities of Springerville and Eagar in east central Arizona were substantially revitalized by the creation of a large coal fired electric power plant in the 1980's. The plant is a substantial employer in its own right and this fact is easily established by a straight forward economic base survey (Gibson and Glenn, 1999). Also revealed by the base survey is the existence of three other substantial employers directly tied to the plant – employers which fit Andrews, (1953) description of primary indirect. These firms provide on-going maintenance services, security services, and general contractor-type services on an "out-source" basis. None of these firms would be in the area if not for the Tucson Electric Power Plant. Together these firms form a convincing cluster which through the multiplier process supports additional local serving jobs. The power plant is clearly more than just a large employer – it is the core of a complex of linked industries.

The notion of clusters took a very different direction in a study of seasonality and tourism in the White Mountains (Gibson and Evans, 2002). The White Mountains are known as a summer vacation area until the White Mountain Apache built the Sunrise Ski area in the 1970's. Whereas it is widely known that the tourism cluster is a major engine in the White Mountain Region, investors needed to have evidence that the ski area was truly a winter season engine before additional funds could be made available for expanded snowmaking capability. An economic base study convincingly described the role of the ski area as the major source of support for winter season tourism. The economic base approach identified the ski area as the "engine" and it identified almost 250 firms in seven key tourist-serving sectors that found their fortunes closely tied to the ski area.

7. LOSS OF A MAJOR EMPLOYER

When any community is faced with the loss of a major employer, there is bound to be widespread concern. Despite the fact that plant closures, military base closures, and a host of other events of this sort are almost common-place in this era of "downsizing", straight-forward approaches to objectively evaluating these changes are in short supply. There are certainly data from published sources that can help answer the question "what happens when a community loses a large, high-wage, export-oriented employer". But economic base surveys done before and after the closure can bring unique perspectives to this problem.

Elected officials and development practitioners want to be able to anticipate the nature and extent of change and they are interested in the experiences of other communities. Economic base studies undertaken in Bisbee, Arizona can shed light on both of these concerns. In the former case we can use the findings of economic base studies to examine such things as:

- changes in multipliers;
- changes in employment;
- changes in full- and part-time employment;
- changes in gender of the workforce; and
- changes in the number and type of business establishments.

In the latter case, we can use Bisbee as a case study of what happened in one community that lost a major employer. Are findings from studies in Bisbee typical of what happens to a community when a large, male-dominated, high-wage employer shuts down? Probably not. But a close look at what happened in Bisbee can nevertheless focus thinking on several key questions that are worth considering in almost any community.

Example. A series of economic base studies were undertaken in Bisbee, Arizona in 1974, 1981, 1986, and 1995. The 1974 study captures the large employment numbers concentrated in the mining sector. Shortly after the 1974 data were collected, Bisbee's copper mining operations were discontinued. The 1981 survey and the two which followed it reflect conditions in Bisbee as the town's employment was reconfigured.

Answers to questions initially asked by Bisbee's leadership are answered by the four economic base surveys conducted since 1974.

Question: *What will happen to overall employment levels once Phelps Dodge Corporation discontinues its open pit and underground mining operations?*

Answer: They will drop sharply and then slowly build over time as Southeastern Arizona grows.

	1974	1981	1986	1995
Full- and Part-time Headcount Employment	3,025	1,929	2,091	2,607
Full-time Equivalent Employment (FTE)	2,603.5	1,452.1	1,637.3	1,685.7

Employment was very robust in 1974. The first post-closure data show a sharp decline in employment. The picture since 1981 is mixed. FTE employment picked-up a bit after 1981 but appears to be flat since 1986. The growth of headcount employment has been solid but gains have come with growth of the part-time and seasonal segments, not the full-time cohort.

Question: *How will multipliers be affected by the Phelps Dodge closure?*

Answer: The multiplier responded as one might expect it moved up and then dropped back down.

1974	1981	1986	1995
1.30	2.17	1.31	1.22

Bisbee was, in many respects, a company town. Its 1974 multiplier was a modest 1.30. Export employment was high given the overall size of Bisbee and its non-basic employment was kept in check by the fact that many local serving goods were provided by efficiently managed company stores. The company stores were maintained into the 1980's. As a result, non-basic employment was propped up at the same time that a huge piece of the export sector had been made redundant. It is worth noting that the Phelps Dodge layoff actually overstates economic loss. Many of those who lost jobs retired or out-commuted to new jobs from their Bisbee residence. These sources of income (retirement and out-commuting) have been accounted for and actually keep the 1981 multiplier from being higher than 2.17.

The 1986 and 1995 multipliers in part reflect the facts that Bisbee has declined in population (and in demands for goods) and that export jobs in public administration and in the synthetic transfer income and out-commuting sectors have grown.

Question: *How, exactly, have full- and part-time employment changed since 1974?*

Answer: The situation is fairly straight forward. Full-time employment dropped substantially and then started the long road to recovery. Part-time employment, often in low wage sectors, has experienced substantial growth.

	Headcount employment			
	1974	1981	1986	1995
Full-time	2,757	1,540	1,595	1,992
Part-time	268	389	496	615
Percent Part-time	8.9%	20.2%	23.7%	23.6%

Question: *Have there been changes in gender in the workforce?*

Answer: In 1974 Bisbee was a mining town. Bisbee had 2,754.3 full-time equivalent employees; 48% of them were in mining. Further, an estimated 95% of the mining employment was male. Times have changed and so has the sectoral makeup of Bisbee's workforce. Over one-half of both full- and part-time employees are female!

Total headcount employment – percent female

	1974	1981	1986	1995
Full-time	27.5%	45.6%	48.4%	50.8%
Part-time	58.2%	57.1%	57.3%	62.0%
Percent Part-time	30.2%	47.9%	50.5%	53.4%

Total headcount employment – number female

	1974	1981	1986	1995
Full-time	757	702	772	1012
Part-time	156	222	284	381

In absolute terms, the number of females in the workforce has shown steady growth, the one exception being the 1974 and 1981 counts for full-time employment.

Question: *Have there been changes in the number and type of business establishments in Bisbee?*

Answer: Changes in the number of businesses have been much less pronounced than might otherwise be expected. In fact, the number of employers has actually increased from 299 in 1974 to 325 in 1995. Much of the increase has been in small tourist-serving establishments such as hotels and motels, eating and drinking places, and gift shops. The biggest changes were in retail trade (+sixteen firms) and services (+twenty-three firms).

The overall structure of employment, however, has experienced massive changes.

Share of total employment

	1974	1981	1986	1995
Primary Sector[4]	44.7%	4.9%	1.9%	1.8%
Secondary Sector[5]	1.0%	6.6%	5.4%	7.3%
Tertiary Sector[6]	54.3%	88.5%	92.7%	90.9%
Total	100.0%	100.0%	100.0%	100.0%

[4] Agriculture and mining
[5] Manufacturing and construction
[6] Transportation, communication and public utilities; wholesale trade; retail
trade; finance, insurance and real estate; services; and public administration

The primary sector, which includes mining, has all but disappeared. The secondary sector has seen some growth mostly in manufacturing. But the real story is in the tertiary sector; Bisbee has obviously shifted away from mining to services.

Changes in employment are highly concentrated the big loss is in mining. Public Administration jobs, especially those associated with county government, have been the largest single source of replacement jobs, although retail trade, manufacturing, and especially services have contributed, too.

CONCLUSION

Economic base analysis has long been recognized as an effective approach for regional problem solving. Further, there is a rich literature that explores a host of technical issues associated with base studies. This paper was written to push economic base solutions into problem areas that have previously relied on other approaches. It has persued this goal in two ways. First, by using findings from regional and sequential base studies to answer economic development practice questions. And second, by taking a best practice perspective when presenting findings. The result is a paper which bridges the applied research-practice interface and, hopefully, one that will be of use to both thoughtful practitioners and implementation-oriented applied researchers.

ACKNOWLEDGMENTS

I would like to thank Erik Glenn for his very substantial data collection and summarization efforts on the 1995 Bisbee study and for his thoughtful comments on drafts of this paper.

LAY JAMES GIBSON
Department of Geography and Regional Development
and
Economic Development Research Program
The University of Arizona
Tucson, Arizona U.S.A.
Phone: 520-621-7899
Email: ljgibson@ag.arizona.edu

REFERENCES

Alexander, J.W. (1953) An economic base study of Madison, Wisconsin, Wisconsin Commerce Papers, Volume I, No. 4 (Madison, Wisconsin, University of Wisconsin).

Andrews, R. (1953) Mechanics of the urban economic base: historical development of the base concept, Land Economics, 29, pp. 161–167.

Bailly, A. (1975) L'organisation urbaine, th,ories et modŠles (Paris, Centre de Recherche d'urbanisme).

Blumenfeld, H. (1955) The economic base of the metropolis: critical remarks on the "basic-nonbasic" concept in: P.D. Spreiregen (ed.), The modern metropolis, its origins, growth, characteristics, and planning, selected essays by Hans Blumenfeld, pp. 331–368 (Cambridge, Massachusetts, The M.I.T. Press, 1971).

Blumentahl, D., Gluck, M., Seashore Louis, K. AND Wise, D. (1986) Industrial support of university research in biotechnology, Science, 231, pp. 242–246.

Blumenthal, D., Epstein, S. AND Maxwell, J. (1986) Commercializing university research, lessons from the experience of the Wisconsin Alumni Research Foundation, The New England Journal of Medicine, 314, pp. 1621–1626.

Claval, P. (1968) La th,orie des valles, Revne g,ographique de l'Est, VIII, pp. 3–56.

Gibson, L.J. (1987) Restructuring the landscape, Yearbook of the Association of Pacific Coast Geographers, 50, pp. 7–20.

Gibson, L.J. (1993) The potential for tourism development in nonmetropolitan areas in: D.L. BARKLEY (Ed.) Economic Adaption: Alternatives for Nonmetropolitan Areas, pp. 145–64 (Boulder, CO, Westview Press).

Gibson, L.J., Barr, J.L. & O'Keefe, T.B. (1975) Measuring the community economic base: a comparative analysis, American Industrial Development Council Journal, 10, pp. 7–35.

Gibson, L.J., & Evans, B. (2002) Regional dependence on tourism: The significance of seasonality, Yearbook of The Association of Pacific Coast Geographers, 64, pp. 112–127.

Gibson, L.J. & Glenn, E. (1999) The round valley region economic base study: A generic case study of three hypothetical communities, Economic Development Review 16, pp. 53–62.

Gibson, L.J. & Reeves, R.W. (1974) The roles of hinterland composition, externalities, and variable spacing as determinants of economic structure in small towns, The Professional Geographer, 26, pp. 152–158.

Gibson, L.J. & Worden, M.A. (1981a) Estimating the economic base multiplier: a test of alternative procedures, Economic Geography, 57, pp. 146–159.

Gibson, L.J. & Worden, M.A. (1981b) A university-based delivery system for applied geographic research, Proceedings of Applied Geography Conferences, 4 pp. 11–16.

Gibson, L.J. & Worden, M.A. (1984) A citizen's handbook for evaluating community impacts: an experiment in community education, Journal of the Community Development Society, 15, pp. 27–43.

Mulligan, G.F. (1987) Employment multipliers and functional types of communities: effects of public transfer payments, Growth and Change, 18, pp. 1–11.

Mulligan, G.F. (1988) A general model for estimating economic base multipliers, The Australian Journal of Regional Studies, 3, pp. 52–68.

Mulligan, G.F. (1994) Multiplier effects and structural change: applying economic base analysis to small economies, Review of Urban and Regional Development Studies, 6, pp. 78–94.

Mulligan, G.F. & Fik, T. (1994) Using dummy variables to estimate economic base multipliers, Professional Geographer, 46, pp. 368–381.

Mulligan, G.F. & Gibson, L.J. (1984a) A note on sectoral multipliers in small communities, Growth and Change, 15, pp. 3–7.

Mulligan, G.F. & Gibson, L.J. (1984b) Regression estimates of economic base multipliers for small communities, Economic Geography, 60, pp. 225–237.

Mulligan, G.F. & Kim, H.H. (1991) Sectoral-level employment multipliers in small urban settlements: a comparison of five models, Urban Geography, 12, pp. 240–259.

Porter, M.E. (1990) The Competitive Advantage of Nations (New York, The Free Press).

Porter, M.E. (1998a) Clusters and the New Economics of Competition, Harvard Business Review, pp. 77–90.

Porter, M.E. (1998b) On Competition (Boston, Harvard Business School Publishing).

Porter, M.E. (2000) Location, Competition, and Economic Development: Local Clusters in a Global Economy, Economic Development Quarterly, 14, pp. 15–34.

Reeves, R.W. & Gibson, L.J. (1974) Town size and functional complexity in a disrupted landscape, Yearbook of the Association of Pacific Coast Geographers, 36, pp. 71–84.

Richardson, H.W. (1985) Input-output and economic base multipliers: looking backward and forward, Journal of Regional Science, 25, pp. 607–661.

Roderique, D.B. (1986) Private sector economic impacts and community responses to large scale energy development. Unpublished MA Thesis. Tucson, AZ, University of Arizona, Department of Geography and Regional Development.

Rusden, S.A. (1988) Management of the community economic base as a strategy for economic development. Unpublished MA Thesis. Tucson, AZ, University of Arizona, Department of Geography and Regional Development.

Smilor, R.W., Dietrich, G.B., AND Gibson, D.V. (1993) The entrepreneurial university: The role of higher education in the United States in technology commercialization and economic development, UNESCO, pp. 1–11.

Sombart, W. (1907) Der Begriff der Stadt und das wesen der st,,dtbildung Archives für sozialurssenschaften und sozial politik, 25 (Berlin).

Tiebout, C.M. (1962) The Community Economic Base Study (New York, Committee for Economic Development).

Vias, A.C. (1995) Specification of economic base multipliers for small Arizona communities. Unpublished MA Thesis. Tucson, AZ, University of Arizona, Department of Geography and Regional Development.

Vias, A.C. & Mulligan, G.F. (1997) Disaggregating economic base multipliers in Arizona communities, Environment and Planning A, 29, pp. 955–974.

Vias, A. (1996) The Arizona community data set: A long-term project for education and research in economic geography, Journal of Geography in Higher Education, 20, pp. 243–258.

Vierck, S.L. (1983) Regional impacts of substantial reduction in basic employment: economic and demographic impacts of the 1974–75 mine closures on Bisbee, Arizona. Unpublished MA Thesis. Tucson, AZ, University of Arizona, Department of Geography and Regional Development.

HARRY TIMMERMANS

CHAPTER 7
RETAIL LOCATION AND CONSUMER SPATIAL
CHOICE BEHAVIOR

ABSTRACT

This chapter reviews progress made in applied geography with respect to analyzing and predicting spatial shopping behavior. Developments in models of store and shopping center choice during the past decades are reviewed. In addition, progress in simulating pedestrian movement is outlined.

INTRODUCTION

Geographers traditionally have made substantial contributions to retail management and retail policy. Retail locations decisions have a major impact on the success of a store. The saying "location, location, location" suggests that bad location decisions are difficult to compensate by other elements of the marketing mix such as pricing, merchandising and promotion. Space and location are the very core of geography; hence many geographical theories, results of spatial analyses and spatial methods are potentially relevant for better informed decisions in retailing.

In addition to contributions to the retail industry (retail organizations, real estate developers, consultants, etc), geographers have made important contributions to retail policy. In many Western societies, governments have attempted to exercise some degree of control on retail development. For example, to avoid empty downtown areas, retail policy in many countries has stimulated retail business in downtown areas and prohibited retail growth in out-of-town or peripheral locations by implementing appropriate planning control. Equity issues (good accessibility and a basic provision of retail stores in every neighborhood) have also been high on the policy agenda, especially in European countries in the 1970s and 1980s. Store opening hours policies and local parking policies are other examples of policies that generated substantial discussion, diverging views, and in many cases frustration and anger among shopkeepers and store employees.

In this chapter, I will review the geographical literature on spatial shopping location decisions in an attempt to demonstrate the applied relevance of this tradition in geographic

A. Bailly and L. J. Gibson (eds.), Applied Geography, 133–147.

research. Restrictions on the length of the chapter prevent me from discussing all relevant work in this area of research. This chapter will focus on selective topics and approaches that can be identified in the literature. It is organized as follows. First, I will discuss developments in modeling spatial shopping behavior at the regional level. Next, progress in simulating pedestrian behavior will be discussed.

MARKET POTENTIAL AND LOCATION DECISIONS

An assessment of market potential is a key issue for the decision to open a new store. Because shopping behavior is distance-sensitive and potential customers and demand are not equally distributed across space, market potential is also location-sensitive. The expected sales in a store will vary by location. Retailers have to collect information about the expected sales in a new store, given the distribution of their potential customers, the strength of their competition and the spatial behavior of their potential customers. Expected sales can then be compared with their expected costs (merchandise, advertisement, lease, etc), and this provides important input to their decision of whether or not to open a new store at a particular location.

A central element in this process concerns the prediction of future sales levels in the new store. This problem has been high on the research agenda of retail geography, and important contributions have been made over the years. It should be emphasized however that most geographical studies have been focused at the level of the shopping center/area as opposed to the level of the individual store. The latter has been more the area of interest of marketing researchers. The approaches that were used, however, are quite similar.

The problem of estimating future sales has typically been addressed by developing a model of spatial shopping behavior. This means that researchers formulate a set of assumptions about shopping behavior, translate these assumptions into a mathematical expression, adjust the model to available data about shopping behavior, test whether the model reproduces these data satisfactory and, if so, use the model to predict future sales by changing the explanatory variables of the model, while fixing the so-called parameters of the model that were used to tune the model to the available data. The most widely used model in retail geography has been the production-constrained spatial interaction model (Wilson, 1971). It assumes that (i) the probability of visiting a store/shopping center is proportional to the attractiveness of the store/shopping center, (ii) the probability of visiting a store/shopping center is inversely proportional to the degree of competition in the area, and (iii) the probability of visiting a store/shopping center is inversely proportional to some measure of distance or travel time. These assumptions can be expressed in the following mathematical expression:

$$p_{ij} = \frac{A_j^\alpha f(d_{ij})}{\sum_{j'} A_{j'}^\alpha f(d_{ij'})} \tag{1}$$

where,

p_{ij} is the probability that a customer living in residential zone $i(i = 1, 2, 3, \ldots)$ will visit store or shopping center j ($j = 1, 2, 3 \ldots$);

A_j is the attractiveness of store or shopping center j;

$f(d_{ij'})$ is a function of the distance between residential zone i and store or shopping center j';

α is a parameter to be estimated.

In most studies, α has been assumed equal to 1.0, consistent with the assumptions articulated above. However, in some cases, it was argued that larger shopping centers attract trade more than proportionally. If this is true, α should be larger than 1.0.

The attractiveness variable has been operationalized in many different ways. In most geographical studies of shopping centers, attractiveness has been operationalized in terms of the amount of retail floorspace. Over the years, however, multi-dimensional measures have increasingly been used. While geographers tended to focus on such variables as number of parking lots, number of anchor stores and the like, marketing researchers have used variables such as store atmosphere, price level, quality of merchandise and service quality. Moreover, different distance decay functions have been used to express the notion that customers are distance sensitive and that the probability of visiting a store, ceteris paribus, decreases with increasing distance. Distance itself has been measured in terms of Euclidean distance, distance over the transportation network, travel time, etc.

The spatial interaction model predicts the probability that a customer will choose a particular store or shopping center. Predictions of sales levels were obtained by multiplying these probabilities with the known population size in each residential zone and expenditure levels per capita, and summing these residential zones. In doing so, the level of aggregation may differ considerably. Sometimes average figures were used, sometimes figures were broken down according to socio-demographics (e.g. income level) and product class. Predicted sales can then be compared with norms regarding for example sales/sq foot that exist in the industry to assess market potential and viability.

The model also allows one to predict the shift in sales across shopping centers. Although this information might not be particularly relevant to the retailer, planning authorities have used this information to decide whether or not to allow development in particular areas of the city. Examples of the classical production-constrained spatial interaction model, also know as the Huff-model (Huff, 1963), and its extensions can be found in Lakshmanan & Hansen (1965), Gibson & Pullen (1972), Jain & Mahajan (1979), Ghosh (1984), and Gonzales-Benito, et al. (2000) to name a few.

Spatial interactions problems have been criticized in terms of their lack of underlying theory about individual spatial behavior. It led to the development of so-called discrete choice models that were based on economic and psychological theories of individual choice behavior, leading to a specific specification of the model. It should be noted that the specific mathematical expression was also applied in the context of spatial interaction models, and in fact some authors simply continued to apply their old models, only using a different terminology. Strictly speaking, however, this is incorrect as far as the underlying theory is concerned.

Like spatial interaction models, discrete choice models allow one to predict the probability that a consumer will choose a particular shopping center or store. Again, this choice is predicted as a function of locational and non-locational attributes of the choice alternatives and socio-economic characteristics of customers. It results in predictions of market share, consumer demand, and when linked with external criteria, provides indicators of feasibility and competitive impact. Thus, the goals and outcomes of these different modeling approaches do not differ; the main difference is their theoretical underpinning.

Discrete choice models can be derived from different theories, such as Lancaster's consumption theory, random utility theory and Luce's psychological choice theory (see e.g., Ben-Akiva and Lerman, 1985). Although one should be aware of the difference between these theories, one of the possible derivations of the model is as follows. It is assumed that an individual i is faced with J shopping alternatives in his choice set A_i. Each shopping alternative j can be described by a bundle of attributes $\mathbf{x}_j$. Further, it is assumed that individuals derive some utility for each attribute, and that these attribute utilities are combined according to some algebraic equation to arrive at an overall utility for shopping alternative j. Thus,

$$U_{ij} = f(\mathbf{x}_j) \forall j \in A_i \tag{2}$$

In most studies, a linear additive function is assumed to describe this integration process (function f). The linear additive function describes a compensatory decision making process in the sense that lower values on some attributes may, at least partially, be compensated by higher evaluation scores on one or more of the remaining attributes. The linear additive function can be expressed as:

$$U_{ij} = \sum_k \beta_k x_{jk} \tag{3}$$

where, x_{jk} describes the level (or value) of attribute k for shopping alternative j and the β's are the parameters to estimate.

In order to predict the choice of shopping alternative, an additional assumption is required to link overall utility to choice. For example, Luce (1959) assumed that the choice probabilities are proportional to these overall utilities. Hence, the probability of individual i choosing shopping alternative j is equal to:

$$p_{ij/j, j' \in A_i} = \frac{U_{ij}}{\sum\limits_{j' \in A_i} U_{ij'}} \qquad (4)$$

The assumption of such deterministic behavior is of course very strict. It does not account for inconsistency in consumer choice behavior. Moreover, it does not account for the fact that the modeler is faced with measurement error. If one assumes that preferences are not fixed but stochastic, then the utility function should be expressed as a random utility, consisting of a systematic utility component (V_{ij}) and an error component (ε_{ij}). Thus,

$$U_{ij} = V_{ij} + \varepsilon_{ij} \qquad (5)$$

If one assumes that an individual will maximize his/her utility, then the probability of shopping alternative j being selected from choice set A_i is given by the following equation:

$$P_{ij}/A_i \, \mathrm{Pr}(V_{ij} + \varepsilon_{ij} \geq V_{ij'} + \varepsilon_{ij'}), \, \forall j, j' \in A_i \qquad (6)$$

To calculate these choice probabilities, additional assumptions have to be made regarding the distribution of the error terms. Over the years, many different choice models have been suggested (see Timmermans and Golledge, 1990 for details). The most commonly used model is the multinomial logit model, which can be derived by assuming that the error terms are identically and independently Gumbel distributed. This model has the following form:

$$p_{ij} = \frac{\exp(U_{ij})}{\sum\limits_{j' \in A_i} \exp(U_{ij'})} \qquad (7)$$

In case of spatial applications, one of the attributes will typically be distance or travel time, resulting in the following spatial choice model (χ being the parameter for distance):

$$p_{ij} = \frac{\exp(\sum\limits_{k} \beta_k x_{ijk} - \chi d_{ij})}{\sum\limits_{j' \in A_i} \exp(\sum\limits_{k} \beta_k x_{ij'k} - \chi d_{ij'})} \qquad (8)$$

It should be noted that we kept in subscript i. In an applied context, however, it is usually dropped. In that case, the error terms are assumed to reflect heterogeneity among customers as well. The above set of rigorous assumptions results in an easy-to-use model but at the cost of ending up with a model specification, which can be rather unrealistic under particular circumstances.

In particular, the multinomial logit model is based on the rather rigorous assumption that the introduction of a new store or shopping center will draw market share from the existing stores or centers in direct proportion to their overall utility. This property does not hold only for the multinomial logit model; it also applies to the Huff/production-constrained spatial interaction model. One might argue that more realistically, the (dis)similarity of stores will influence choice probabilities in the sense that more similar stores will compete more for the same market share.

There are several alternatives to avoid this IIA-property of the multinomial logit model. The assumption of identically and independently distributed error terms may be relaxed. Alternatively, the existence and/or attributes of competing choice alternatives may be included in the specification of the utility function of a particular choice alternative, leading to the mother or universal logit model (e.g., Timmermans et al., 1991). However, the best-known way of developing non-IIA choice models is to specify a nested logit model. The idea behind the nested logit model is that the modeler constructs a hierarchical structure representing the order in which alternatives are assumed to share unobserved attributes. By placing similar alternatives in the same nest, the expected shifts in market shares can be approximated (Ben-Akiva and Lerman, 1985).

Unlike many other application contexts, the nested logit model did not receive that much attention in retailing research, especially not in geography. Although they were never positioned that way, other non-IIA models dominated in geography. Fotheringham (1983) introduced the so-called competing destinations model, originally to study migration, but later it was also applied to retail location problems (Fotheringham, 1988). When applied to spatial shopping behavior, the choice behavior is conceptualized as a hierarchical choice process in which individuals first select a part of their environment or a shopping center, and then select the shopping center within that environment or a store within a shopping center. This hierarchical choice process is modeled by incorporating the accessibility of a shopping alternative to all the other potential shopping alternatives in the attractiveness term of the production-constrained spatial interaction model. The simplest approach to the specification of this additional term is the use of a Hansen-type accessibility measure. The competing destinations model was applied with mixed results by Guy (1987) to data on food and grocery shopping behavior in the Cardiff area. The improvement in fit over a conventional spatial interaction model was small but led to improvements in the aggregated predictions of shopping flows. However, the sign of the parameter was counterintuitive in that shopping centers facing greater competition tended to attract more trade. Thus, agglomeration effects appeared to be present, which is unexpected for food and grocery shopping.

Whereas Fortheringham elaborated the spatial interaction model, Borgers and Timmermans developed several models that incorporated both the similarity among shopping centers and spatial structure effects into the multinomial logit model. The model they suggested (Borgers and Timmermans, 1987, 1988) can be considered as the multinomial logit equivalent of Fotheringham's competing destinations model, although it also included substitution effects. Measures of spatial structure and dissimilarity were introduced in the utility function. A problem of this model, shared with the competing destination model, is that it does not behave as theoretically expected when the choice alternatives are located at more than three different locations. Following an idea introduced in the marketing literature by Kamakura and Srivastava (1984), Borgers and Timmermans (1985a, 1985b) developed a second model that allows the estimation of substitution and spatial structure effects. Similarity and spatial structure effects are incorporated into the variance-covariance matrix of the random disturbance terms of the utility function of the logit model. The covariance structure is modeled explicitly in terms of distances in the attribute space between the shopping centers as a measure of substitution. An analysis of consumer choice data for hypothetical structures indicated that when two shopping centers are located close to each other, choice probabilities of the remaining shopping centers decrease (agglomeration effects). When, in addition, both centers differ in terms of their attractiveness, the choice probability of one shopping center may increase while the choice probability of the other center decreases (redistribution effects). The predictive ability of both models was tested and compared to that of other substitution models using data on the choice of 34 shopping centers in the Maastricht region, The Netherlands. The substitution/spatial structure effects models performed only slightly better than the conventional MNL model: a finding consistent with the results obtained for simulated data sets (Borgers & Timmermans, 1987). The models that accounted for spatial structure effects produced the best results. This study also led to the conclusion that the spatial transferability of the substitution and spatial structure models is slightly better than those obtained for the MNL model.

Most of these models of spatial choice behavior were derived from data about actual shopping behavior in real markets. However, a new shopping center may have characteristics that are beyond the domain of experience. In such cases, the predictive validity of the model may be limited. In other cases, such as teleshopping and e-commerce, the new service may not even exist, implying that data on actual shopping behavior are not available. Geographers have been in the front seat to develop so-called conjoint, decompositional preference/choice or stated preference models for such situations (Timmermans, 1984). The mathematical expression of these models is the same as those derived from revealed preference data, but the data collection differs. It is assumed that shopping centers or services can be characterized in terms of a set of attributes. Each attribute has a series of levels. One then creates an experimental design to generate a set of hypothetical shopping centers that varies the positions of the attributes in a controlled manner. The specifics of the model determine the creation of an appropriate design. Once the design is created, respondents are asked to rank or rate the resulting attribute profiles in terms of their overall preference. Alternatively, the profiles can be placed into choice sets, and respondents are asked to choose from each choice set, the profile they like best. The overall evaluations of the designed shopping centers are then decomposed into a set of part-worth utilities associated with every level or position of each attribute. The ability of the estimated part-worth measures to recover an individual' s observed evaluation responses is assessed by a goodness-of-fit measure. The best fitting conjoint model for a given decision task is identified by comparing these goodness-of-fit measures for alternative models. Once part-worth utilities have been estimated, a choice rule is postulated to predict the choices that an individual is likely to make given the estimated preference function. A simple, commonly used, postulate is that an individual will choose the shopping center with the highest overall evaluation. Alternatively, different probabilistic rules can be postulated (e.g., Timmermans & van der Heijden, 1984). Shopping behavior can then be simulated by defining real-world shopping centers in terms of their positions or levels on each attribute varied in the experimental task, calculating each individual's overall utility score for each shopping center on the basis of the estimated preference function, and applying the postulated choice or decision rule to map an individual's estimated overall preference or utility into choices. Applications of this approach to retailing can be found in SchuIer (1979), Timmermans (1980, 1982), Moore (1990), Louviere, *et al.* (1994), Oppewal, Louviere and Timmermans (1997). If the experimental task is a choice as opposed to a ranking or rating task, typically a multinomial logit model is assumed to represent the choice process. More recently, however, non-IIA models have been suggested for experimental design data (Timmermans and van der Waerden, 1992; Timmermans, 1996).

The spatial interaction and discrete choice models have received ample application in retailing both in academic and applied research. Nevertheless, a critical reflection of the assumptions underlying these models indicates that they relate to very specific shopping occasions. First, these models assume that the shopping trip is made from home: there are no mechanisms to capture multi-stops or shopping trips originating from other locations than home. Secondly, there is rarely an explicit consideration of different items that need to be purchased or multiple purposes for the trip. At best, the competing destination model and its equivalents allow one to incorporate agglomeration and/or competition effects, but only in an indirect manner. Thirdly, it is assumed that the attractiveness or utility of shopping alternatives is constant over time. In reality, however, it may depend on the time of the day, the day of the week, the time elapsed since the last shopping trip, the available transport mode, etc. Finally, the emergence of new retail formats, such as box stores, power centers, factory outlet, retail parks etc (see e.g., Wrigley, 1994, Cliquet, 2000, Morganosky and Cude, 2000, Hahn, 2000) may have an impact on spatial shopping behavior. In other words, the shopping activity may be considerably more complex than is assumed in these models. The last decade therefore has witnessed increasing attention to multi-stop, multi-purpose trips and even to the modeling of comprehensive daily activity-travel patterns. An overview of this recent stream of research in given in Timmermans, Arentze and Joh (2002).

Originally, models of multi-stop, multi-purpose behavior relied on semi-Markov process models or Monte Carlo simulations (e.g., Horowitz, 1980; O'Kelly, 1981) A review of these early approaches and their problems is provided in Thill and Thomas (1987). More recently, utility-maximizing models and optimizing models have dominated the field (e.g., Bacon, 1995, Krider and Weinberg, 2000). An important contribution in this regard was made by Kitamura (1984), who introduced the concept of prospective utility. It states that the utility of a destination is not only a function of its inherent attributes and the distance to that destination, but also of the utility of continuing the trip from that destination. Based on this notion, Arentze, Borgers and Timmermans (1993) developed a model of multi-purpose shopping trip behavior. It assumed a list of items that need to be purchased in different frequency. The choice of a shopping center to purchase a particular item is predicted as a multinomial logit function. The probability of being any other good during the same trip is then the choice between either buying this item during the same trip or buying it during another trip. A recursive equation is derived from this premise, which allows one to predict the frequency of purchasing different goods and the distribution of visits across shopping centers. Dellaert et al. (1998) generalized this approach to account for both multi-purpose and multi-stop aspects of the trip chain. Arentze and Timmermans (2001) showed how store performance indicators can be derived from such models.

More recently, trip chaining is studied as part of more comprehensive activity-travel patterns. Models such as DAS (Bowman, et al. (1998), PCATS (Kitamura and Fujii (1998), CATGW (Bhat, 1999) and Albatross (Arentze and Timmermans, 2000), to name a few operational models, predict which activities are conducted where, when, for how long, the transport mode involved and sometimes yet other facets of such patterns. Shopping behavior is considered time and space dependent, and is influenced by considerations of time, distance, attractiveness, various constraints and the activity schedule itself. Some of these models are still based on the nested logit model, other models are based on decision heuristics. Baker (1994, 1996, 2000) developed a space-time model of aggregate consumer shopping behavior using differential equations, defining distributions of when and where consumers enact aggregate shopping behavior. It should be evident that such activity-based models are no longer based on assumptions such as home-based shopping trips and time-invariant utilities. It should however also be realized that such models are considerably more complex, less easy to use, and require more detailed data.

PEDESTRIAN MOVEMENT

The models discussed in the previous section have been used to assess the feasibility of new stores and shopping centers. They typically examine spatial shopping behavior at the regional scale. In addition, there is a stream of literature in applied geography that has looked at pedestrian movements. Such studies are relevant to evaluate retail developments and plans within city centers and shopping centers and to predict the likely impact of new local transportation plans.

Already in the early 1970s, research has shown that the commercial viability of inner-city shopping streets is highly influenced by pedestrian movement and that the impact of new retail developments is closely related to the locational patterns of magnet stores and the distribution of the transport termini (Johnston and Kissling, 1971; Pacione, 1980; Walmsley and Lewis, 1989; Lorch and Smith, 1993). More recent transport policies such as the reduction of car use and mobility in city centers, restricted parking, increased parking fees, one-way streets and other policy measures may have a dramatic impact on retail viability. In general, pedestrian shopping behaviour in a shopping center depends on their knowledge about shops, the street network, the distribution of shops, and the choice mechanisms that are involved in deciding where to shop, in what order, and which route to take. Traditionally (Sandahl and Percivall, 1972; Borgers and Timmermans, 1986c,d; Hagishima *et al.*, 1987) gravity models, sometimes embedded in Markov chains, have been used to predict pedestrian movement. Thus, either explicitly or implicitly, researchers have assumed that pedestrian destination and route choice can be viewed as the result of utility-maximising behaviour, in which pedestrians trade-off the attractiveness of stores or shopping streets and the distance or time it takes to visit that store. Later, the assumption of utility-maximizing behavior has been replaced by studies examining choice heuristics of pedestrians (Hayes-Roth and Hayes-Roth, 1979; Gärling, 1987; Van der Hagen, Borgers and Timmermans, 1991; Kurose, Borgers and Timmermans, 2000).

Much of this research is empirical in nature. An interesting, rapidly growing field of research concerns the development of cellular automata and multi-agent models of pedestrian flows. Stimulated by complexity theory, these models attempt to show that simple behavioral rules can induce emergent patterns in pedestrian movement. Following the success of cellular automata models in other areas of geography, urban planning, regional science and transportation, the late 1990s have seen the appearance of cellular automata models to simulate pedestrian movement (Blue and Adler, 1998; 1999). The challenge of cellular automata modeling therefore is to find the set of rules that would validly generate seemingly chaotic emergent patterns in pedestrian movement. One way of introducing this greater variability and complexity is to introduce agents in the simulated environment. PEDFLOW (Kukla *et al.*, 2001), STREETS (Schelhorn *et al.*, 1999) and Amanda (Dijkstra, Timmermans and Jessurun (2000) are all agent-based approaches for the modeling of pedestrians in urban environments. Every pedestrian is conceptualised as an agent, each with its own goals, information, agenda, etc. A set of rules determines how these agent behave in the environment. To date, these models have only showed how typical patterns can be simulated. Further work is required to develop fully operational models, which can be used to assess how pedestrians react to changing retail land use patterns and a changing environment.

CONCLUSIONS

The aim of this chapter has been the discuss progress in some areas of retail geography to indicate the kind of tools that have been developed and improved over the years and that have been applied in practice. The choice of topics was restricted to cross-sectional models of spatial shopping behavior at the regional level and models of pedestrian movement at the local level. Together, they provide practitioners with a rich set of tools to make better informed decisions. Progress in the past has been triggered by an understanding in practice of the limitations of these models. It emphasizes the importance of the link between academic and applied research. The dissemination of these tools would benefit from their inclusion in geographical information and decision support systems Kohsaka (1997), Benoit and Clarke (1997) and Arentze, Borgers and Timmermans (1996, 2000) represents examples of recent progress in this area of research.

REFERENCES

Arentze, T.A., Borgers, A.W.J., and Timmermans, H.J.P. 1993. A Model of Multi-Purpose Shopping Trip Behavior. *Papers in Regional Science* 72: 239–256.

Arentze, T.A., Borgers, A.W.J., and Timmermans, H.J.P. 1996. Design of a View-Based DSS for Location Planning. *International Journal of Geographical Information Systems* 10: 219–236.

Arentze, T.A., Borgers, A.W.J., and Timmermans, H.J.P. 2000. A Knowledge-Based System for Developing Retail Location Strategies. *Computers, Environment and Urban Systems* 24: 1–20.

Arentze, T.A., Borgers, A.W.J., and Timmermans, H.J.P. 2000. *Albatross: A Learning-Based Transportation Oriented Simulation System*. European Institute of Retailing and Services Studies, Eindhoven.

Arentze, T.A., Borgers, A.W.J., and Timmermans, H.J.P. 2001. Deriving Performance Indicators from Models of Multipurpose Shopping Behavior. *Journal of Retailing and Consumer Services* 8: 325–334.

Bacon, R.W. 1995. Combined Trips and the Frequency of Shopping. *Journal of Retailing and Consumer Services* 2: 175–184.

Baker, R.G.V. 1994. An Assessment of the Space-Time Differential Model for Aggregate Trip Behaviour to Planned Suburban Shopping Centres. *Geographical Analysis* 26: 341–362.

Baker, R.G.V. 1996. Multi-Purpose Shopping Behaviour at Planned Suburban Shopping Centres: A Space-Time Analysis. *Environment and Planning A* 28: 611–630.

Baker, R.G.V. 2000. Towards a Dynamic Aggregate Shopping Model and its Application to Retail Trading Hour and Market Area Analysis. *Papers in Regional Science* 79: 413–434.

Ben-Akiva, M, and Lerman, S. 1985. *Discrete Choice Analysis: Theory and Application to Travel Demand*. Cambridge, Mass.: MIT Press.

Benoit, D., and Clarke, G.P. 1997. Assessing GIS for Retail Location Planning. *Journal of Retailing and Consumer Services* 4: 239–258.

Bhat, C. 1999. A Comprehensive and Operational Analysis Framework for Generating the Daily Activity-Travel Pattern of Workers, Paper presented at the 78th Annual Meeting of the Transportation Research Board, Washington, January 10–14, 2001.

Blue, V.J., and Adler, J.L. 1998. Emergent Fundamental Pedestrian Flows from Cellular Automata Microsimulation. *Transportation Research Record* 1644: 29–36.

Blue, V.J., and Adler, J.L. 1999. Cellular Automata Microsimulation of Bi-Directional Pedestrian Flows. *Journal of the Transportation Research Board*, 135–141.

Borgers, A.W.J., and Timmermans, H.J.P. 1985a. Context Effects and Spatial Choice Models. Paper presented at the 25th Regional Science Association Conference, Budapest.

Borgers, A.W.J., and Timmermans, H.J.P. 1985b. Effects of Spatial Arrangement and Similarity on Spatial Choice Behavior. Paper presented at 4th Colloquium on Theoretical and Quantitative Geography, Veldhoven.

Borgers, A.W.J., and Timmermans, H.J.P. 1986c. City Centre Entry Points, Store Location Patterns and Pedestrian Route Choice Behaviour: A Micro-Level Simulation Model. *Socio-Economic Planning Sciences* 20: 25–31.

Borgers, A.W.J., and Timmermans, H.J.P. 1986d. A Model of Pedestrian Route Choice and the Demand for Retail Facilities within Inner-City Sopping Areas. *Geographical Analysis* 18: 115–128.

Borgers, A.W.J., and Timmermans, H.J.P. 1987. Choice Model Specification, Substitution and Spatial Structure Effects: A Simulation Experiment. *Regional Science and Urban Economics* 17: 29–47.

Borgers, A.W.J., and Timmermans, H.J.P. 1988. A Context-Sensitive Model of Spatial Shopping Behavior. In *Behavioural Modelling in Geography and Planning*, eds. R.G. Golledge, and H.J.P. Timmermans, pp. 159–179. London: Croom Helm.

Bowman, J.L., Bradley, M., Shiftan, Y., Lawton, T.K. and, Ben-Akiva, M.E. 1998. Demonstration of an Activity-Based Model System for Portland. Paper presented at the 8th World Conference on Transport Research, Antwerp.

Cliquet, G. 2000. Large Format Retailers. *Journal of Retailing and Consumer Services* 7: 183–196.

Dellaert B., Arentze, T.A., Borgers, A.W.J., Timmermans, H.J.P., and Bierlaire, M. 1998. A Conjoint Based Multi-Purpose Multi-Stop Model of Consumer Shopping Centre Choice. *Journal of Marketing Research* 35: 177–188.

Dijkstra, J., Timmermans, H.J.P., and Jessurun, J. 2000. A Multi-Agent Cellular Automata System for Visualizing Simulated Pedestrian Activity. In *Theoretical and Practical Issues on Cellular Automata – Proceedings of the 4th International Conference on cellular Automat for Research and Industry*. Pp. 29–36. Karlsruhe.

Fotheringham, A.S. 1983. A New Set of Spatial Interaction Models: The Theory of Competing Destinations. *Environment and Planning A* 15: 15–36.

Fotheringham, A.S. 1988. Consumer Store Choice and Choice Set Definition. *Marketing Science* 7: 299–310.

Gärling T. 1987. The Role of Cognitive Maps in Spatial Decisions and Choices. Unpublished paper, Department of Psychology, University of Umeå.

Gibson, M., and Pullen, M. 1972. Retail Turnover in the East-Midlands: A Regional Application of a Gravity Model. *Regional Studies* 6: 183–196.

Ghosh, A. 1984. Parameter Nonstationarity in Retail Choice Models. *Journal of Business Research* 12: 425–436.

Gonzales-Benito, O., Greatorex, M, and Munoz-Gallego, P.A. 2000. Assessment of Potential Retail Segmentation Variables: An Approach Based on a Subjective MCI Resource Allocation Model. *Journal of Retailing and Consumer Services* 7: 171–179.

Guy, C.M. 1987. Recent Advances in Spatial Interaction Modelling: An Application to the Forecasting of Shopping Travel. *Environment and Planning A* 19: 173–186.

Hagishima, S., Mitsuyoshi, K., and Kurose, S. 1987. Estimation of Pedestrian Shopping Trips in a Neighbourhood by Using a Spatial Interaction Model. *Environment and Planning A* 19: 1139–1153.

Hagen, X. van der, Borgers, A.W.J., and Timmermans, H.J.P. 1991. Spatiotemporal Sequencing Processes of Pedestrians in Urban Retail Environments. *Papers in Regional Science* 70: 37–52.

Hahn, B. 2000. Power Centres: A New Retail Format in the United States of America. *Journal of Retailing and Consumer Services* 7: 223–232.

Hays-Roth, B., and Hays-Roth, E. 1979. A Cognitive Model of Planning. *Cognitive Science* 3: 275–310

Horowitz, J. 1980. A Utility-Maximizing Model of the Demand for Multi-Destination Non-Work Travel. *Transportation Research B* 14: 369–386.

Huff, D.L. 1963. A Probabilistic Analysis of Shopping Center Trade Areas. *Land Economics* 38: 64–66.

Jain, A.K., and Mahajan, V. 1979. Evaluating the Competitive Environment in Retailing Using Multiplicative Competitive Interactive Models. In *Research in Marketing* ed. J. Sheth. Greenwich: JAI.

Johnston, R.J., and Kissling, C.C. 1971. Establishment Use Patterns within Central Places. *Australian Geographical Studies* 9: 166–132.

Kamakura, W.A., and Srivastava, R.K. 1984. Predicting Choice Shares Under Conditions of Brand Interdependence. *Journal of Marketing Research* 21: 420–432.

Kohsaka, H. 1997. Monitoring and Analysis of a Retail Trading Area by a Card Information/GIS Approach. *Journal of Retailing and Consumer Services* 4: 109–116.

Kitamura, R. 1984. Incorporating Trip Chaining into Analysis of Destination Choice. *Transportation Research* 18B: 67–81.

Kitamura, R. and Fujii, S. 1998. Two Computational Process Models of Activity-Travel Choice. In *Theoretical Foundations of Travel Choice Modelling*, eds. T. Gärling, T. Laitila, and K. Westin. pp. 251–279. Oxford: Elsevier.

Krider, R.E., and Weinberg, C.B. 2000. Product Perishability and Multistore Grocery Shopping. *Journal of Retailing and Consumer Services* 7: 1–18.

Kukla, R., Kerridge, J., Willis, A., and Hine, J. 2001. PEDFLOW: Development of an Autonomous Agent Model of Pedestrian Flow, Paper presented at the 80th Annual Meeting of the Transportation Research Board, Washington, D.C., January 7–11.

Kurose, S., Borgers, A.W.J., and Timmermans, H.J.P. 2001. Classifying Pedestrian Shopping Behaviour According To Implied Heuristic Choice Rules. *Planning and Design* 28: 405–418.

Lakshmanan, T.R., and Hansen, W.A. 1965. A Retail Market Potential Model. *Journal of American Institute of Planners* 31: 134–143.

Lorch, B.J., and Smith, M.J. 1993. Pedestrian Movement and the Downtown Enclosed Shopping Centre. *Journal of the American Planning Association* 51: 75–86.

Louviere, J.J., and Gaeth, G.J. 1987. Decomposing the Determinants of Retail Facility Choice using the Method of Hierarchical Information Integration: A Supermarket Illustration. *Journal of Retailing* 63: 25–48.

Louviere, J.J., and Gaeth, G.J. Johnson, R, and Chrzan. 1994. Accommodating Ideal Brands and Testing the Predictive Validity of Brand-Anchored Conjoint Analysis: A Multiple Choice experiment Approach. *Journal of Retailing and Consumer Services* 1: 21–30.

Luce, R.D. 1959. *Individual Choice Behaviour: A Theoretical Analysis.* New York: Wiley and Sons.

Moore, L. 1990. Segmentation of Store Choice Models using Stated Preferences. *Papers of the Regional Science Association* 69: 121–131.

Morganosky, M. and Cude, B.J. 2000. Large Format Retailing in the US. *Journal of Retailing and Consumer Services* 7: 215–222.

O'Kelly, M. 1981. A Model of Demand for Retail Facilities Incorporating Multistop, Multipurpose Trips. *Geographical Analysis* 13: 134–148.

Oppewal, H., Timmermans, H.J.P., and Louviere, J.J. 1997. Modelling the Effects of Shopping Centre Size and Store Variety on Consumer Choice Behaviour. *Environment and Planning A* 29: 1073–1090.

Pacione, M. 1980. Redevelopment of a Medium-Sized Central Shopping Area. *Tijdschrift voor Economische en Sociale Geografie* 71: 159–168.

Sandahl, J., and Percivall, M. 1972. A Pedestrian Traffic Model for Town Centres. *Traffic Quarterly* 26: 359–372

Schelhorn, T., O'Sullivan, D., Haklay, M., and Thurstain-Goodwin, M. 1999. STREETS: An Agent-Based Pedestrian Model. Presented at Computers in Urban Planning and Urban Management, Venice, 8–10 September, 1999.

Schuler, H.J. 1979. A Disaggregate Store-Choice Model of Spatial Decision-Making. *The Professional Geographer* 31: 146–156.

Thill, J.C., and Thomas, I. 1991. Towards Conceptualizing Trip-Chaining Behavior: A Review. *Geographical Analysis* 19: 1–17.

Timmermans, H.J.P. 1980. Unidimensional Conjoint Measurement Models and Consumer Decision-Making. *Area* 12: 291–300.

Timmermans, H.J.P. 1982. Consumer Choice of Shopping Centre: An Information Integration Approach. *Regional Studies* 16: 171–182.

Timmermans, H.J.P. 1984. Decompositional Multiattribute Preference Models in Spatial Choice Analysis: A Review of Some Recent Developments. *Progress in Human Geography* 8: 189–221.

Timmermans, H.J.P. 1996. A Stated Choice Model of Sequential Mode and Destination Choice Behaviour for Shopping Trips. *Environment and Planning A* 28: 173–184.

Timmermans, H.J.P., Arentze, T.A., and Joh, C.-H. 2002. Analyzing Space-time Behavior: New Approaches to Old Problems, *Progress in Human Geography* 26: 175–190.

Timmermans, H.J.P., and van der Waerden, P.J.H.J. 1991. Mother Logit Analysis of Consumer Shopping Destination Choice. *Journal of Business Research* 23: 311–323.

Timmermans, H.J.P., and Golledge, R.G. 1990. Applications of Behavioural Research on Spatial Choice Problems II: Preference and Choice. *Progress in Human Geography* 14: 311–354.

Timmermans, H.J.P., and van der Heijden, R.E.C.M. 1984. The Predictive Ability of Alternative Decision Rules in Decompositional Multiattribute Preference Models. *Sistemi Urbani* 5: 89–101.

Timmermans, H.J.P., and van der Waerden, P.J.H.J. 1992, Modelling Sequential Choice Processes: The Case of Two-Stop Trip Chaining. *Environment and Planning A*, 24: 1483–1490.

Walmsley, D.J., and Lewis, G.J. 1989. The Pace of Pedestrian Flows in Cities. *Environment and Behavior* 21: 123–150.

Wilson, A.G. 1971. A Family of Spatial Interaction Models and Associated Developments. *Environment and Planning A* 3: 1–32.

Wrigley, N. 1994. After the Store Wars: Towards a New Era of Competition in UK Food Retailing. *Journal of Retailing and Consumer Services* 1: 5–20.

PART II

A WORLD PERSPECTIVE

JORGE GASPAR

CHAPTER 8
APPLIED GEOGRAPHY IN WESTERN AND
SOUTHERN EUROPE

I. INTRODUCTION AND OVERVIEW

In Western and Southern Europe there is a substantial literature dealing with the application of theoretical concepts, methodologies and techniques developed by university-based geographers. Two of the most recent were authored by contributors to this book – Phlipponeau 1999 and Pacione 1999. These two works are complementary and cover topics ranging from local and regional planning to geography's contributions in a variety of public and private sector settings.

Whereas "applied geography" is well established in Western and Southern Europe there is still controversy – especially when the discussion turns to the place of applied work in the university curriculum. Even strong advocates of applied geography debate its place in undergraduate education and especially if it should be in the first levels of the geography curriculum.

The concept of applied geography is not new. According to Phlipponeau H.J. Herberston who was the first geographer to use the term "applied geography" around 1880 or so (Phlipponeau 1969, 22; Haure 1965). But even earlier than 1880 there were clear indications that geography was an applied science. A textbook published in Coimbra, Portugal in 1830 as part of the university's curriculum reform begins with definitions: "geography is the science that deals with the description of the planet we inhabit" and it is divided into two parts ... "theoretical geography which indicates in general the objects that must be described" and "applied or practical geography which, by applying the principles of theoretical geography, describes effectively the earth in its former and present states" (Liçes Elementares de Geografia e Chronologia). Despite these early references, however, some will still argue that "real" applied geography did not emerge until after World War Two. Post-war applied work was driven by the demand for knowledge which would be useful to the work of spatial, social, and economic restructuring of a war-torn continent. Whereas the contributions of geographers to town and country planning since the late 19th Century in Great Britain are acknowledged it is argued that it is during the Post-World War Two period that geography really came into its own as a planning-oriented discipline in Europe.

151

A. Bailly and L. J. Gibson (eds.), Applied Geography, 151–168.
© 2004 Kluwer Academic Publishers. Printed in the Netherlands.

The general potential of geography as an applied discipline was discussed by Omer Tulippe in his 1956 paper "Geographie Appliquée." A few years later two important books followed. In the United Kingdom and France, respectively, Applied Geography and Geographie et Action: Introduction à la Geographie Appliquée (Stamp 1960 and Phlipponeau 1960) were issued.

As noted previously, the geography – planning connection became more intense during the 1950's. The first, post-war Congress of the IGU (International Geographical Union) was held in Lisbon in 1949. This meeting saw the creation of a Commission for Regional Planning Studies chaired by Jean Gottman (1952). Additionally in the 1950's articles and manuals referring to the geography and planning theme were produced by Freeman (1958).

The post-World War Two environment encouraged geographers to become engaged in real-world problem solving. It is interesting to speculate about the role that the European Union might play in creating the next wave of interest in applied geography. The socioeconomic and political realities of the European Union are apparent with the present 15 member European community. There is little consensus among academics on many of the issues facing the 15 member EU that exists today (Huntington 1997). Imagine the prospect for consensus when the European Union expands to 25 members in 2005. Speculation about a 25 country European Union is well beyond the scope of this paper. Our focus here will be an applied geography in the current 15 European Union countries and in the "could-be" countries of Norway and Switzerland.

Potentially, the European Union itself will support a "new" applied geography inasmuch as it is a new regional space. The EU by its presence redefines scale economies and notions of complementarity, regional strategies, and the need for interventions of various sorts.

The move toward a more multi-cultural and multinational European Region has certainly had an influence on geographic thinking in general and in the way that geography is taught in particular. The new geography curriculum in the European Union recognizes the emerging diversity of the new map of Europe and the increased importance of flows and linkages in integrating this new region. It may be too much to say that geography was saved as a teaching subject in the primary and secondary levels in many of Europe's countries. But it does seem clear that geography's standing as a distinctive discipline was enhanced by the need to understand the new geographical region called the European Union.

Additionally the European Union has created demand for other "geographical products." Throughout Europe educators, public administrators, and business people have found new needs for those skilled in cartography, GIS, and physical, social, and political planning. Educational institutions have sometimes reacted effectively to the demands of the new European market place but more often then not the most aggressive leadership in the movement to meet demand for "geographic products" has come from various professional associations.

Sometimes geographers are represented by their own scholarly societies in geography, regional science, and regional studies but more frequently the capabilities of applied geographers will be advocated by professional associations which serve practitioners in urban planning, regional planning, environmental engineering etc.

In many European universities applied geography is more-or-less synonymous with regional planning both because its content is appropriate and because so many geographic techniques have clear benefits. At Humboldt University in Berlin, to name one example, Applied Geography and Regional Planning is a clearly identified unit within the Human Geography Section.

How broad, or narrow, should our definition of applied geography be? Phlipponeau (1999) takes the position that applied geography is not a specific, narrowly defined field. He argues that all geography, directly or indirectly is applied inasmuch as plans and policies can be enhanced when informed by geographic perspectives.

The fact that applied geography is a term that covers a variety of topics is supported by a quick review of two web sites. (See the applied geography and planning link in Cybergeo's site (www.cybergeo.prese.fr) or author key words in (www.cf.wos.isiglobalnet2.com) for the term *applied geography*.

Easily the two most comprehensive contemporary discussions of applied geography in Europe are those offered by Pacione (1999) and Phlipponeau (1999). The latter author provides a revised and updated discussion that draws on his highly regarded earlier work to launch a *plaidoyer pour la geographie appliquée*. Pacione, on the other hand, presents a highly structured discussion of the roles of the applied geographer with special emphasis on policy issues. Despite the very substantial contributions of Phlipponeau and Pacione a great deal of work remains to be done. Europe's cultural, social, and economic institutions are undergoing profound and rapid change. This change presents applied geographers with new challenges and opportunities.

II. GEOGRAPHY AND POLITICS – AN OLD MARRIAGE

The term *political geography* emerged a century ago in association with F. Ratzel's classic work but the connection between geography and politics goes back to the works of Plato and Aristotle. The practical application of geography sooner or later involves the design and implementation of policies – and one way or another "the art of politics."

As early as the Renaissance there was a clear relationship between geography and politics: the discovery of new seas and new lands raised questions of political control, resource ownership, and trading rights. Geographers were responsible for mapping and otherwise evaluating trade routes and describing Europe's discoveries. And implicitly it was geographers who made evident Europe's "civilizing mission" which was based on their "superior culture." Ever since this age of exploration and discovery these notions of geographical determinism have been used as an instrument of power.

The Berlin Conference and the sharing of Africa represent a defining moment in the history of geography and political action. In the 19th Century geographical societies established themselves as key actors in the business of influencing policy aimed at managing and enlarging colonial empires. This, of course, was in addition to their more objective roles as cartographers and database managers.

A perhaps exagerated view of the roles of geographical societies was found in Portugal at the end of the 19th Century. Ramalho Ortigão, one of Portugal's leading writers wrote an open letter to the Prince Regent and future King of Portugal urging him to be careful of geographers and calling Lisbon's Geographical Society the "ante chamber of power;" he even suggested that many Ministers were, in fact "dangerous undercover geographers."

The proactive position of geographers through World War One produced a backlash and it encouraged geographers to redirect their efforts. The years between the First and Second World Wars saw growth of both a regionalist approach and a neopositivist focus. As it turned out both of these "new" approaches made geographers extremely useful during the post-World War Two restructuring and reconstruction process.

Geography's move away from political activism after World War One was actually the result of both push and pull forces. The push was the decline in determinism. The pull was planning. Sir Patrick Geddes (1898, 1915) established the basis for the application of geographic principles to regional and urban planning thus building the bridge between Britain's geographical tradition and the emergent French regional school (Livingstone 1992).

Despite these shifts and changes the marriage between geography and politics has not been lost. Since World War Two geography has played an important role in helping the developed world better understand the less developed world. In Europe there was a shift to new scales, especially the regional and the local (Pinchemel 1999). There also has been growth in the area of electoral studies which have followed the early work of French geographer André Siegfried (1913). Geographic studies of voter behavior and the deliniation of districts to unite (or divide) constituencies are two important examples.

The growth of quantitative geography in the 1950's and 1960's did a great deal to raise the profile of electoral geography by making it more objective and less partisan. The media, political parties, and public administrators all find these studies of voting behavior and preferences. An appropriate example comes from Portugal. In 2000 the government through the Parlimentary Affairs and Science Ministries commissioned three university-based units (two geography departments and a demography and statistics department) to independently redesign the countries electoral districts to fit the requirements of new electoral laws.

III. GEOGRAPHY AND GENERAL EDUCATION: TEACHING IS STILL THE PRIMARY APPLICATION!

From its earliest beginnings geography has had two faces. On the one hand geography is an essential element in every person's general education. It helps us understand better the world that we live in. As Sack (1997) says ... "we humans are geographical beings ... " on the other hand, geography has a long history of offering effective and efficient solutions to a wide range of very practical and often critical problems.

Basic geography is important both as a part of a solid general education but also because it is the foundation for more advanced applications. Historically Europe in general and Southern and Western Europe in particular have maintained a strong tradition in geographic education. Unfortunately, since 1960 geography has been in decline at least in some European countries. In extreme cases it has disappeared from the curriculum. This situation has negative repercussions at several levels – citizens are less well informed about issues and opportunities at local, regional, national, and supra-national levels, the flow of students from basic education to more advance geographic education is reduced, and finally the supply of professional geographers available to make contributions to commerce and society is diminished.

In Europe (and in much of Australasia and the United States too) human geography per se has been replaced by "social studies" and physical geography by "environmental science" in the primary and secondary schools. Whereas these courses are not without value they do not present the same integrated vision that is the hallmark of a good geographical education.

Another issue, one tied to the notion of "integrated vision" mentioned above, has to do with the trend to separate physical and human geography. There is a strong case to be made for a single geography that considers both the human and physical environments. It can be argued that this fully integrated geography best serves the general education needs of informed citizens and leads to advanced applications in everything from GIS and remote sensing to regional and local planning. Given the sweep of the geographers interests from land-use management issues at a local scale to global change a traditional and fully integrated geography would seem to be appropriate.

Given widespread interest today in the so called global-local continuum and a renewed interest in spatial management of large systems a return to a more integrated approach in geography would be welcomed in many quarters. Similarly the growing popularity of sustainable development and the expanded attention given to finding solutions for environmental problems at the local scale also reinforces the attractiveness of combining the human and physical dimensions. Gloucester University (www.glos.ac.uk) has developed an elaborate and perhaps model curriculum that stresses the importance of geographic scale when defining applied geography as "... the application of geographic knowledge and skills to identify the nature and cause of environmental, economical, and social problems and inform policies which lead to their resolution." They go on to point out that whereas their programs are based in human geography they teach courses in a way that allows a full range of environmental elements to be considered.

Perhaps human geographers are more concerned about the lack of physical geography in their programs than vice versa. It is not easy to find physical geography courses that pay much attention to the human environment and even more difficult to find entire programs that do so. It is not impossible to find curricula that deal with environmental hazards and risk in a sophisticated way but such courses are surprisingly few and far between.

Unfortunately, the trend in Portugal mirrors that of much of Europe. Physical geographers have developed strong programs in geomorphology, climatology and coastal dynamics but with very limited application to the real problems encountered in the human environment. There are a few exceptions – mainly in the management of coastal areas – but these exceptions are not significant enough to invalidate the generalization above.

IV. POST WORLD WAR TWO RECONSTRUCTION: THE EMERGENCE OF THE *New Geography* AS THE FOUNDATION FOR SPATIAL AND LAND-USE PLANNING

Following the Second World War geography experienced a "scientific revolution" that produced a *New Geography* – one that emphasized quantitative approaches to modeling and evaluation, and perhaps best of all modeling. This shift in approach put geography on the same level as the most rigorous and respected social sciences and it positioned geographers to assume positions of responsibility on integrated multidisciplinary research teams with both natural scientists and social scientists concerned with the spatial management of economic activities and residential areas.

It was the legacy of Sir Patrick Geddes that positioned geographers to make enormous contributions to the process of rebuilding Europe following the Second World War. But it has been geography's move to statistical analysis and quantitative models that has given geographers credibility in territorial planning since the 1950's.

Geography's position as a preferred theoretical foundation for spatial planning has been enhanced by its role in building and refining location theories and by the wide spread adaption of computer-based applications in geographical information systems and other domains of this nature.

Two final factors also proved critical in geography's rise as a major foundation for the planning profession. First, highly regarded scholars such as T. Hägerstrand, P. Gould, and P. Hall established a trend when they effectively reconciled high quality academic work with the demand of those engaged in practice and policy. Their work had "demonstration value" that effectively communicated geography's usefulness as both an academic subject and as an applied subject for business and government. Second, geography departments effectively placed graduates in public agencies and private companies and in doing so, "educated" employers about the value of hiring those with geographic training.

V. FROM MACRO REGIONALISATION TO LOCALIZATION: APPLIED GEOGRAPHY IN THE EUROPEAN UNION

A report presented to the third meeting of the Ministers responsible for EU politics and regional planning[1] establishes the important role of applied geography in the scientific and technical support of the European community. The commission of the European communities report was entitled *Europe 2000 – The Development: Perspective of the Community Territory*. This report is a synthesis of policy oriented case studies dealing with the human occupation and organization of places as well as their evaluation over time.

Although some terminology is over used or misused (the word "zone" is used in a variety of ways, for example) the report does a fine job of presenting applied geography *per se* in a flattering light. Further, it reflects several different theoretical orientations. The French geographical school is present with its emphasis on the regionalist paradigm but the report also reflects the approaches of the "new geography" when it deals with the European urban system. The *Executive Summary* provides an overview of the text and an overarching theme – Europe's Economic Geography and Emerging Trends. This discussion is followed by summaries of the major themes: demography; infrastructure and spatial networks (transport, information technology and telecommunications, energy, environment); the future of specific regions (urban zones, rural zones, frontier regions, coastal regions and islands) and lastly; cooperation between cities and regions.

The report's introduction features both sector and regional studies. The emphasis is on socioeconomic disparities and on the discriminatory policies that have created and perpetuated these disparities. Chapter one deals with demography. It is a pragmatic

---

1. CEC 1991, Hagul, 18–19 December.

chapter that defines terms, diagnoses problems, and offers solutions. The chapter looks at demography through the eyes of a geographer. The fact that the chapter pays attention to the importance of scale (local, regional, national, and international) helps assure that migration issues get appropriate attention.

The next section focuses on the geography of production and the space economy. It discusses traditional patterns and the role of new localization factors in creating an evolving map of European production activity.

The discussion of transport deals both with modes and with shifts of activity that follow changes in systems and networks. The issue of transport modes is effectively illustrated by a case study of the cross-channel tunnel and its regional impacts. Discussions focusing on the centre-periphery model, illustrate the regional shifts that can occur when there are system improvements that facilitate the flow of people, goods, and ideas within and between regions. A similar discussion of telecommunication and energy networks shows the role of system enhancement in these areas in producing a more even development and the reduction of economic asymmetry. This section also underscores the role of, and need for, interregional cooperation and international cooperation among neighbouring states.

Environmental issues and the conservation of natural resources are covered next. Among the critical issues covered are the requirements for truly sustainable development and the wisdom of preserving Europe's natural heritage.

The final section deals with the overarching concern for "community," i.e., European community. It is a geography-based discussion that deals with the big ideas that geographers can and do deal with effectively – urban systems and systems of cities, the future of rural regions, new and emerging roles for cities and regions, the special problems associated with border regions and coastal regions and islands. The concluding chapter deals with policy issues. It is cautiously written inasmuch as spatial development policies are the domain of the member states – not the EU *per se*. Discussions focus on three main themes – the need for more data and for harmonization of all data; creation of a spatial development committee; and recommendations for various interventions in the specific domains of frontier areas, city-region cooperation and centre-periphery connections.

When considering this report it is worth commenting on the Commission's emphasis on the importance of the principle of subsidiarity when applied to the responsibilities of the states, regions and local authorities in the areas of spatial and land planning. The report makes clear that the role of the Commission is squarely in the area of coordinated planning. The Commission provides technical assistance to promote balanced and sustainable economic development and it offers the member states "big picture" perspectives. But the responsibility for functional planning and for the articulation of specific policies lies with the member states.

The report is unlike many other reports in tone – it recognizes that compliance is voluntary and that approaches must be sensitive to cultural differences. In the document *Europe 2000+* the regional approach of the French School is in evidence but so too are other approaches such as the abundant use of cartographic images favored by the Dutch and the solidly Germanic focus on the "Raumondnung."

Europe 2000+ was officially presented at the ministerial meeting in Leipzig in September of 1994 during the German presidency. It was intended as a basis for the more comprehensive European Spatial Development Perspective (ESDP).

Europe 2000+ claims a dynamic spatial perspective. But it has come under fire for dividing Europe into eight regions. Their goal was to produce "trans regional studies" but given that there were 12 EU member states at the time it is easy to imagine that any regionalization scheme would have its critics. Still, the regionalization is interesting: the North Sea Area; the Centre Capitals; Atlantic Arc; Alpine Arc; Continental Diagonal; New Länder; Mediterranean Area. This approach was applied geography but it was offered in marketing language. The intent was to show Europe as a dynamic area where regional boundaries cut across national boundaries and to convince the reader that the new Europe should be focused on functional regions with contemporary significance, not just nation states with historical significance. But some readers found the approach more off-putting than fresh. In the end, the model offered in *Europe 2000+* ran out of steam and did not seem to have a significant impact on ESDP's results.

The final version of the ESDP was presented at the Spatial Planning Minister's Council meeting on May 15, 1999. This version offered country-specific elements for each of the 15 countries which by then had joined the EU. The fact is that spatial planning and mangement is still the exclusive prerogative of each member state. Recognition of this fact, however, does not diminish the need for more general documents designed to have the members "think regionally" and to help set directions for EU policies developed by the Commission.

The ESDP has established the following goals for its member states: 1. a balanced and polycentric urban system and a new and more effective urban-rural partnership; 2. parity of access to infrastructure and knowledge; and 3. sustainable development that includes thoughtful management and protection of natural environments and cultural heritage.

As part of ESDP's charter, a European Spatial Planning Network (ESPON) was initiated. Each of the 15 member states formed a team to represent its interests. Geographers were represented on all teams and in some cases geographers were the dominant professional group. The Nordic Centre for Spatial Development in Sweden served as project manager.

During the first phase which occurred during the period 1998–2000 the overview document (Study Programme on European Spatial Planning or SPESP) was prepared as were several thematic reports from working groups. Since 2000 various specific themes identified by the SPESP have been explored. The actual work of preparing these studies has been completed by a variety of organizations, public and private, which deal with development planning and research. Organizations come from the existing 15 EU countries and from the 10 countries approved to join the EU in the near future. This program offers Europe's applied geography community an exceptional opportunity and it has willingly responded to the challenge.

VI. A BRIEF COUNTRY BY COUNTRY OVERVIEW OF EUROPE'S APPLIED GEOGRAPHY COMMUNITY

All European Geography has an applied dimension; the focus is often in spatial and land planning. The differences from country to country can be substantial. Further, given different "schools" and paradigms differences within countries can be substantial too.

France. France has a long and rich tradition in applied geography, especially in the years following World War Two. The list of well known names includes Jean Gottman who founded the IGU's Regional Planning Commission in 1949 and J. Tricart and E. Juillard who created human and physical geography laboratories in 1947 at Strasbourg, with the clear intention of becoming "players" in the fields of regional development and land-use planning. (Phlipponeau, 1999; Gottman, 1952; Juillard, 1948.)

Other major names associated with French-style applied geography include Ph. Pinchemel, J. Labasse, M. Rochefort, G. Chabot, and more recently, R. Brunet, T. St. Julien, D. Pumain, and A. Bailly. Those listed above are well known for their work in France and in some cases, for work done in Africa and South America through university-based projects, government funded technical assistance projects, faculty-student networks, or simply through direct contacts with high-level public and private decision makers. Work might be tied to regional planning and development assignments or project evaluation for investment projects of various sorts. While his standing in geography is sometimes overlooked the substantial influence of Jean François Gravier on the political sphere is worthy of mention. (Gravier. 1947).

Similarly, and without naming names, we can note that many prominent urban and regional planners in both the private and public sectors have geography backgrounds. The same can be said about planners with NGO's; international consultants, and certainly the European Commission(s).

Belgium. The Belgium applied geography landscape is, perhaps, similar to the French case but even more impressive given the country's small size and large non-Francophone population. The so-called Liège school with names such as O. Tulippe and J. Sporck is certainly distinguished. Liège has counter parts at Louvain and at the Free University in Brussels. Both of these institutions have strong planning units that are major players in the profession at the national level and at the EU level. (Mérenne-Schoumaker, 1996).

Germany. There is a long tradition of applied geography in Germany; geographers are found in a variety of corporate and government settings. If there was any question about applied geography it was clearly answered after the Second World War by the active role played by geographers in spatial planning. But post-war reconstruction was only the start. The creation of the EU, the on-going expansion of the EU and, since the 1990's the reunification of Western and Eastern Germany have all been initiatives that have called on applied geographers. Additionally, applied geographers have found roles in the massive urban and industrial restructuring initiatives that have been ongoing since the 1970's.

Germany's federal structure and its preference for decentralized decision making favours regional, sub-regional and local governments and this has been good for applied geographers. All levels need professionals for functional planning; coordinative planners are vital to assure that the various administrative scales are smoothly integrated into one federal system.

Germany has a history of contracting with its universities for both spatial and sectoral applied research. Additionally there are private consultancies that compete for studies of these sorts. Both types of organizations employ applied geographers. Add to this the even more important federal, regional, and local government research institutes and it is clear that the demand for applied geographers is enormous.

Austria. In Austria applied geographers work in a variety of topic areas but the emphasis seems to be more on physical geography and environmental studies and on geographic information systems, (GIS), and on related techniques, than in more traditional regional planning. Departments of geography tend to emphasize physical geography and techniques which seems appropriate given the Alpine landscapes that typify this nation.

Austrian geographers, like their neighbours in Germany, are active in EU initiatives as consultants or through university departments.

Switzerland. Despite Switzerland not being an EU member, its applied geography is still strongly influenced by its three close neighbours, Germany, France and Italy. This means that there is a strong tradition in applied physical geography and spatial planning, with an emphasis on the Alps in the German-speaking geography departments and more human geography and regional planning in the Francophone and Italian speaking departments. Due to its federal structure and the local power of the cantons, there is a strong demand for applied geography in terms of land-use, urban and regional planning, and spatial development. Not only faculty members are involved in cantonal consultancies but also many current and past students are employed in local governments. Professional geographers have an extremely strong influence on local matters.

The Netherlands. The Dutch have a long tradition of applied research in both planning and environmental management issues. Given this it is hardly surprising that both theoretical and applied geography curricula reflect these areas.

Geography as a discipline is exceptionally strong in the Netherlands; university departments are typically strong in both human geography (economy, society, and culture) and physical geography. The fact that the Netherlands hosted the 1996 Congress of the IGU in the Hague speaks well for geography's clout in this country. Geography's standing at top universities such as Utrech, Amsterdam and Groningen also suggests that the Dutch consider geography to be a fundamental discipline.

The Nordic Countries. Early on geography became well established in the Nordic Countries because of their active roles in exploration and discovery in many parts of the world and because of there roles in the settlement of places such as Lapland, Greenland, and Iceland.

Geography grew during the Post World War Two period and strengthened itself by adapting "best practices" from geographers in the U.K., North America, France, and Germany. The fact that the IGU Congress was hosted by Sweden in 1960 suggest that Nordic geography in general and Swedish geography in particular was held in high regard by the world geographical community. The 1960's certainly represent a high point in the history of both theoretical and applied Nordic geography. Torsten Hägerstrand for one had a tremendous influence on the progress of geography in both Europe and North America.

The 1950's and 1960's were decades of massive economic and social transformation in Sweden, Norway, Finland, and Denmark. Geographers played key roles by finding creative and innovative solutions to regional, urban, and land-use planning problems.

In more recent years there has been a decline in the influence of Nordic geography. This is at least in part the result of a shift away from geography in K-12 education which has lead to a "shrinking" of academic departments. The downward trend was slowed and possibly reversed by the entry of Sweden and Finland into the EU. The role of geographers elsewhere in Europe in the planning and development of European space through programs such as ESDP and ESPON provided a "demonstration effect" that spoke well for the practical applications of geographic training.

The United Kingdom. The U.K. has long been a European powerhouse in geography, both theoretical and applied. Academic geographers have a distinguished history of responding effectively and publicly to issues having social, economic, and political implications. Applied geography has existed since the 19th Century. Today some universities grant academic degrees that are explicitly in applied geography.

Geographers clout in the U.K. has been enhanced by at least two things. First, geography has managed to hold on to its position as a fundamental element in the K-12 curriculum. Second, it has established itself as a viable theoretical and technical "foundation science" for those seeking professional degrees in planning and management. Many of the highest profile departments do not have applied geographer *per se!*

Whereas much of the research carried out by U.K. geography departments is applied it is, more often than not, "background" research rather than research aimed at immediate implementation. Data gathered as part of the evaluation process for geography departments in the U.K. during the 2001 year show that funding comes from a limited range of entities that are often closely aligned with the academic world, e.g., Scottish Natural Heritage and the Ordnance Survey.

Many of the big names in British geography are prominent as consultants both as individuals and through university departments. However, some such as Sir Peter Hall are better known for their work in other fields such as urban and regional planning than they are for their work as geographers. Despite this observation British geographers have a high profile in applied geography. Perhaps largely because of their work on EU projects in environmental management, public policy articulation, territorial planning, economic restructuring, education, and project evaluation they possibly are better known beyond the U.K.'s borders that with in them.

As in other European countries the contributions of geographers are often made in degree programs in economic development, urban planning and regional planning. As a result, the discipline of geography sometimes gets less credit for its contributions than it deserves. The same can be said about geographic techniques that are in great demand; automated cartography and GIS capabilities are sometimes housed in other units such as planning and civil engineering.

In conclusion, the applications of geography in the U.K. show great diversity and teaching and research are convincingly articulated. But because geography is sometimes "buried" in other domains, e.g., planning, its full utility may be underappreciated.

Italy. Italy has a long history in cartography and more recently it has found itself on the cutting edge in GIS. Geographic capabilities are largely concentrated in universities or research centers with ties to universities. Geography might be a freestanding discipline or it might emerge as a part of a department of urban and regional economy or architecture and urbanism. Other units that may house those doing applied geography are those focusing on urban and regional planning, local studies, sustainable development; tourism development, regional management, and coastal landscape protection.

Whereas Italian geography does not have a long and rich history of contributions to spatial and landscape planning, individual contributions have emerged from time to time (Sestini, 1945; Dematteis, 1995).

Finally, and on a somewhat related note, it is worth mentioning that in the past several years the Società Geografica Italiana has been revitalized. Because the Società and its international journal suggest a rededication to geography in Italy, the IGU selected the Society's headquarters in Rome as the home for it archives.

Spain. The Institute of Applied Geography was founded in Saragossa in 1954 under the direction of J.M. Casa-Torres. But it was not until democracy was reestablished in the 1970's that geographic teaching and research really came into their own in Spain. An extra boost came in 1978 with a new constitution that encouraged regional autonomy (and increased demand for planners and regional analysts).

Since the 70's there has been substantial growth in the number of students taking geography courses and in the number of departments offering them in both the older, more established universities and in the new universities too. And as noted above, the proliferation of autonomous entities stimulated demand for the graduates of these programs. Because geographers have performed well in high-profile positions it is anticipated that many will emerge in influential political positions. Those working at the regional level are already showing up in regional organizations; some of these are sure to make their way into the central administration. Similarly, at the municipal level they are increasingly showing up in politically significant positions, e.g., as municipal council members and as mayors. Additionally, applied geographers are appearing in the private sector as consultants. This cohort is smaller and less influential than the public sector cohort described above (many are university professors) but it is significant nevertheless.

Spanish applied geographers have established a solid track record for producing high-quality work on a great variety of topics – regional planning and development including tourism, transport, historic preservation, environmental management and preservation, management of coastal areas including marine environments etc.; feasibility studies for infrastructure investments and for facility location; land-use studies; and strategic planning studies including area opportunity analyses.

At the university level, Spain has several post-graduate, masters and Ph.D. programs with a strong applied geography component. At the K-12 or pre-university level, however, geography has lost some ground in part because some of its content has been "repackaged" as social studies or as environmental science. Further damage has been done by the decline in school-age population as Spain ages.

On a more positive note, applied geography has been helped by the strengthening of geographic societies. The "Colégio de Geógrafas Professionales" was created in May of 1999 when the King and Prime Minister signed a proclamation giving it the same standing as other organizations for academic professionals. Similar organizations worthy of mention include a professional geographers organization covering Andalesia, Catalonia, and Cantabria, the well-established Spanish Geographers Association, regional and sub-regional associations, and even a Federation of Young Geographers Associations.

Portugal. Geography has a long and rich history in Portugal but it was truly revitalized in the mid-to-late 1970's as a result of the country's democratization and modernization process. The past 30 years or so have seen a growth in geography in general and the emergance of applied geography and the increased demand for geographers in urban and regional planning and in private industry too. But geography's real strong hold is still primary education. Further, the demand for geographers here continues to be strong because of policies which have strengthened compulsory primary education.

Most of the increase in demand for applied geographers in Portugal is tied to Portugal's entry to the EU in 1986. As is the case elsewhere in the EU, geographers are well-suited too the work of social, economic, and territorial restructuring. Additionally, there is substantial demand for those with geographic training in local, regional, and "central" public administration organizations.

Geographers have emerged as important contributors to studies of the socioeconomic and physical environments. They have proved to be effective as analysts, diagnosticians, and strategists. As noted above they have also, found their way into the upper levels of public administration in a variety of roles. Similarly, geographers have risen to positions of responsibility and authority in a broad range of sectors from agriculture to trade to infrastructure development and management.

Portugese applied geographers have also done well as a result of cooperative ventures involving the EU and Portugal's public administration sector. There has been a special place for geographers in the production of plans at several different scales and in program evaluation in the Community Support Frameworks (CSF) and community initiative program areas.

The EU connection has had another benefit for applied geographers. Because the EU depends on public and private networks that join member nations, there is mobility for consultants and technicians. Portugese geographers now have improved access to opportunities elsewhere in the EU. The third round of regional development plans for Portugal which were supported by the European Commission saw widespread involvement by geographers from both the academic and applied sectors.

At the local or municipal scale the demand for geographers was largely driven by expanded requirements for municipal plans. Now that most communities have established basic municipal plans geographers are moving on to "second generation" strategic and regional planning assignments.

Finally, the National Spatial Development Plan which was initiated in 2003 is the most recent initiative which promises employment for substantial numbers from Portugal's applied geography community.

VII. A CONCLUDING NOTE: FROM GENERAL EDUCATION TO FOCUSED APPLICATION

The importance of applied geography in the fields of local and regional planning and more recently the contributions by geographers to the business of creating a vision for a unified Europe have mirrored the theoretical and methodological evolution of geography itself. During that period geography moved from being a big-picture generalist discipline to one more likely to find practical solutions to real-world problems where analyses are based on real data and on more sophisticated understandings of models and of spatial dynamics. This does not mean that geographers have given up on the work of producing visions for the future or of anticipating the eventual outcomes related to specific interventions.

In the process of this evolution geographers seem to be getting more pragmatic. The imaginative can deal with the ideal world but the pragmatic planner realizes that his world is far from perfect.

The European Union has provided abundant opportunities for geographers to participate directly in the planning for the 15 members now in the EU and the total of 25 members anticipated in early 2005. The initial process of forming the EU and subsequent exercises in enlarging it have given applied geography a real rallying point. It has brought together individuals, university departments, public institutions, and private firms. Effectively, the construction of the European Community has lead to the creation of a new geography which has been made richer by confronting new challenges and opportunities and stronger by its exposure to new roles and new professional associations.

Globalization, regionalization, and localization are aspects of the same processes at different scales. Understanding how these processes play-out at different scales may be the greatest challenge ever faced by geographers.

JORGE GASPAR
CEG, University of Lisbon

REFERENCES

Alcoforado, M.J., Lopes, A., and A. Andrade, H. (1999) Cartes thematiques et cartes du "risque" d'occurrence de basses tempéatures en milieu urbaine à Lisbonne. *Publications de l'Association Internationale de Climatologie,* Volume 12, pp. 433–441.

Bailly, A., Pumain, D., and et Ferras, R. (1993) *Encyclopedie de la Geographie,* Paris, Economica.

BBR (2001) *Study Programme on European Spatial Planning* – Final Report + CD with thematic reports. Bonn: Bundesant für Bauwesen und Raumordnung.

Casas-Torres, J.M. (1967) *Applied Geography in Spain,* Commission of applied Geography, IGU, pp. 75–84.

Dematteis, G. (1995) *Progetto implicito – il contributo della geografia urbana alle scienze del territoriso,* Milano, Franco Angeli.

uropean Commission (199) ESDP *European Spatial Development Perspective.* Luxembourg: OOPEC.

Ferreira, D.B (2000) Environmental impact of land use change in the inner Alentego (Portugal) in the 20th century. In Slaymaker, O. (Ed) *Geomorphology, Human Activity and Global Environmental Change,* John Wiley & Sons Lda., pp. 249–267.

Freeman, T.W. (1958) *Geography and Planning,* London, Hutchinson & Company.

Garcia-Ramon, M.D. and Nogue-Font, J. (1990) "Professional Geography and Institutionalisation of Academic Geography in Spain", *Geography in Spain*, Barcelona, R.S.G. y A.G.E.

Geddes, P. (1915) *Cities in Evolution* London: William and Norgate.

Geddes, P. (1898) "The Influence of Geographical Conditions on Social Development", *Geographical Journal,* Volume 12, pp. 580–587.

Gottman, J. (1952) *L'Amenagement de l'Espace – Planification Regionale et Geographie*, Paris, Armand Colin.

Gravier, J.F. (1947) *Paris et le désert français*, Paris: Flammarion.

Hall, P. (1992) *Urban and Regional Planning*, London: Routledge.

House, J.W. (1965) "Great Britain. Origin and Evolution of Applied Geography", in Géographie appliquée dans le monde, Prague, CGA.

Huntington, S.P. (1997) *The Clash of Civilizations and the Remaking of World Order*, Simon and Schuster.

Labasse, J. (1966) *L'organisation de l'espace*, Paris, Hermann.

Lições Elementares de Geografia e Chronologia, com seu atlas appropriado, Real Imprensa da Universidade de Coimbra: 1830.

Livingstone, D.N. (1992) *The Geographical Tradition*, Oxford: Brackwell.

Mérenne-Schoumaker, B. (1996) "Léxperience du SEGEFA en Geographie appliquée", *Géographes associés*", Volume 19, pp. 25–29.

Pacione, M. (1999) *Applied Geography*, London: Routledge.

Pereira, A.R., Laranjeiro, M.M., and Neves, M. (2000) A resilende checklist to evaluate coastal dune vulnerability. *Littoral 2000 Responsible Coastal Zone Management: The Challenge of the 21st Century (Zagreb), Periodicum Biologorum*, Volume 102, Supplement 1, pp. 309–318.

Philipponneau, M. (1999) *La Geographie appliquée*, Paris: Armand Colin.

Pinchemel, G, Pinchemel, P. (1988) *La face de la Terre*, Paris, A. Colin.

Sestini, A. (1948) "II presaggio antropogeográfico come forma d'equilibrio", *Boll Soc. Geografica Italiana*, Volumes XII, s. VII, Numbers 6–7, pp. 1–8.

Sack, R.D. (1997) *"Homo Geographicus. A Framework for Action, Awareness and Moral Concern"*. London, Johns Hopkins.

Siegfried, A. (1913) *Tableau politique de la France de l'Quest*, Paris, A. Coling.

Stamp, L.D. (1960) *Applied Geography*, London, Penguin.

Tulippe, O. (1956) "La geographie appliquée", *Bulleting de la Societé Belge d'etudes Geographiques*, pp. 59–113.

Van Wessep, J. Ternwindt, J.H., and Augustinues Pieter, G. (1996) *The State of Dutch Geography*, Utrecht.

Williams, R.H (1996) *European Union Spatial Policy and Planning*. London: Paul Chapman Publishing.

Zêzere, J.L., Ferreira, A.B., and Rodrigues, M.L. (1999) The role of conditioning and triggering in the occurrence of landslides; a case study in the area north of Lisbon (Portugal). *Geomorphology*, Volume 30, Number 1–2, pp. 133–146, Elsevier.

GYÖRGY ENYEDI

CHAPTER 9
APPLIED GEOGRAPHY IN CENTRAL EUROPE

1. INTRODUCTION

This paper is not concerned with the definition of applied geography, the topic of another study in the present volume. Nevertheless, we need to state a few starting points. First of all, we shall not discuss the application of *geographical knowledge* as such, since it reaches far back to the beginnings of agriculture, the building of towns and the beginnings of sailing. We speak about applied geography only if it makes use of the body of scientifically established knowledge of geography as a codified discipline in the interest of the successful operation, and attainment of the aims, of the economic, political and social user distinct from scientific life. The link to this utilisation is the specialised geographer well acquainted with the conditions and aims of application, whether working within the academic world or at an applying organisation. Generally, researchers speak about applied geography, while users require knowledge necessary for town planning, the location of industry, flood control, etc. and they are unconcerned with disciplinary origins. The majority of the researchers do not consider applied geography as a branch of geography but as a special approach. I have narrowed down Pacione's comprehensive definition ("... applied geography may be defined as the application of geographical knowledge and skills to the resolution of social, economic and environmental problems" Pacione 2001, p. 3) by concentrating on its social/economic application, disregarding its countless technical applications from afforestation to flood control. Second, aside from one or two examples, I shall forgo referring to historical antecedents and review only the past 50 years, in the course of which the methodology facilitating the applicability of geography (quantitative geography), the training of geographers in application and a number of areas of application developed. During this period, the development of Eastern Central Europe (comprising today's Czech Republic, Slovakia, Poland and Hungary)* and Western Central Europe (Austria, Germany and Switzerland) differed in a number of respects, which also determined the modes and directions of the application of geography. This is why Central Europe deserves a separate chapter in the volume.

---

* I included only these four countries in the investigation, though Slovenia and Croatia are also part of the same historical region.

A. Bailly and L. J. Gibson (eds.), Applied Geography, 169–185.
© 2004 *Kluwer Academic Publishers. Printed in the Netherlands.*

First, the paper summarises Central Europe's developmental characteristics, then it traces the stages of the application and the theoretical-methodological development of geography. Finally it outlines the main areas of application, thus giving a thematic, rather than a country by country or chronological, description.

2. THE CHARACTERISTICS OF CENTRAL EUROPE'S DEVELOPMENT IN THE 20TH CENTURY

2.1. *The development of nation-states.* The nation-states in Central Europe developed later and as a result of a different type of process than in Western and Southern Europe. Put simply, most Western and Southern European nation-states came about through the integration, or unification, of historical regions (principalities, counties or shires, provinces) that possessed a degree of autonomy earlier. The form of this integration was dictated by internal power relations. In Central Europe the modern nation-states were created, after the First World War, with the partitioning of the Austrian-Hungarian Monarchy, and partly in the western areas of the Russian Empire, as provided for by the Versailles peace treaties. The boundaries of the new nation-states were drawn in accordance with the strategic interests of the victorious big powers, without regard to internal power relations, nationality/ethnic and historical regions. These artificial boundaries did not prove stable, changing a number of times in the 1930s and 1940s. After the Second World War, Poland regained its independence within considerably changed borders. In 1992, Czechoslovakia separated into two states. In the neighbouring South-Eastern and Eastern Europe the formation of new states has, perhaps, not ended yet. As a result of all this, the application of *geography for geopolitical purposes* in the Central European region has remained important to date. When legal rule over a geographical spatial unit passed from one state to another, it necessitated the recording of geographical place names in the new language on a mass scale, and the modification of the territorial subdivision for public administration.

2.2. The past 50 years encompassed *three economic historical periods,* namely, industrial take-off, the change of industrial structure, and de-industrialisation. In Western Europe industrial take-off characterised the 19th century, which means that the other two periods had longer to develop, thus becoming more embedded in society. In Central Europe take-off also began in the 19th century, but reached general development only in the area that constitutes the present-day Czech Republic. In the other countries take-off was insular and its sectoral structure followed modern trends only in the industry of the Budapest agglomeration. The agrarian character remained strong for a long time; in 1949, 52% of the Hungarian population and 60% of the Polish and Slovakian population were employed in agriculture. This structural lag was the reason for the widespread application of geography in optimising the utilisation of the ecological potential of agriculture and, in general, also in classical studies on the location of industry. It was relatively late (only in the second half of the 1970s) that attention began to turn toward environmental protection, service type economic activities and town planning conserving historical values.

2.3. The third Eastern European peculiarity is the existence of the state socialist political system between 1948 and 1989/1990, a one-party system which was characterised by the totalitarian rule of the communist party and the predominance of state ownership, or, at least, direct state control of the economy. All in all, the state socialist economy was a peculiar experiment in catching up in the Central European semi-periphery, which, with the support of the totalitarian state power, led to rapid growth until the close of the industrial take-off. But the absence of a market economy arrested technological development, and deteriorating economic efficiency finally led to a prolonged economic crisis, ending with the collapse of the communist political system based on command economy. The experiment in catching up thus failed (Berend, 1996).

During this period, applied geography could but serve primarily economic growth. Geographers were widely employed in economic and regional planning. Even regional science was developed in economic geography workshops (Enyedi, 2002). In the beginning, applied geographical research directed at the amelioration of regional disparities was encouraged by the ideologically egalitarian communist regime, but not the study of poverty or social discrimination, nor, from the 1980s, the research on environmental damages–for as 'bearers of bad tidings' they were unwelcome. In Czechoslovakia research focussing on these problems was either prohibited or restricted to technical questions; in Poland and Hungary though research was not impeded, its results were not implemented. The development of applied geography was influenced by the fact that between 1968 and 1988 Czechoslovakian geography was unilaterally linked to Soviet geography, and that Poland (after 1956) and Hungary (after 1962) were actively involved in international scientific co-operation (in 1968–76 International Geographic Union had a Polish president, in 1976–84 a Polish vice-president and in 1984–92 a Hungarian vice-president; IGU held its very first regional conference in 1971 in Budapest). Put simply, after 1945, three periods may be distinguished in the application of geography. The *first* is the period of post-war reconstruction and rapid industrialisation when the state power urged the inclusion of geography in state planning, and its big plans of "transforming nature". Going out of academia was expected, especially at universities where, by tradition, secondary school teachers were trained. Research institutes, founded in the 1950s, that belonged to the national academies of science and were not involved in teaching, played an important role in the application of geography. During the *second* period, the universities adjusted to the requirements of practical application (as regards their training system, the research programmes at the various departments), but—during the prolonged crisis of the state socialist economy and growing social discontent from the 1970s on—the state power no longer required the services of geography.

During this time, applied geography played the role of a critical social science. It did not "retreat" to the tranquillity of basic research, but prepared regional and town development plans, proposed environmental protection programmes, etc. It is a curious paradox that although the results of applied geography were not implemented, they were built into the system of social knowledge of the period. But since the turn in 1989 they have frequently been implemented. Human geography is often indirectly applied, and its actual impact is hard to measure because decision making falls outside the sphere of geography.

The *third* period began in 1989. In 1989, the deviation from the mainstream European development came to an end. The decade beginning with 1990 brought about extraordinarily swift changes of remarkable dimensions, specifically, the reinstitution of parliamentary democracy and democratic local governments, the restoration of market economy, the privatisation of the state economy, the emergence of international competition, and the effects of globalisation. At the time I am writing this paper, the Central European countries have already signed the accession treaty with the European Union. Correcting the 50-year detour in slightly more than a decade required extraordinary energy.

A number of new areas opened up for applied geography, for instance, in researching international competitiveness of cities and regional relations extending across borders, in local development plans, environmental protection projects, etc. much like Krugman, Central European economists also discovered that the economy is space dependent, consequently, a growing number of them take part in the application of geographical knowledge.

A considerable part of the areas of application are connected with state or local-governmental institutions. The strong organisations of the private economy belong to the transnational big business sphere, and these do not rely on local experts, for instance, in selecting an industrial location. The small and medium-size firms of the national capital seldom commission the preparation of strategic studies.

2.4. *Institutional background.* The existence and acceptance of applied geography also depends on the institutional background. Besides the geographical institutes of universities, in every country there are research and planning institutes acting as the workshops of applied geography. These institutes were maintained by the national academies of science, ministries connected with urban and regional planning, regional public administrative organs—and sometimes by state companies. The hidden forms of private enterprise appeared in the 1970s and 1980s when research institutes contracted planning and development studies for state organisations for a fee in Poland and mostly in Hungary where private consulting firms were also given permission to operate in 1982. In this respect, the situation is the least favourable in the Czech Republic where geographical institutes and others employing geographers have been losing ground—after considerable earlier expansion (Haviarová and Kučera, 2003). Between 1949 and 1959 in

Czechoslovakia, thirteen new geography departments were set up, among which the economic geography departments were the first to become involved in regional planning. In 1943 in Bratislava, the Research Institute of Geography of the Slovakian Academy of Sciences was established (more or less symbolically at the time), which has gained considerable weight since (and is still very active). The Czechoslovak Academy of Sciences established its institute of geography in Brno in 1963, which was closed in 1993. Today, there are only two small geographical research teams, one at the Institute of Geonics in Brno, the other at the Institute of Sociology of the Czech Academy of Sciences in Prague. As regards Czech universities, only the Charles University in Prague and the University of Brno maintain a geographical institute each comprising several departments, and seven universities have a geography department. The corresponding data in Slovakia are: one university in Bratislava with its own geographical institute and four universities with a geography department. State economic and urban planning institutions also dissolved. In Czechoslovakia the collapse of the reform experiment in 1968 was followed by a hard-line political regime which led to the rigidification of the institutions that became manifestly anachronistic after 1990. Many identified these institutions with the unpopular communist regime and, therefore, were eager to see them dissolved. The establishment of the Ministry of Regional Development, also serving as the information and methodological centre for the regions, was indicative of the gradual restoration of these institutions.

There were institutional changes in Poland and Hungary, too, affecting mainly in-house institutes based in ministries; many of these were dissolved, but, in general, the institutions applying geography expanded. By maintaining close relations with international organisations, the institutions carrying out regional research continuously adjusted to the new tasks of applied geography and took over its research methods. Already in the 1970s, the Institute of Geography of the Polish Academy of Sciences changed its name to Institute of Geography and Spatial Economy and the Hungarian Academy of Sciences—keeping the Geographical Research Institute—established, in 1984, the interdisciplinary Centre for Regional Studies, which still employs a number of geographers.

2.5.   Applied geography in Central Europe profits from the fact that it combines the influence of *a number of international trends*. Besides the Anglo-Saxon schools that dominate the international literature, the French school is also well-known and the German influence is, by tradition, considerable. Furthermore, during most of the past half a century Soviet (mainly Russian) geography was also influential. Just by way of illustration, British influence was strong in Polish geography (the topic of the first British–Polish geography seminar in 1959 was applied geography); in Czechoslovakia and Hungary regular work relations were established with the French and French-language school, thus with E. Juillard, M. Phlipponeau, O. Tulippe and J. Tricart (Leszczycki, 1964, Strida, 1968). In Central Europe, lying on the boundary of the various European cultural zones, the synthesis of various theoretical concepts and models is traditional, which lends diversity to the interpretation of applied geography as well.

In summary, the application of geography began more or less simultaneously in the advanced parts of the world. Its forms and methods developed partly as a result of its development as a discipline, and partly depending on the changes in the areas of application (e.g. location of industry). The Central European characteristics mentioned above were also reflected in Czech, Polish, Hungarian and Slovakian applied geography. We may also add—and it is not just a Central European problem—that it is hard to judge the position of applied geography on the basis of publications. Most of the results of applied geography are never published. However, the demand for application influences (through its financing potential as well) basic research, thus—beyond our personal experiences—we can also infer the trends in application from scientific publications.

3. GEOPOLITICS, TERRITORIAL SUBDIVISION OF PUBLIC ADMINISTRATION

3.1. As mentioned, the First World War was followed by significant *border changes* in Central Europe, especially after the partitioning of the Austrian–Hungarian Monarchy. It was logical for geography and cartography to move beyond basic research and try to provide the scientific basis for drawing the new state boundaries (by the logic of war, i.e. unsuccessfully in the case of defeated Hungary and successfully in the case of the victors). In 1918, the Institute of Geography of the Charles University in Prague was commissioned to participate in the preparations of the peace treaty. This task mobilised the whole Czech geography profession headed by V. Svambera. This was the first and for many decades the single most important joint venture of Czech geography. V. Dvorsky was the key figure of the team, who was not only party to the preparations, but also a member of the official delegation participating in the peace negotiations in Paris, who also took part in the field inspections (mainly between Poland and Czechoslovakia) marking out the future borders. Subsequently, several members of the team moved to Bratislava and participated in the organisation of Slovakian geography. The team used primarily geopolitical arguments to justify the new boundaries which enclosed 3.5 million Germans, 1.5 million Hungarians and 0.5 million Ruthenians in Czechoslovakia. For a few years after the war, the Prague Institute of Geography was occasionally consulted on the spatial organisation of the new Czechoslovak state and the foundation of the (geostrategic) Military Institute of Geography—following which the application of geography was no longer called for (Haviarová – Kucera, 2003). Count Pál Teleki, the leading personality of Hungarian geography and cartography, also participated in the preparation of the peace negotiations, where, using a series of ethnic maps and referring to the unified ethnic nature of nation-states, he argued (unsuccessfully) for a more favourable boundary delimitation for Hungary. (It is interesting to note that after the partitioning of the Ottoman Empire Teleki participated as an expert in the demarcation of the boundaries of the new Near Eastern states, in which case he followed British geopolitical interests, see Tímár, 2001).

Polish geography played a role mainly in connection with the territorial changes after the Second World War. After the war, 54.7% (178,000 sq. km) of Poland's area in 1939 was annexed to the Soviet Union; at the same time, an area of 101,000 sq. km in the north and mostly in the west that previously belonged to Germany was annexed to Poland. Geography was applied only on a modest scale in boundary delimitations, all the greater was its task in replacing German place names (including the names of the natural scenery) with Polish place names (Leszczycki, 1945).

The Czech Republic's and Slovakia's—peaceful—separation represented the main change in recent history. It meant only that the internal boundary between the two federative states became an external boundary.

3.2. *The territorial subdivision of public administration* in Central European countries also changed. This was partly caused by the above-mentioned boundary changes; the modifications (or proposals for modification) were also justified by internal policy factors; furthermore, in the 1990s, conforming to the European Union's NUTS system (territorial units for statistics) strengthened the endeavour to create decentralised regions. Due to the way in which the nation-states were formed and to the boundary changes, regionalism in Central European countries is weak. Historical regions, which remained within the boundaries of one country, are few. Public administration in the respective states is centralised and unitarian, and there are no autonomy aims in the subnational territorial units. The strengthening of decentralisation and regionalism in the European Union is exerting an influence on Central Europe too. The four (Central European) countries represent four different cases, but in each country they are working on public administration reform, and, without exception, in co-operation with public administration lawyers and geographers (Barlow, Lengyel and Welch, 1998; Regulska, 1993).

Hungary is the only post-socialist country where the territorial subdivision of public administration did not change (aside from small corrections) under communism. The county, a one thousand-year old institution, represents the basic subnational territorial administration unit, and there is very strong public identification with the county. (In addition, political parties are also built from county organisations.) Economic and public administration geography has been proposing for decades the reduction of the number of subnational units, that is, the introduction of ten, seven or six regions instead of the current 19 counties (Hajdú, 2001). The 1971 law on regional planning delimited six economic regions for long-term economic planning. The 1996 law on regional policy defined seven regions, each with regional development councils. But only in 2002 did work begin on public administration reform pointing in the direction of real regionalism. According to plans, the seven regions would have elected regional governments and their own budget, and the counties would cease to exist. I would not be surprised if the county survived this reform, too. In principle, the Hungarian local-government system is highly decentralised, since the 1990 law on local governments granted important authority to the elected bodies of the basic municipal units, the communes. However, exercising this authority depends on the reallocation of centralised budget resources. Locally generated financial resources are insignificant (except larger cities and tourist areas).

In the other three countries territorial subdivision was modified a number of times over the past 50 years, and there have been changes since 1990 as well. Polish regions came closest to conforming to the European Union's regional criteria (as a result of the public administration reform of 1999). Czech reform experiments are much more modest, while Slovakian public administration reforms were accompanied by fierce ethnic debates (Slavik, 1998; Bakker, 1998).

4. REGIONAL AND LOCAL PLANNING

This is perhaps the broadest area of the application of geography. The concept of regional and local planning includes regional planning, town planning, the uncovering of regional disparities and regional policy aimed at diminishing disparities. Inevitably, we must also distinguish between the socialist and post-socialist periods; not primarily from a political point of view, but because of the essential differences in the areas and methods of the application of geography during these two periods. The two economic systems differ not only in the way they operate, but the initial decades of the two systems—the 1950s and 1990s, respectively—also show great dissimilarities in economic structure, world economic processes and the system of economic-social aims. Obviously, the areas and methods of application of geography changed greatly in fifty years. During the socialist period, the main areas of application were the promotion of the rapid growth of the economy and the uncovering and reducing of regional disparities. The still ongoing research on regionalisation was also begun. Its aim was to assist public administration reform, on the one hand, and to outline the regional frames of economic development programmes, on the other. The new elements of the post-socialist period include the international competitiveness of regions and cities, the treatment of the new type of regional disparities, localisation of the new economy and the preparation for the spatial integration into the European Union.

Over the years, applied geography has played several different roles. In the 1950s, in consequence of the nationalisation of most of the economy, it became a state task to select the sites of industrial companies, to supply labour and to develop the agricultural production structure utilising geographical–ecological features. In other words, the state did everything that under the conditions of a market economy is accomplished by companies and market elements, such as prices, demand, and costs. State economic management suddenly found itself in need of a large body of applicable geographical knowledge. As much as circumstances allowed, geography departments, specialising in teacher training and working with only a small staff, refused application, or they focussed on physical geographical (e.g. hydrogeographical) application. But the socialist society needed the application of geography. "Socialist society defines very concrete requirements and creative tasks for geography... geography has acquired new content. It has a new user, the socialist state, and its new goal is to serve the economic and regional planning of the new society (Hruska, 1953, p. 163).

By the 1990s, training in applied geography and a comprehensive research agenda were in place. But the institutional environment had changed. The new institutions, e.g., regional development councils—and the private firms involved in urban planning and regional development, as well as multiplant space-dependent firms (e.g. banks building a branch-network) have created this new demand for applied geographers.

4.1. *Regional disparities* appeared in various forms, and were always given special political attention. Given the egalitarian nature of the socialist society, the sudden increase in social-regional disparities after 1989, shocked the public and led to political tensions. Consequently, the governments have always intervened in the correction of disparity promoting processes.

   4.1.1. Some disparities were the consequence of *the Second World War*. They came as a result of the flight and displacement of masses of people; certain regions became depopulated (e.g. Sudetenland in today's Czech Republic). Elsewhere a population exchange took place, that is, the population settled in an unfamiliar environment. War damages also differed in magnitude; they were very extensive in Poland and slight in the Czech Republic. Reconstruction reproduced the significant regional disparities of the 1930s.

   4.1.2. During the first decades of the socialist period, attention was paid mainly to economic structural differences. Regional development meant fundamentally the industrialisation of agrarian regions and the expansion of the urban network. Public utilities and services were relegated to the background. Heavy industry dominated the economic agenda; locations were driven by raw material production, energy resources and transport routes. Processing industrial plants were often located in former agricultural market towns. Large industrial investments were made at the expense quality of life and agriculture production. Even in the 1960s and the 1970s, the growth of cities was characterised mostly by industrialisation and the related population migration and housing. In every country a few new, so-called socialist cities were built where egalitarian socialist urbanist principles could be tested. Studies have showed how these cities gradually returned to the traditional Central European urban structure and life style (Szirmai, 1987).

Two big changes of ownership shook agriculture: land reform, i.e. the division of the large estates among the family farms in 1945–46, and the collectivisation of family farms between 1949 and 1962. In Poland family farms were not collectivised, but their size and technical development were restricted. Large-scale farming was modernised most successfully in Hungary and, as a result the country became an important net exporter.

Regional development policy has been aimed at removing regional differences. Accordingly, the main areas of the application of geography were (1) geographical location of big investments, (2) surveying the ecological potential of agriculture, and (3) geomorphological surveying for physical planning.

From the 1970s the malfunctioning of the socialist economy intensified. It became clear that decreasing regional differences did not lead automatically to the regional equalisation of living conditions. For instance, the rural/urban dichotomy remained. Backward infrastructure made accessing of health, educational and administrative centres difficult for rural people. Housing shortage in cities became permanent. The attention of applied geographers turned increasingly toward social–regional disparities. But the state seldom supported such investigations and publication also met with difficulties, especially in Czechoslovakia. Applied social geography, which appeared as a new trend and was strongly influenced by the Munich school in Czechoslovakia and Hungary, became stronger (Hartke, 1959; Ruppert, 1968). During this second period, research serving the protection of the natural environment also intensified; the times when applied geography was exclusively used to locate production facilities were over!

4.1.3. *The return to a market economy* increased significantly the regional disparities of both the economic structure and living conditions. This had two sources: (a) the transformation crisis and (b) the rapid penetration of Western European regional processes in just a few years (which was facilitated by the common boundary with the European Union and the strong tradition of German–Austrian relations). A few elements of the *transformation crisis* that strengthen regional differences are: (1) total economic output dropped significantly and reached the 1989 level only after 10 years; (2) state economy collapsed; (3) the locational disadvantages that the non-market type location of the state economy and social protectionism had concealed, came to the surface, large numbers of uneconomical industrial plants, which were established mainly to provide employment, were closed down; (4) both domestic investors and outside investors favored investments in already prosperous regions – mainly in metropolitan regions (e.g. Western Hungary, the urban regions of Budapest, Bratislava, Prague). Differences between developed and underdeveloped regions grew substantially. Accession to Western European regional processes (1) destroyed the old industrial zones in short order (but without the simultaneous rise of new economic branches), (2) accelerated the concentration of financial and business services and economic decision making in **big cities**, and (3) abruptly exposed the Central European regions and towns to strong continental and global competition, for which they were unprepared.

Over a short period of time development levels became more uneven; personal income declined and unemployment increased. Worse yet, the government's institutional system and the legal system were also not prepared to respond to the economic down-turn.

The transition to a market economy took place at different rates and using different methods in the four countries. It was carried out most rapidly in Hungary through radical privatisation, whereby significant foreign direct investments were attracted before the onset of the world economic recession, and, since 1996, economic growth has been rapid. Hungary was also the first to adopt the law on regional development (in 1996). All in all, the slower transformation of the other countries was also successful.

In the 1990s, the application of geography extended to new areas and new commissioners appeared. The old central government institutions weakened, but new regional and local governments and, development agencies stepped in their place. Studies done for private firms were not made public (e.g. the branch-office location plan that the Centre for Regional Studies of the Hungarian Academy of Sciences prepared for the Austrian Creditanstalt). The geographical research institutions often used the fees they received for applied research to cover the expenses of a part of their basic research. Bicik and Hampl (2000) summarised the applied research trends of Czech geography as follows: (1) questions of regional development, including the geography of the new economy, the transformation of the settlement system and the preparation for integration with the EU; (2) participation in transdisciplinary projects, such as, for instance, the utilisation of former Soviet military bases; (3) land-use studies on the transformation of agriculture and the village; and (4) urban geographical and population-geographical studies involving mainly demographic and migration analyses. Hungarian and Polish investigations analysed the geographical location of the new economy (e.g. the bank system, Gál, 2000; Stryjakiewicz and Potrzebowski, 1995), showing little interest in industry (Barta, 1998). Regional and urban competitiveness have been frequent topics. Social disparities (except in respect of employment and unemployment) have not been included among the topics of applied geography. However, much attention has been given to cross-border economic co-operation.

4.2. *Regional and town planning*

Regional planning resembles the levelling of regional disparities discussed in paragraph 4.1 insofar as the planning process is the same, but its aim is different. The aim concerns the optimal operation of the object of the plan—an industrial plant, a housing estate, the economy of a city, or a resort district. (Optimalisation may extend to the output or profitability of the economy, the number of visitors of a resort centre, etc.) It involves not so much regional policy aims as technical or management tasks.

During the 1950s, '60s and '70s, geographers participated in the location of big industrial complexes. Geographers were well suited to this work; they could think not only in terms of questions concerning the location of industry, but also in terms of the entire affected area (i.e. the location of labour, protection of the natural environment and connected services). Big projects such as the East-Slovakian Steelworks in Kosice, the location of the Czech, Slovakian and Hungarian nuclear power plants, the steelworks in Nowa Huta near Cracow, Poland etc. are all examples. Some of these projects came to form a rust belt in the 1990s.

The 1990s gave rise to new tasks in town planning as well. The internal functional structure of cities changed; earlier metropolitan agglomerations became metropolitan regions based on a more complex internal network. Large-scale housing privatisation created a special challenge. New business functions had to be included in master plans. The real estate market revived and gradually resumed normal operation. Towns were given substantial attention and were guaranteed extensive autonomy by the new democratic laws on local governments. Eastern European towns needed to transform themselves to match the Western European model.

Applied geographers took part (and continue to do so) in the strategic planning of Central European metropolises; a special challenge is trying to define the international role of the capital cities. One thing is certain, that Central Europe does not have a single centre. Vienna continues to hold its earlier advantage in attracting international organisations; Budapest may fill the role of a gateway city toward Southeast Europe; Prague emphasises its role as cultural and tourist centre; Warsaw's geographical position makes it very suitable for mediatory roles between Central Europe, Eastern Europe and the Baltic states. The Slovakian capital, Bratislava, because of its close proximity to Vienna, acts as its complement rather than as an autonomous macro-regional organisational force.

As opposed to the earlier housing-estate constructions, attention focussed on the rehabilitation of old districts. After all, Central Europe is full of aesthetically very valuable old towns. Understandably, the participation of geographers was extensive in the preparation of town-development programmes and the new master plans of big cities (Barta, 1998; Cséfalvay and Rohn, 1992; Enyedi, 1998; Tosics and Hegedüs, 1998; Sykora, 1993).

4.3. Czech geography plays an outstanding role in applied demographic research. Czech demography was developed at the Institute of Geography of the Charles University back in the 1930s. Distribution of the population according to national origin and migrations aroused great political interest. These phenomena can only be interpreted in geographical context. Czech geographers also took part in the organisation of population censuses.

Elsewhere in Central Europe demography was connected more with statistics, but the geographical study of the population played an important role in regional planning in every country. The reasons for this were: (1) After the Second World War, in every country there were extensive population displacements, changing the population size and composition of whole regions. (2) Industrialisation-driven urbanisation induced large-scale migration from villages to cities, the surveying and forecasting of which was necessary for urban planning. (3) The location of industry—particularly in the 1970s—sought free labour already in very short supply. This industrialisation was of a processing industrial nature and established new industrial plants mostly in rural zones. This required knowledge of the geographical distribution of labour of working age available for new employment (young people and dependent women). (4) Long-term planning demanded regional population prognosis. (5) Aside from Poland, population decline and ageing characterised the three countries for a long time. Everywhere state measures were adopted to boost the birth-rate with only moderate success; the high death rate compared to Western Europe is also a cause of population decline. This required the regional investigation of the demographic structure and the dynamics of the population. The availability of good-quality statistical databases providing regional breakdowns made population geographical research attractive for many geographic scientists.

In the 1990s, analysis of international migrations represented a new element in the research agenda. It contains many uncertainties, for example, concerning the separation of temporary migration and illegal migration, or distinguishing citizens with permanent residence abroad from emigrants. This is a new phenomenon in the Central European region since under communism emigration (with a few exceptions) was considered illegal and there was no possibility of return. At the same time, the socialist countries were not important immigration target countries. Both the regulation of migrations and research working out the basis of regulation are as yet undeveloped.

Large scale migration is not an issue in Central Europe. The main concern is brain drain, which is the subject of various studies. Brain drain within a country is given less attention. Aside from the Czech Republic, there are important university centres in the underdeveloped regions of each country (e.g. Debrecen in Hungary,

Kosice in Slovakia and Lublin in Poland), which, in principle, could lay the foundations of the knowledge-based economy of the respective regions, but hardly any graduates remain there.

5. ENVIRONMENTAL STUDIES

Physical geography played a traditionally important role in Central European geography. In Hungary the first geography department grew out of the geology department in the 1870s. In every country geography departments are part of the faculty of natural science. (Multivarious development of human geography took place only in Poland.) This would suggest that among the topics of applied geography environment utilisation and environmental protection played an important role. But this was not the case. Geomorphology, which dominates physical geography, has only limited applicability (e.g. in demarcating the areas of danger of landslides). Nevertheless, geographers participated in the exploration of natural resource use, e.g. in agro-ecological surveying, in discovering the gravel deposit of alluvial fans, etc. Separation became more widespread with the increasing importance of solving environmental problems (from the end of the 1960s). This also required co-operation between physical and economic geographers, which did not go smoothly (Demek, 1985). Czechoslovak geography's prestige was greatly enhanced by the fact that the Research Institute of Geography in Brno of the Czechoslovak Academy of Sciences directed the long-term programme of the Comecon countries, entitled "The Development of Research Methods for Examining Man's Impact on Nature" (1971–1980). It was during this period that regular relations were established between physical geography on the one hand and planning institutes and local authorities on the other.

The situation of research supporting environmental protection policy shows a special picture in Central Europe. Increasing damage to the environment was not denied officially (as it was in the Soviet Union and Romania), but it was deemed a relatively easily solvable problem and exclusively the state's task. The illegal political opposition also regarded environmental problems as safer grounds for confrontation than overt opposition of the political system, which struck a more sensitive chord in the political leadership. For this reason many monitoring and research findings were classified state secret. The socialist leadership was helpless in the face of the rapidly growing environmental problems (notwithstanding that the main polluting institutions and firms were state-owned) since there were no funds for environmental protection investments during the deepening and protracted economic crisis (Enyedi, Gijswijt and Rhode, 1986).

After 1989, environmental protection research greatly intensified, but the relative positions of geography deteriorated. The other earth sciences—from climatology to geology—used their better equipment and greater field experiences to expand the scope of their environmental protection activity. But new applied geographical topics were also formulated, for instance, the role of environmental quality in regional competitiveness, conservation of the environment or organic agricultural production as new rural activities. This fact also included social geography in environmental research (Fodor, 2001; Szirmai, 1999). International projects, e.g. the International Geosphere–Biosphere Programme in which Polish geographers were highly involved, contributed greatly to the resolution of applied research tasks.

6. THE TRAINING AND EMPLOYMENT OF APPLIED GEOGRAPHERS

Universities are the centres for training applied geographers. In some cities (Budapest, Warsaw, Poznan) the work of the economic geography departments of universities of economics is also considerable. By tradition, the training of geographers meant secondary-school teacher training, where geography was paired with another discipline (mostly history or biology). Applied geography as a profession spread gradually. New courses, such as regional planning, applied geomorphology and location theory, were introduced (as a result, many students did not subsequently pursue a teaching career). Geographers received a good natural scientific basic training, but their knowledge of social sciences was sketchy, which was a disadvantage in economic planning. Declining demand for geography teachers propelled university departments in the direction of applied geography. In the 1960s, the separation of geography teacher training and geographer training began in different ways in the various countries. The main occupational areas of specialised geographers were in regional and town planning, tourism, the application of GIS and environment management. For this reason, geography and economics (and sometimes technical knowledge and public administration) are linked in both training and work. In Central Europe regional science developed mostly at the geography departments.

The number of geography students increased considerably in the 1990s; specialised geography graduates find jobs readily and the areas of employment are expanding. Often there is also a demand for the (non-geography related) application of techniques learned at the university (e.g. the use of GIS in real-estate market analyses).

However, employers today are not looking for geographers since they generally do not know what geographers actually do. It is therefore helpful if the specialised qualification, like 'regional planner' or 'environmental manager', is added to 'geographer' in diplomas (as it is done in Hungary). It was mentioned in the interviews with Slovakian applied geographers in 2003 (Havariová and Kučera, 2003, p. 25) that "the marketing of geographers in the labour market is poor, the representatives of better known professions are hired more readily" ((B. Divinsky); "small private planning firms or architectural offices employ engineers to do everything, even the job of geographers, in order to save on labour" (S. Smiesková); or "geographers specialised in GIS are in great demand in Slovakia, but usually for programming or computer-network management jobs." (P. Trenbos). It is also arguable whether or not the performance of field work by someone with a degree in geography is applied geography *per se*. Gardavsky

(1986) proposed that the application of geography be directed at solving the conceptual problems of regional development, rather than counselling on segments of a problem. The practical strength of geography lies explicitly in the comprehensive investigation of spatial systems, in the study of the reciprocal effects of the physical and social elements of space, which it does not always take advantage of (Postolka, 2002). The social perception of applied geography would be greatly enhanced if geographers (geographical workshops) undertook to direct, or manage, important regional development or environmental protection projects—which, naturally, also require the ability to direct relatively big, multidisciplinary teams. There are many examples that show that the same person can successfully carry out basic research and solve practical tasks. According to I. Zemko, a respondent of the above cited Slovakian interview series, "instead of polemics about the limits of geography, geographers should be concerned with whether or not they want to stay competitive". This also requires a clear understanding on our part of the strong points of applied geography, and that we should not only try to find out what practice demands, but also to formulate, and take to the market, our own offering.

BIBLIOGRAPHY

Applied geography has quite a long list of publications in East-Central Europe. I have had no intention to give a substantial overview of this literature; I put into the bibliography those what was quoted in the text. Special credit due to Ms Eva Havariova (Slovakia) and Tomas Kucera (Czech Republic) who prepared a manuscript for me titled "Applied Geography in Czechia and Slovakia")

Bakker, E. 1998 Local self-government and ethnic minorities: local political power of Slovakia's ethnic minority. In= Barlow, M.- Lengyel, I.- and R. Welch: *Local Development and Public Administration in Transition*. Szeged: JATE Press pp.103–113

Barta, Gy.(ed) 1998a *Budapest – nemzetközi város* Budapest: MTA

Barta, Gy 1998b Industrial restructuring or deindustrialization? In Enyedi, Gy (ed.) *Social Change and Urban Restructuring in East-Central Europe* Budapest: Akadémiai Kiadó pp. 189–209

Berend, I.T. 1996 *Central and EasternEurope 1944–1993. Detour from the Periphery to the Periphery*. Cambridge-New York-Melbourne: Cambridge Univ. Press

Berényi, I. 1992 *Az alkalmazott szociálgeográfia elméleti és módszertani kérdései*. Budapest: Akadémiai Kiadó

Bicik, I. and Hampl, M. 2000 Czech human geography: research and problems *Sbornik CGS* vol 105,no 2. pp 118–128

Cséfalvay, Z. und Rohn, W. 1992 *Die Transition des ungarischen und Budapester Wohnungsmarket* Wien: Verlag der Österreichischen Akademie der Wissenschaften

Demek, J. 1985 Cesti geografove a studium zivotniho postredi (1945–1985) *Sbornik CSG* vol.90, no 2. P.108–119

Enyedi, Gy.- Gijswijt, A.J.- Rhode, B.(eds) 1987 *Environmental Policies in East and West* London: Taylor Graham

Enyedi, Gy. 1996 *Regionális folyamatok Magyarországon* Budapest: ELTE

Enyedi, Gy. (ed.) 1998 *Social Change and Urban Restructuring in Central Europe* Budapest: Akadémiai Kiadó

Fodor, I. 2001 *Környezetvédelem és regionalitás Magyarországon* Pécs-Budapest: Dialóg Campus

Gardavsky, V. 1986 Poznávaci a praktická funkce geografie *Geograficky Casopis* vol. 38, no 2–3. p136–141

Hajdú, Z. 2001 *Magyarország közigazgatási földrajza.* Budapest-Pécs: Dialóg Campus

Hartke, W. 1959 Gedanke über die Bestimmung von Raumen gleichen sozialgeographischen Verhaltens. *Erdkunde* vol. 13, no 4. p. 426–436

Haviarová, E. – Kucera, T. 2003 Applied geography in Czechia and Slovakia. Prague:Manuscript

Häufler, V. 1967 *Dejiny geografie na Universite Karlove 1348–1967* Praha: Universita Karlova

Hruska, E. 1953 Geografie a územni planováni *Sbornik CSZ* vol. 58, no 2. p.163–164

Leszczycki, S. 1960 The application of geography in Poland *Geographical Journal*, vol. 126, no. 4, pp. 418–426.

Postolka, V. 2002 Ceska geografie versus zivotni prostredi (reflexe po roce 1989) *Sbornik CGS* vol. 107, no 1. p. 50–62

Regulska, J. 1993 Local government reform in Central and Eastern Europe. In= Bennett, R.J. (ed.) *Local Government in the New Europe.* London: Belhaven p.183–197

Ruppert, K. 1968 *Zum Standort der Sozialgeographie* Kallmünz/Regensburg: Verlag Lassleben

Slavik, V. 1998 National minorities and the transformation of public administration in Slovakia In= Barlow, M.-Lengyel, I. and Welch, R (eds.) *Local Development and Public Administration in Transition* p. 95–103

Stryjakiewicz, T. – Potrzebowski, G. 1995 The newly emerging banking system in Poland and its spatial organization *Geographische Zeitschrift* vol 83, no 2 p. 87–99

Strida, M. 1968 Applied geography in regional planning *Sbornik CSZ* vol 73, no 3 p. 283–284

Sykora, L. 1998 Commercial property development in Budapest, Prague and Warsaw. In= Enyedi, Gy. (ed.) *Social Change...* op.cit. p 109–136

Szirmai, V. 1987 *Csinált városok* (Gyorsuló idö) Budapest: Magvetö

Szirmai, V. 1999 *A környezeti érdekek Magyarországon* Budapest: Pallas Stúdió

Tímár, E. 2001 Teleki Pál egy kevéssé ismert munkája, a moszuli jelentés. *Földrajzi Közlemények* vol. 125, no 3–4 p. 65–85

JOHN W. FRAZIER

CHAPTER 10
APPLIED GEOGRAPHY IN 20TH CENTURY NORTH AMERICA:
A PERSPECTIVE

INTRODUCTION

Applied geography in North America, especially in the United States, experienced a resurgence during the last quarter of the 20th century. After being well established pre-1955, its visibility within academic geography declined after 1955, resulting in confusion over its purpose and practice. Prominent North American academic geographers practiced geography during the 1920s–1950s as government employees and consultants. Applied geography was recognized within the formal literature of the discipline as a key dimension of the subject. Also, prior to 1955, applied geography had become institutionalized as worthwhile practice within the private and non-academic settings.

Professional academic geography underwent significant rapid changes in the 1950s and 1960s that, among other things, turned the discipline's focus toward spatial theory, while seriously questioning previous human-land, regional frameworks that not only characterized teaching and research agendas of the past, but also guided applied geography practiced by disciplinary leadership for more than a generation. North American academic geography also "busied" itself with preparation of the professoriate and became less interested in other issues. Together, the "revolutionary" new thinking and teacher-training foci, turned professional academic geography not only away from the non-academic world but led to the devaluation of applied geography as a dimension of the discipline. Despite these trends, some of the discipline's leading theorists practiced geography unnoticed by their colleagues because their work had no formal outlet in academic geography.

By the 1970s, calls came from both non-academic and academic sources to revisit the applied dimension of the discipline. Leading North American journals carried articles that clarified the roles of applied geography and directed its future. Geography's leading North American professional organization also rediscovered applied geography. A national Applied Geography Conference, now in its 26th year, began in 1977 in Binghamton, N. Y. An "Applied Geography Specialty Group" was initiated within the

A. Bailly and L. J. Gibson (eds.), Applied Geography, 187–210.
© 2004 Kluwer Academic Publishers. Printed in the Netherlands.

Association of American Geographers and the James R. Anderson Medal of Applied Geography was created to honor one of the leading applied geographers in North America by annual recognition of a distinguished career in applied geography. By the close of the 20th century, applied geography had once again found a comfortable position within the discipline in North America.

The purpose of the remainder of this chapter is to provide a more detailed perspective on the evolution, recession and re-emergence of North American applied geography within these two periods, with primary examples largely drawn from the United States. Before discussing the periods in question, it is useful to define some terms.

APPLIED AND APPLICABLE GEOGRAPHY

Unfortunately applied geography, although a relatively straight forward concept, suffered from confusing perceptions of its meaning after 1955. Also, it has been confused by somewhat different connotations inside and outside of academia. A clarification of the terms applied and applicable geography also is useful for discussing trends.

Applied geography as science: The academic connotation

Applied geography was linked to science by early professional academic geographers. In 1936, Colby argued for regional composite planning and claimed the "close association of geographic science and geographic application "was central to American geography" (Colby, 1936, p. 4). His notion of geographic science included the study of areal units with the utilization of classification, mapping and analytical comparisons. In addition, applied geography, like all applied science, extended the scientific method to include design and effectuation stages. The link was re-established by Frazier in 1982:

> ... "Although the questions asked make geography distinctive, as science it shares the goals of explanation and prediction through the use of the scientific method while it seeks to improve the human condition.
>
> Applied geography, as a science, likewise shares the goals of general science, but as an applied science it is an extension of pure science. Applied geography, like other applied sciences, has several attributes: it is user-oriented; (2) is action-oriented; and (3) extends the experimental method to include evaluation and implementation stages to achieve the first two attributes.
>
> Although pure geographic research includes observation, hypothesis elaboration, deduction of the effects based on the hypothesis, and new observation for verification, applied geography continues with the evaluation and includes an implementation stage." (Frazier, 1982, p. 14).

As perspectives from inside academia, these definitions emphasize the importance of science. The academic practitioner controls the application. Methological and other decisions are typically the responsibility of the academic applied geographer, who often is serving as a consultant while employed by a college or university. The academic environment of the applied researcher typically provides a great deal of independence and stresses the importance of the scientific method.

Applied geography as client-driven tasks: The non-university setting

By contrast, applied geography in a non-university setting typically involves an applied geographer performing a set of tasks that are repetitive and occur within a reporting hierarchy. A team member often performs repetitive tasks, such as directed fieldwork, map production, writing, or decision-making. Each member is accountable for only a small part of a final product, such as a piece of geographic software, atlas, store location, or some other product.

The non-academic, private setting constitutes a different work place than the academic one (Frazier, 1994). Teamwork is more the norm and free market principles (e.g. competition, efficiency, product quality control, etc.) play important roles in highly competitive market places. Similarly, government settings, where product-orientation is stressed (e.g. census products), differ from academe. This is not to suggest that non-academic agencies ignore science or that universities have no accountability. Rather, it is to emphasize the different daily environments (Frazier, 1994). Universities stress science; non-academic institutions emphasize competitive products and employ teams of people to produce them.

One of the results of these different work environments is the connotation assigned to applied geography. In academia, applied geography is a process that is deeply rooted in science but, for practical purposes, must provide directed actions in pursuit of a client's goal. In non-academic settings, applied geography is a more inclusive concept. An applied geographer is one cog in a production wheel that results in applied geography. Given narrow tasking, even though science may guide the production process, non-academic applied geographers may be unaware of the scientific principles guiding the process. Any single participant may be performing research, producing some aspect of the final product, supervising others, or implementing findings through specific actions. Each assignment is crucial to the quality of the end product. As a result, those in the non-academic setting are more inclusive when discussing applied geography as a concept. In their case, the entire process is considered applied geography and all who participate in any single task are applied geographers.

Where in academia the final product is likely a consulting report, in non-academic settings a more diverse set of applied geography products is produced. In addition to reports that inform and direct decisions, products can include new digital data bases, any GIS product, an atlas, the opening of a retail outlet, the location of a noxious facility, or the creation of a routing scheme. These are but a few examples of applied geography products.

Useful knowledge: The tie that binds

Although workplace imperatives dictate different operational styles and result in somewhat dissimilar perspectives, academic and non-academic geographers share the common goal of serving a client by producing a product with a user orientation. Michael Pacione made the useful observation that the glue between applied geographers in different settings is the emphasis placed on useful geographic knowledge (Pacione, 1999). Whether emphasizing science or product, applied geography operates from the perspective that directed actions require useful knowledge that guides decisions. This binding perspective is reflected in Barry Wellar's differentiation of theoretical and applied geography:

> "...Pure (theoretical) geography being driven by a researcher's own curiosity about a geographic problem, and applied geography being driven by a clients needs" (Wellar, 1998).

As Pacione has noted, this sometimes requires that an applied geographer use knowledge to support an unpopular position, rather than merely supporting an institutional decision (Pacione, 1999). No matter the situation, useful knowledge guides direct actions for a client.

This clear distinction between applied and theoretical research begs the question: How much academically-based research meets the criteria (client-driven, useful information with a user-orientation) of applied geography? The answer is some does and some does not. However, a close relative of applied geography, applicable geography, has been an important part of the applied science perspective of geography.

Differentiating applied and applicable geography

To understand the difference between applied and applicable, one only needs to consult a dictionary. Webster's defines applicable as "capable of being applied; relevant, pertinent, or useful" (McKechnie, 1983). There are a fair number of academics who are interested in performing relevant contemporary research that could inform a prospective audience (relevant constituency) but are unable to develop a client relationship. For example, one might be vitally interested in a local pollution problem or the potential location of a noxious facility but have no means of becoming a consultant on the project. That researcher might consider the broader public a client but not engage them directly. The desire to provide useful geographic information relevant to the decisions or actions related to such an issue can be extremely strong. The same is true for many issues. For this reason, many geographers, who consider themselves applied geographers, conduct research with the intention that it provide useful information in the decision-making process surrounding a specific issue. In short, applicable geography differs from applied geography because it is not client-driven. However, the two concepts share the intention of producing useful geographic information to be consumed by decision makers, rather than solely contributing to theory. Thus, unlike curiosity-driven research, which informs geographic theory, applied and applicable geography are intended for a non-academic audience.

NORTH AMERICAN APPLIED GEOGRAPHY IN THE PRE-1955 PERIOD

By the beginning of the 20th century, useful geographic knowledge already had been employed in various ways, including in cost-benefit analyses for the Isthmian Canal that was later built in Panama (James, 1972). Among the most significant and voluminous applied geography efforts, however, were those associated with two world wars and their aftermath.

Early applied geography: War-related activities

Professional geographers serving in both civilian and military officer capacities conducted applied geography. Preston E. James provided a comprehensive overview of applied geographer's contributions during both world wars. He also summarized their contributions in post-war capacities, such as the "commodity studies," which were required because of shipping overloads created by trade and war supplies (James, 1972).

Even prior to World War I, the American Geographical Society had adopted applied geography as central to its mission. J. K. Wright quoted the inaugural issue of the *Geographical Review*:

" It is the essence of the modern ideal that knowledge is of value only when transformed into action that tends to realize the aspirations of humanity" (Wright, 1952, p. 195).

In this spirit, the Society supported field expeditions that "would lead to important discoveries and human development," sought data important to government and business decision-making, and attempted to influence policy. For example, the Society attempted to influence census taking and even a new federal tax system, which Wright characterized as its "most ambitious "undertaking (Wright, 1952, p. 50).

Other national and international Society activities are well documented. Among those most prominent are those associated with the Congressional Act that established the International Meridian Conference in Washington, D. C. and the leadership of Society Director, Isaiah Bowman in an enterprise headquartered at the Society. Bowman was among the most prominent geographers of the wartime era. Not only did Bowan play a prominent role as "Chief Territorial Specialist" at the Versailles Peace Conference, he was responsible for settling many European boundary issues and was credited by Secretary of State Acheson as "one of the architects of the United Nations" (Ogilvie, 1950, p. 229). He also has been termed "Wilson's" and "Roosevelt's Geographer" by

Neil Smith, who argued that American globalism has roots in the intellectual ideas of the past, including those of Bowman (Smith, 2003).

World War II, however, stood out as the period of unmatched demand for useful geographical knowledge, which took the forms of terrain analysis, the preparation of intelligence reports and other formats in all branches of the military. Joseph A. Russell noted that applied geography was carried on in host of federal agencies and that geography was a fundamental course for military forces (Russell, 1954). He also reported that applied geographers supervised the work of geologists, meteorologists, and others, while directing, coordinating and editing all geographic studies for the war effort. At the conclusion of war-related activities, most applied geographers returned to their university posts. However, many remained in federal agencies and new opportunities for newly university-trained geographers appeared.

Applied geography and the university: The case of the university of Chicago

American geographers were actively involved in addressing a range of other national, regional and local geographic problems in the pre-1955 period. Much of this work reflected a dialectic between academic and non-academic geography. Leading academic geographers of national repute converted their academic interests in human-environmental and urban locational problems to government employment. Still, others served as consultants and advocates. Some participated directly in the applied geography process; others produced applicable results. The purpose here is not to detail all the types of applied geography that occurred in this period. Rather, it is to focus on a small set of academic geographers of the Chicago School of Geography, whose efforts exemplify that dialectic and whose works are readily identifiable. For two academic generations, important Chicago-affiliated geographers played leadership roles.

Examples from the 1st generation

Long before Colby identified applied geography as one of the two major dimensions of the discipline, Chicago geographers were busy applying the concepts of their discipline. In Chicago's first generation, two geographers serve as very good examples.

Harlan Barrows, one of the leaders of American Geography between 1919-1942, served as Chair of Chicago's Department and worked for the government. Between 1933 and 1941, Barrows worked on a variety of water resource projects that involved formulating policy, served on the Mississippi Valley Committee of the WPA, and became directly involved in water project assessments and comprehensive plan design for water resource development. Colby and White later noted Barrows policy-making roles and his ability to settle difficult resource issues, such as those associated with the allocation of the Upper Rio Grande waters (Colby and White, 1961).

Carl Sauer, a Chicago trained cultural-historical geographer, applied geographic knowledge for the improvement of human-environmental conditions. In the Michigan Land Economic Survey, his academic theme, "human as agent," was applied to land assessment and classification. Sauer applied well-established academic methods to land use analysis, dealing directly with 'destructive exploitation' and its specific impacts on the land (Sauer, 1919;1921). His operational scheme included a model of potential future uses. He argued that future standards be set to establish land use based on permanence rather than short-term gain. Sauer's efforts were utilized in land planning and published in the academic literature. He was researcher and activist, lobbying for the implementation of this plan, although long distance travel was required (Leighly, 1976).

Examples from the 2nd generation

A second generation of notable applied geographers emerged from Chicago, notably Harold Mayer and Gilbert White. Both received PhDs from Chicago and joined the faculty.

Harold Mayer thought of himself as a practitioner and theorist. He provided details on various individuals who were active in applied urban geographic research and the preparation of city plans. Under the leadership of Hoyt (sector model), geographers played "especially prominent roles" in both applied and applicable research, especially in 1930s city planning efforts during his years at Chicago (Mayer, 1982). Mayer directly participated in these efforts, as well as in port planning in Chicago and other U.S. cities such as Milwaukee and Seattle.

Gilbert White is perhaps the 20th Century's most respected international expert on water resources and natural hazards. White is the premier example of the dialectic between academic and applied geography for more than a half-century. His early applied government service (encouraged by Barrows) on water projects and flood victims led to his work that generated policies for the Roosevelt Administration. White's emphasis on environmental variability and human perception helped the federal government find ways to mitigate the impacts of natural hazards. He held crucial positions on the National Resources Committee and Natural Resources Planning Board, and served in the Bureau of the Budget, Executive Office of the President. FEMA administrator Frank Thomas explained the importance of his work in the 1960s:

"(White prepared... (a report), entitled 'A Unified National Program for Managing Flood Losses' (House Document 465). This report documented the need for non-structural approaches to flood loss reduction. It recommended establishment of programs for delineating flood hazards, for flood proofing or relocation of repetitively damaged properties, and for undertaking a flood insurance feasibility study, all of which subsequently became elements of the National Flood Insurance Act of 1968 Program" (Thomas, 1986).

The government, the public, and the deprived – are the clients of White's applied geography. In addition to performing as civil servant, White served as faculty member at Chicago from 1956–69, before moving to the University of Colorado to establish the Natural Hazards Institute.

In summary, the works of leading geographers of the period pre-1955 period included applied and applicable geography. Practice was not separated from theory. Civic duty in the form of applied science was a normal extension of academic geography. It provided useful geographic knowledge—that which could contribute to social improvement through social policy. Applied geography was valued inside and outside of academic institutions.

During the same period, applied geography was being practiced in non-university settings. Space does not permit elaboration, but a few brief examples clarify another important form of applied geography as practiced in institutional settings.

Institutional applied geography in non-academic settings

The important institutional development of applied geography at the American Geographical Society was reported earlier. Applied geography was also institutionalized in the federal government and in the private sector in this period. Perhaps more than any other federal agency, the Census Bureau has been employed in the acquisition, reporting, and analysis of data for national, state and local programs. Since 1890 under the direction of Henry Gannett, census geography and census data products of the applied geography process, have become key tools for business and government planners. The applications of geography evolved and expanded during the 20th century to include the development of digital national geographic reference files and digital mapping techniques, forerunners to the TIGER files in wide use today. Applied geography, then, also has a long tradition in government.

A final example of applied geography in a non-university setting in this period involves business geography. In the 1930s, William Applebaum accepted a position with the Kroger Company. His introduction of very basic geographic concepts and methods for analyzing market regions proved highly successful, launching careers for others who followed. Initially, applied geography in business focused on measurement techniques, market delimitation and competitor analyses. However, Applebaum maintained a close relationship with a small group of academic geographers and hired new geography graduates. Slowly, applied business geography expanded to include other roles, including location analysis, site selection, and the application of various models. As a result of the dialectic between Applebaum's practice and the theories of a group of academically based geographers, a new academic sub-specialty, business/marketing geography, emerged in the U.S. (Epstein, 1978). Applebaum's legacy continues.

These examples of applied geography in academic and non-academic work places clarify the important role of the applied geography in 20th Century America prior to 1955. Most illustrate the symbiotic relationship between the two worlds. Despite this strong tradition, applied geography experienced a dramatic recession by the 1960s. Applied leaders like Gilbert White did not stop practicing geography. But most of the new academic leadership would turn to purely theoretical pursuits and to preparation of the professorate. Academic geography was consumed by a new vision and by internal controversy until the resurgence of applied geography in the 1970s.

APPLIED GEOGRAPHY AFTER 1955: RECISSION AND RESURGENCE

Applied geography not only was pushed to the discipline's periphery after 1955, it became a confusing concept and was denigrated by academics. By the 1970s, a more tolerant, pluralistic environment and harsh economic conditions, including for the professoriate, led to the re-emergence of applied geography.

Recission: The rise of spatial science and training of the professoriate

Any period of revolutionary change requires a "start date." The seed of change in academic geography for this period is sometimes linked to the 1955 seminar of William L. Garrison at the University of Washington, where the discovery of "spatial thinking" emerged. There are at least two distinctive features of the 1955–1970 sub-period. One is the new emphasis placed on spatial theory. The second involved the demographic and accompanying economic changes in North America, particularly the United States.

The power of spatial thinking and spatial theory

The actual advocacy for spatial thinking can be traced to Schaefer's criticism of geography's claim of exceptionalism among the sciences (Schaefer, 1953). His support for the development of spatial theories and laws was supported by Ullman's plea for a focus on spatial interaction (Ullman, 1953). However, it was the seminar of Garrison in 1955 that appeared to produce some of the leading spatial thinkers of a new generation of geographers, including Berry, Bunge, Dacey, Getis, Huff, Marble, Nystuen and Tobler. These graduates enthusiastically carried the torch for development of spatial theories advocated by Garrison (1959) and others. Their scholarship and influence expanded and spread spatial thinking for decades, transforming departments at Chicago, Northwestern, Iowa and Michigan, among others. Of course, there were other leaders as well.

The spatial theme not only had to attack the existing paradigm as inadequate, it had to be methodologically and conceptually different, including provision of new forms of analysis. Michael R. Hill (1981), among others, noted the links between the spatial theme and the philosophy, logical positivism. A series of publications provided the challenge and new framework for the spatial theory theme. In the 1960s Ackerman spoke of the "research frontier" and the "science of geography" (1963 and 1965). In the same decade, European and North American geographers addressed the need to develop spatial theory, incorporating process explanations for spatial patterns using statistical and other types of models (Bunge, 1966; Haggett, 1966; Hagerstrand, 1967; Chorley and Haggett, 1967; Harvey, 1969). A new wave of geography professors emerged and spread spatial thinking. From Northwestern came Brown, Gould and others (e.g. Brown, 1968; Abler, Adams and Gould, 1971) and, from Iowa emerged King, Golledge and others (King, 1969; Golledge, 1967). By 1971, a text by Abler, Adams and Gould presented the "new" geography and its scientific foundations:

> "... we no longer see the foundation of our science as description of the spatial organization of the world. Our view now is that the explanation of classes of events by demonstrating that they are instances of widely applicable laws and theories is the function of geography" (Abler, Adams, and Gould, 1971, p. 87).

Post-war growth and expansion: preparation of the professoriate
After World War II, and especially in the 1960s, the U.S. economy expanded rapidly and population growth rates were high. The post-war GI Bill fueled university attendance. However, it was the ripple effect of the post-war baby boom that changed the American educational system. Not only were the numbers of school-age children growing rapidly, Americans could better afford to send children to college due to an expanding economy, which further increased college enrollments. Beginning in the 1950s, baby boomers swelled elementary and secondary schools, demanding more classrooms. By the 1960s, Geography enrollments increased and graduate departments became focused on filling the ranks of the expanding professoriate. In fact, during the 1949–1963 period, North American Geography expanded at an unprecedented rate.

The demands on the professoriate and the new scientific approach to geography occurred simultaneously and busied most departments. The resulting emphasis on teacher preparation in this period was so great that U.S. State Department Geographer, Daniel P. Beard argued that this teaching-teachers syndrome had become a troublesome tradition with potentially negative consequences for graduates in a changing economy (Beard, 1976).

The purge mentality and some of its consequences
By 1973 Bunge suggested that a purge mentality had crept into Geography (Bunge, 1973), one that purged the gains of the past for fashionable substitutes. While this is a debatable contention, there can be little doubt that "spatial science" leaders attacked the previous paradigm, resulting in an air of frustration and intolerance. Applied geography associated with the previous generation, including the dialectic of academic geography that had been so prevalent in the pre-1955 period seemed to evaporate thereafter for nearly two decades.

In a retrospective of this period, Gould noted that it was a time of both "intellectual excitement," and a time of "viciousness" and "ugliness" by the old guard, such as Sauer, who spoke of the newer geographer as "competent amateur" (Gould, 1979, p. 140). Gould claimed that the previous paradigm blocked publication efforts of the newer breed. Gould's judgment was that Sauer had been:

> "faced with a new generation, one that was both sick and ashamed of the bumbling amateurism and anti-quarianism (study of antiques) that had spent nearly half a century of opportunity in the university piling up a tip-heap of unstructured factual accounts" (Gould, p. 141).

Gould went on to suggest "it was practically impossible to find a book in the field that one could put in the hands of a scholar in another discipline without feeling ashamed." It was the works of the new breed that now made us proud. Thus, the "revolution" resulted in a true separation of disciplinary viewpoints—one scientific and unforgiving of the past, and another, that was proud, conservative and resentful of the new geography.

Like true revolutions, this one was swift and damaging. This attempt to dispose of traditions in favor of promising new approaches Bunge termed a "purge mentality." To be sure, the differences between traditional and spatial geographies were real and each would gladly have signed the other's death warrant. This tension of the 1960s, resulted in the efforts of the former school to defend and justify their academic integrity and for the latter to zealously pursue scientific proofs.

This occurred at an inopportune time because the so-called "environmental mood" was at the same time evolving in the sciences. One result was the evolution of "environmental geomorphology" as an applied geological science. Geologist, Don Coates, described it "discoveries"

> "...it was recognized that geomorphology could and should play an important role in this new emergence, whereby mankind is studied as one of the agents of sculpturing and changing the landscape... environmental geomorphology is the practical use of geomorphology for the solution of problems concerning human plans to modify landforms or use and change surficial materials and processes" (Coates, 1982, p. 139–140).

Geography's human-environmental themes—the use of "human as agent" and destructive use of the land – applied in the Michigan Land Economic Survey and to water resource planning of the 1920s and 1930s, was being "discovered" with fanfare by other applied scientists of the 1960s.

In short, it appears the academic tensions of the 1960s did not permit the dialectic of academic and applied geography so apparent in the previous generation. Thus, a distinguishing feature of this period was the recission of the applied geography from its previous position in American Geography. Among the casualties were loss of turf and loss of perspective. So much so, that by the 1970s, Harrison reported confusion over the term "applied geography." While some still saw the term as associated with problem solving, *1 in 4* respondents to a survey equated the term with "employment" (Harrison, 1977, p. 298).

Given the 1960s geography focus on "spatial science" and the clear goal of preparing the future professoriate, many geographers felt unhappy and uncomfortable with the new "Manpower" mandate that geography should look outside academia for job placement. Bhardwaj suggested that "non-academic employment" was the reason that academics "looked down upon" and found applied geography "impure" (Bhardwaj, 1977, p. 320). He suggested that a possible explanation was the contrast in workplaces, specifically the academic freedom of the university vs. the "ideological constraints" present in non-academic environments. Others, including Berry, also noted the classical tendency of "purer" colleagues to denigrate the "applied." In speaking of this period, Berry noted the "radical assaults in the 1960s and 1970s on applied geography" and suggested that "most waves of change overshoot and subsequently are rationalized." He believed this characterized the reactions to applied geography (Berry, 1980; 1981).

Several forces of the 1970s began to counter-balance the forces that had pushed applied geography to the periphery of the discipline. Generally their occurrence led to a period of pluralistic thought in Geography. It is in this context that applied geography would re-emerge in several formats.

Relevance, dissent, and pluralism in the resurgence: The 1970s and beyond

The 1970s, less than a generation after "spatial science" had taken hold of the discipline, was a period of challenges to this controlling framework. Criticisms were couched in a variety of formats: social relevance, philosophical positions, and the call for tolerance in a maturing discipline.

A positivistic geography provided too rigid a framework for most geographers but perhaps the loudest critics were among the radical dissidents who challenged the status quo and reported the irrelevance of much of 1960s research. "Establishment geography" had become irrelevant and catered to the elite. Peet described in detail the evolution of radical geography into Marxist geography of the 1970s. Dissident geography at first meant "taking the side of the oppressed, advocating their causes, pressing for fundamental social change." The radical critique early focused on the fact that "spatial patterns could be described in the most minute mathematical detail without disclosing anything significant about causal social processes" (Peet, 1998, p. 74).

Other radical viewpoints emerged as well. Just as Harvey had turned on positivistic science, Bunge (***Theoretical Geography***) experienced a self-reported awakening in 1966 Chicago, when experiential learning with ghetto dwellers taught him more about life's geography than library books (Bunge, 1979). Blaut claimed that dissenting geography constituted a type of "inverted" applied geography, where activities such as Bunge's Detroit Geographical Expedition amounted to taking tools and valuable knowledge into "communities across the U.S. and Canada" to struggle against the status quo results of capitalism (1979).

Other forces also were at work in the 1960s and 1970s that challenged the controlling paradigm. Buttimer (1974) and Tuan (1971), for example, argued that geography was not value free. Still, others from inside the establishment, cried out for change that would embrace social relevance in geography. Morrill, for example, was a persistent advocate of altering the direction of geographic research, proposing geographic strategies, including, "organization," "persuasion," and "action" to facilitate needed societal transformations (Morrill, 1969 and 1970). Zelinsky outlined other roles as "diagnostician," "prophet," and "architect" of the needed societal transformation (Zelinsky, 1973). This combined critique, rooted in social relevance, led to "disillusionment" of many of those formerly promoting spatial analysis. By 1974 Taaffe reported that it was impossible to afford confidence in theoretical approaches without social utility (Taaffe, 1974).

In addition to Beard's criticism, other non-academics also felt alienated from the discipline and believed that existing curricula were inadequate to prepare geographers for non-academic employment opportunities. Some even argued that non-academic research is often "far more prolific or superior in quality" than that which appears in academic journals (Beard, 1976).

This environment of challenge and change led to a pluralistic geography, one in which geographers would become more tolerant of opposing viewpoints, both philosophically and methodologically. Applied geography re-emerged in this environment "as a breath of fresh air" (Anderson, 1977).

Applied geography since 1955

Applied geography continued during the recission period. In fact, prominent spatial thinkers of the "Washington School" provided important applications. However, because it was dwarfed by theoretical work in spatial analysis, it was obscured and fugitive. Thus, most applied geography during the rescission was produced outside of academia or overlooked by it. By the mid-1970s, after self-criticism of theoretical geography, cries for social relevance and the realities of declining demand for academic geographers, a number of institutionally based actions strengthened the resurgence of applied geography in North America.

From rescission to resurgence: Threads of continuity linked the periods
Despite the rescission explained in the previous section, a number of prominent academic geographers associated with spatial theory practiced applied geography and influenced regional, national and international policy, although these works were not

published nor discussed in academic geography. Two excellent examples are Garrison and Berry.

Garrison, early champion of spatial thinking and spatial analysis, performed applied spatial analysis in transportation planning and land use. Funded by the Federal Highway Administration, Garrison demonstrated some of the earliest quantitative relationships between accessibility and land values. His work was influential in the primary planning of the freeway system. His work on land-use developments adjacent to freeway interchanges and the impacts of highway relocations on the spatial structure of commerce was client-driven and informed assessment procedures and the policies of local and federal governments. Began at the University of Washington and continued at Northwestern, this type of research also influenced future U.S. and Canadian applied geographers, such as Duane Marble and Barry Wellar (Wellar, 2003). Garrison also received funding in the 1960s from the Office of Naval Research for pioneer applied GIS work.

Berry, undoubtedly one of the most cited geographers of all time, became an internationally known urban theorist while at the University of Chicago in the 1960s. Berry's numerous academic publications contributed greatly to the evolution of the Chicago School of Urban Geography. Given the theoretical emphasis of the period, it is not surprising that Berry's applied research went largely unnoticed by the discipline during that era. In 2003, Berry commented that:

> "... the geographic community is far less aware of my applied research and the tension between my theoretical and applied research over four decades" (Berry, 2003).

Hired at Chicago in 1958 by Gilbert White, Berry almost immediately undertook applied urban research. White received permission from his Dean to place Berry in a "second" office, one inside Chicago City Planning (Berry, 2003), where he worked two days per week guiding and performing applied studies.

Actually, a number of Berry's applied research activities ran in parallel and cut across the decades of the rescission and resurgence of applied geography. One involved commercial redevelopment studies that informed Chicago's urban renewal program. At the same time, he examined relationships between urban land use, environmental quality and urban planning. Perhaps even less well known than these activities is his applied work in developing countries in urban geography and urban planning. Berry spoke recently of the impact of direct experience on his applied work, such as in Calcutta, which:

> "began to open my eyes to other realities, and to the realization that some portion of my life would have to be devoted to helping alleviate ... poverty and degradation that I saw" (Berry, 2003).

What followed was policy-related regional development work, not just for India, but for Indonesia, Chile and Brazil.

In the U.S. and Canada, Berry performed other applied urban research that benefited from the tension between urban theory and practice. For example, his work on metropolitan growth resulted in better urban definitions, the selection of specific corridors for highway development, and new and useful concepts, such as "counter urbanization."

These brief examples illustrate the continuity in the applied geography tradition among academics. At the same time, useful geographic knowledge was being applied outside academia. Bart J. Epstein chronicled the leadership role of William Applebaum in the development of a new applied geographic field, marketing-business geography during the 1950s (Epstein, 1978). The original emphasis on methods of market delimitation and measurement evolved to include impact and locational analyses includes the application and modification of the Huff model developed by geographer David Huff at the University of Washington in the 1950s. Huff and Epstein worked as academic consultants for corporations during the rescission. More importantly, increasingly, applied geographers were employed in the business field (Epstein and Schell, 1982). Unfortunately, because of the emphasis on theoretical geography and even the negative connotations given to applied geography during the rescission, many practicing business geographers withdrew from academic geography meetings. They would be reunited during the 1970s resurgence of applied geography.

Actions within academic geography: The re-awakening after 1970
A number of events took place in the 1970s, while Geography was becoming more tolerant of non-positivistic viewpoints and the demand for the professoriate was diminishing. In addition to the awareness brought by criticism, cries for social relevance, and curricular change, a few additional, significant events occurred that stimulated a reinstitutionalization of applied geography. The first came from geography departments and, then, from the Association of American Geographers.

Geography departments and applied geography
Some small non-Ph.D. granting geography departments had long emphasized the practicality of geographic training for employment. Indeed, during the 1960s, planning employment had become one of the chief sources of jobs for geographers (Mayer, 1982). However, action that was intended to place applied geography on a higher plane within the discipline occurred in a number of ways in the 1970s. Two examples illustrate this.

In the spring of 1977, the Department of Geography, Michigan State University held a symposium entitled "Applications of Geographic Research." Lawrence Sommers, then Chair of the Department, received financial support from the University and challenged the Geography faculty to consider the applied implications of their research. The result was a two-day symposium with presentations and fifteen faculty and discussants. A monography, *Applications of Geographic Research* (Winters and Winters, 1977) resulted. A number of the faculty and graduate student participants at this Symposium became leaders in applied geography of the period.

The second example originated at Binghamton University. John W. Frazier conceived of the Applied Geography Conference. Long-time practitioner and academic geographer, Bart J. Epstein of Kent State University, joined Frazier as co-planner and co-director of the conferences. The early conferences were rooted in the strong tradition of applied geography, awareness of recent pluralism, the practical experiences of the founders and the belief that applied and applicable geography deserved formal recognition in the discipline. Broad leadership and guidance were necessary for the Conferences to be successful. Early guidance from key, well-respected leaders, including James R. Anderson, USGS, John Lounsbury, Arizona State University, Jacob Silver, U.S. Census, Lawrence Sommers, Michigan State University, and Thomas J. Wilbanks, Oak Ridge National Laboratory, later evolved into a Steering Committee that represented business and government interests (e.g. Dayton-Hudson Corporation, The May Stores, K-Mart, JC Penney, Thomson Associates, United States Military Academy, USGS, U.S. Bureau of the Census, among others).

The first conference was held in 1978. A "Papers and Proceedings" captured the major contents of the Conference and the publication has continued since. The early Conferences were by invitation only to provide quality control. The openness to applicable and applied geography attracted a wide range of academic and practicing geographers and resulted in several major themes in the early conferences, including the application of methods and technology, economic development, business and retailing studies, and issues associated with curricular change. The Conferences attracted traditionally trained academic geographers, as well as practitioners, such as those working on the edge of GIS and remote sensing applications. They attracted a mix of the young, including students, and the established that interacted freely with the newcomers. In summary, the Applied Geography Conferences were established out of the needs of the discipline. That fact that Geography was emerging from a period of rescission likely enhanced interest. Bringing together those seeking a common bond in a useful forum resulted in success.

Meanwhile, the Association of American Geographers began to realize it was in the best interests of its memberships to foster applied geography. When opportunities arose, the AAG took advantage.

Actions by the association of American geographers to institutionalize
applied geography
In support of applied geography, the AAG leadership attended the annual AGCs. In fact, occasionally, they made an effort to schedule the Fall AAG Council Meeting

at the Conference site. Two opportunities arose shortly after the AGCs began and provided opportunities for direct support of the resurgence of applied geography in North America.

The first opportunity arose in 1980 when the AAG created specialty group memberships (Khater, 2003). The Association included applied geography as one of the initial options for group affiliation. The success was immediate. The Applied Geography Specialty Group (AGSG) was among the five most popular groups based on enrollments. It attracted non-academic geographers employed in the private and public sectors, as well as university members, including students. Non-academic geographers were elected to leadership positions and played active roles in establishing future efforts to promote applied geography. Although it had no formal role in the Applied Geography Conferences, there was a great deal of interaction between the two and eventually the Specialty Group contributed to the Conferences as a co-sponsor.

The second opportunity emerged from the unfortunate death of James R. Anderson, who was nominated for the vice-presidency of the AAG in 1979 and also was involved the planning of the Applied Geography Conferences. Shortly after Anderson's death, his family made a donation to the AAG in his name and the AAG Executive Director, Patricia McWethy, charged the Specialty Group with the creation of criteria and a jury for a proposed annual award, The James R. Anderson Medal of Honor in Applied Geography. The award itself was important in the recognition of a distinguished career. However, the elevated status given applied geography in the AAG was unprecedented. First awarded in 1983 at the AAG's annual banquet meeting, it shared the stage with all other annual distinguished honors granted by the Association. Since 1983, twenty Medals have been awarded for distinguished careers in applied geography in government, private enterprise and the university, including two to Canadians, Roger F. Tomlinson and Barry Wellar.

The Applied Geography Specialty Group and Anderson Medal are two important means by which applied geography became institutionalized by the AAG in the recent era.

The source of applied geography: Demand for applied geography
outside the academy
It should be clear form the preceding sections that restored academic interest in applied geography has been broadly beneficial to geography as a discipline and society as a whole. Driven by self-criticism and the pressure for funding and non-academic job placements, many colleges and universities reached outside the academy to establish an applied geography interface. In addition, some geographers made new connections with academic colleagues to perform interdisciplinary applied research. These efforts enriched both academic and non-academic work places by bringing additional intellectual strength to pressing real-world problems. What should not be overlooked, however, is that the non-academic world provides the demand for most applied geography

problems. This led Torrieri and Radcliffe to argue that client-driven research arises
where cost-effective and time-based solutions are prerequisites, namely in the government and private sectors, which fund most applied research (Torrieri and Ratcliffe,
forthcoming). The discussion of various applied geographic research activities in this
chapter, including the wartime efforts, commodity studies, The Inquiry, New Deal
projects and policies, and business geography, support that contention. The same argument can be made for the recent past and the current period. Perhaps the best example
is the rapidly changing and expanding geographic technologies, themselves products
of applied geography, that have greatly stimulated research applications among applied geographers. These include, of course, GIS, remote sensing, image processing,
and global positioning systems. The result has been an increasing volume of applied
geographic research on a variety of themes (Torrieri and Ratcliffe, forthcoming).

THE FUTURE OF APPLIED GEOGRAPHY IN NORTH AMERICA

To avoid Bunge's warning that academia separates itself from practice and becomes
irrelevant (Bunge, 1979), academic geography must continue to recognize the importance of applied geography as an equal partner. Only then does it benefit from the
tensions that accompany theory placed in practice. As we begin the 21st century, a professional infrastructure that supports both these dimensions seems in place. However,
given experiences of the 20th century, we should vigilantly guard this balance. Among
the challenges of the 21st century is the necessary leadership to go beyond the role of
social criticism and include directed action to change society and solve its problems.
Hoping our work is applicable is a far cry from immersion in problem solving. Whether
paying customers or disenfranchised groups requiring empowerment, geography needs
to identify and serve a clientele in order to close the gap between applicable and applied geography. This will require desire and commitment to undertake controversial
problems and participate in active ways in the world of practice. When this occurs, institutional leadership can and should assist practicing geographers in important ways.
First and foremost, it should foster focused working groups that are able to contribute
to the solution of specific problems in the way the AAG responded to NSF's call for
proposals for Homeland Security. An effort at the national level that recognizes a few
good leaders surrounded by competent supporting researchers, including practitioners,
is a useful model, especially if it considers regional linkages that support the overall
mission. Such a structure could be similar to the one that connects the use of applied
water resource and emergency service (hazards) research by flood plain managers and
others.

Implicit in this suggestion is the need for connecting the strengths of theory and practice
between academia and the other sectors. Utilizing traditional strengths is a logical
pursuit, while seeking new useful applied geography is also a priority. A model for this
continuing tension between theory and practice is the U. S. Census Bureau. Berry's
contribution of new and useful urban geographic concepts was noted earlier. He and

other geographers have contributed concepts that appear as organizing concepts for census geography. The continuing technological innovations of the Bureau, which are routinely put to the touchstone by university and non-university researchers and generate new and useful applied and theoretical research, were also mentioned. This interactive process creates a tension that is mutually beneficial and continues to produce even more useful geographic knowledge.

Similarly, continuous, long-term research foci provide a strong basis for collaboration between academic and applied geographers and their clients. There are many well-established research traditions that should be strengthened further in the 21st century, while taking advantage of new geographic technologies. Among those that are readily identifiable from discussion in this chapter is transportation geography. Transportation networks, their purpose, structure, flows and impacts on people and areas, have long been a subject studied by geographers in theoretical and applied contexts. The commodity studies and aspects of military geography mentioned earlier in this chapter are representative of this research theme, as are the studies of the Isthmian Canal Commission at the turn of the 20th century (James, 1972) and the 1950s and 1960s work of Garrison, Marble and Wellar on Interstate Highways and their interchanges previously reported in the chapter. This research focus continued in the 1970s and 1980s, including in the evaluative and operations planning work performed at Oak Ridge National Laboratory (e.g. Sorenson, 1986; Southworth, et al., 1986, Oak Ridge Staff, 2002). Finally, the recent applied research by Wellar and his colleagues on the "Walking Security Index" has provided methods for the identification and remediation of dangerous highway intersections (Wellar, 1998). This is but one example of noteworthy research foci that has combined geographic theory and practice in constructive ways for more than a century. Such long-term accomplishments should be continued in the 21st century.

At the same time, applied geography will be asked to meet new challenges that require new useful geographic knowledge, as well as being informed by existing theory. Applied geography must bring new and innovative methods and information to the solution of society's most vexing problems. The mention of a few of the likely issues that will require applied geography expertise and insight helps clarify the challenges of a new century in North America.

The first involves terrorism and homeland security. The events of 9–11 not only raised our consciousness as a nation, they resulted in coordinated action to prevent future terrorist events. Among those actions was a request from the National Science Foundation for academic disciplines and their practitioners to provide frameworks for assisting in homeland security. Applied geography's role throughout the 20th century in wartime studies, including the direct employment of geographers in the federal service, and since is very clear. Geographic concepts and useful geographic knowledge are vital to the protection of the United States. Not surprisingly, then, Geography was among the first disciplines to provide a coordinated and prompt response to this call. The AAG co-ordinated a funded project with a tight deadline for response. The project, entitled "The Geographic Dimensions of Terrorism: A Research Agenda," required coordination of a wide variety of geographical assets directed at this national priority. A volume fol-lowed and included the perspectives of thirty leading scholars on issues of geographic technologies and geo-spatial data, as well as thoughtful contributions on hazards and ter-rorism (Cutter, Richardson, and Wilbanks, 2003). Contributors from geography include both academic and private presentations and are joined by public policy perspectives. This is an excellent example of the provision of necessary institutional coordination at the national level mentioned above. However, the real test for 21st century applied geography will be provision of continuity in this vital topic. Surely some of the authors of this volume will find individual consulting roles but the coordinated effort should not stop with the production of a single volume. The best for the discipline would be to move forward from suggesting the applicability of the geography provided in this volume to detailed action plans that guide their implementation. Also, Canada is not exempt from this problem and both nations would benefit from shared information and research on this topic.

The second example includes the application of GIS and other geographic technologies in responsible ways that improve the human condition. Obviously, geography must continue to be involved in the evolution of the concepts and theories of this rapidly evolving set of geo-technologies. However, the greater social responsibility lies where geographers likely have far greater expertise. That is in the identification of human spa-tial problems, caused by an array of environmental and institutional decisions, whose study and solution can be enhanced by spatial analysis that employs geo-spatial data and the application of geo-technologies as part of the methodological solution. In 2002, Jack Dangermond speculated about the future of GIS analysis on the Internet (Danger-mond, 2002). His vision is one of an interconnected world, one that can instantly access numerous databases and GIS tools that can assist the solution of global, regional and local problems. His vision for geography is that it will provide leadership in problem resolution by applying its knowledge of environmental and spatial processes and prob-lems in conjunction with GIS and other geo-technologies available on the Internet. His challenge is that geography should use its best minds and technologies to perform criti-cal analysis that better informs policy. For Dangermond, this is the connection between geo-technologies and applied geography.

The final example involves one of North America's most vexing problems, persistent racism and ethnocentrism in a rapidly changing racial/ethnic environment. Immigra-tion in the 1990s contributed to increasingly multiculturally diverse populations in the United States and Canada. Yet, vexing racial/ethnic problems continue to plague both nations. Some of these persistent problems that periodically boil over into conflict and violence beg geographic analysis. Changing cultural landscapes and spatial and power relationships, which are important geographic topics, are never far below the urban-behavioral surface. Canada's "French problem" took on a more complex meaning in the 1990s when immigrants were blamed for blocking French separatism. In the U. S., Latinos became the largest minority group in the country by 2000 and Asians continued to enter the U. S. in record numbers. Yet, African Americans remained the most disenfranchised group in the nation (Frazier, Margai, and Tettey-Fio, 2003). The disparities between white Americans and non-white groups have become an increasingly important topic among geographers. In fact, the 25th annual Applied Geography Conference held a special "Race/Ethnicity and Place" Mini-Conference in 2002 to examine a number of issues associated with this topic. Perhaps one of the most important trends noted in 2000 was the increasing segregation among the nation's white and non-white children. Surely applied geography has something to offer in the recognition and solutions related to American and Canadian racial/ethnic problems as expressed as cultural and iniquitous landscapes.

Applied geography has a long and proud heritage. In the last generation, after a brief recession, applied geography has emerged even stronger than before. There has been no more an exciting and challenging time for applied geography than is available at the beginning of the 21st century. Exciting and trying issues beg geographic analysis. The discipline must select, develop, and support key, focused initiatives rooted in tradition and supported by its new technologies and theories. Such initiatives should draw upon our best minds and take full advantage of the eager and committed, similar to the one that placed geography back in many pre-collegiate classrooms in America by the close of the 20th century. Even stronger efforts will be needed to place geography in vital institutions, such as in the creation of a Geography Division in the Office of Homeland Security. Leadership is required and means providing service in the form of vision, lobbying, and delivery of a product. It also means inspiring talent to participate, having good judgment, and insisting on quality performance. Such initiatives elevate geographic practice to a position of equity with geographic theory. Leadership must come from our associations and institutions, which must commit resources to the enterprise. However, it must also come from individual desire to engage clients and, potentially, confrontational adversaries. It is necessary to influence decisions by addressing clients and adversaries in clear unwavering voices that speak from informed positions.

Most of the problems mentioned above are national or regional by nature. However, many also have local dimensions and geographic expression. What better place to commit one's expertise and effort than at home? Who knows, such activity and commitment may result in national ramifications. Whatever the case, there is no shortage of pressing global, regional, and local problems that require an applied geography approach. As long as geographers are willing to contribute to the solutions of problems that are inherently geographic or have geographic dimensions, there will be applied geography. Where geography is put into action and provides useful knowledge to the non-academic world, applied geography will thrive.

REFERENCES

Abler, R., Adams, J. S., and Gould, P. (1971). *Spatial organization: The geographer's view of the world*. Englewood Cliffs: Prentice Hall, Inc.

Ackerman, E. A. (1963). Where is the research frontier? *Annals of the Association of American Geographers*, 53(4), 429–440.

Ackerman, E. A. (1965). *The science of geography*. Washington, D. C. : National Academy of Sciences, Natural Research Council, Publication No. 1277.

Anderson, J. R. (1977). Personal communication. Letter related to planning First Geography Conference, December 12, 1977.

Beard, D. P. (1976). Professional problems on non-academic geographers. *The Professional Geographer* 28, 127–31.

Berry, B. J. L. (1980). The role of applied geography in creating future geographies. Keynote presentation to third annual applied Geography Conference, Kent State University, October, 1980.

Berry, B. J. L. (1980). Creating future geographies. *Annals of the Association of American Geographers*, 70 (4), 446–958.

Berry, B. J. L. (1981). Applied geography at the public-private interface. *Papers and Proceedings of the Applied Geography Conferences*, 4, 3–5.

Berry, B. J. L. (2003). Confessions for an applied theoretician. James R. Anderson lecture presented to 99th AAO Annual Meeting, March 6, New Orleans, La.

Bhardwaj. S. M. (1977). Some directions in applied medical geography: a third world perspective. *Papers and Proceedings of Applied Geography Conferences*, 1, 318–330.

Brown, L. A. (1968). *Diffusion process and location, a conceptual framework and bibliography*. Philadelphia: Regional Science Research Institute.

Bunge, W. W. (1966). *Theoretical geography*. Lund: Lund Studies in geography.

Bunge, W. W. (1973). The geography. *The Professional Geographer* 25 (4), 331–37.

Bunge, W. W. (1979). Perspectives on *theoretical geography. Annals of the Association of American Geography* 69(1), 169–74.

Buttimer, A. (1974). *Values in geography*. Washington, D.C.: AAG, Commission on College geography, Resource Paper. No. 24.

Chorley, R. J. and Haggett, P. (ed.) (1967). Models in geography. London: Methuen.

Coates, D. R. (1982). Environmental geomorphology perspectives. In J. W. Frazier *Applied Geography. Selected perspectives*. Englewood Cliffs: Prentice Hall, Inc., 139–65.

Colby, C. C. (1936). Changing currents of geographic thought in America. *Annals of the Association of American Geographers* 26, 1–37.

Colby, C. C. and White, G. F. (1961). Harlan Barrows, 1877–1960. *Annals of the Association of American Geographers* 51: 395–400.

Cutter, S. L., Richardson, D. B., and Wilbanks, T. J. (eds) (2003). *The geographical dimensions of terrorism*. New York: Routledge Press.

Dangermond. J. (2002). Applied geography and GIS. James R. Anderson lecture presented to the 98th AAG Annual Meeting, Los Angeles.

Epstein, B. J. (1978). Marketing geography: a chronology of 45 years. *Papers and Proceedings of Applied Geography Conferences* 1: 372–79.

Epstein, B. J. and Schell, E. (1982). Marketing geography: problems and prospects. In J. W. Frazier (ed.) *Applied geography. Selected perspectives*. Englewood Cliffs: Prentice-Hall, Inc., 263–82.

Frazier, J. W. (1982). *Applied geography. Selected perspectives*, 3–22.

Frazier, J. W. (1994). Geography in the workplace: a personal assessment with a look to the future. *Journal of Geography* 93(1), 29–36.

Frazier, J. W., Margai, F. M., and Tettey-Fio, E. (2003). *Race and Place. Equity issues in urban America*. Boulder: Westview Press.

Garrison, W. L. (1959). Spatial structure of the economy. *Annals of the Association of American Geographers* 49, 232–39.

Garrison, W. L. (1959). Spatial structure of the economy II. *Annals of the Association of American Geographers* 49, 471–82.

Golledge, R. G. (1967). Conceptualizing the market decision process. *Journal of Regional Science* 7, 239–58.

Gould, P. (1979). Geography 1957–1997: the augean period. *Annals of the Association of American Geographers* 69(1), 139–50.

Haggett, P. (1966). *Locational analysis in human geography*. New York: St. Martins Press.

Hagerstrand, T. (1966). Aspects of the spatial structure of social communication and the diffusion of information. *Papers and Proceedings of the Regional Science Association* 16, 27–42.

Harrison, J. D. (1977). What is applied geography? *The Professional Geographer* 29(3), 297–300.

Harvey, D. (1969). *Explanation in geography*. London: St. Martins Press.

Hill, M. R. (1981). Positivism. A hidden philosophy in geography. In M. E. Harvey and B. P. Holly (eds.). *Themes in geographic thought*. London: St. Martins Press, 38–60.

James, P. E. (1972). *All possible worlds*. New York, St. Martins Press.

King, L. J. (1969). *Spatial analysis in geography*. Englewood Cliffs, Prentice-Hall, Inc.

Khater, E. (2003). Personal communication, August 25, 2003.

Leighly, J. (1976). Carl Ortwin Sauer, 1889–1975. *Annals of the Association of American Geographers* 66(33), 337–48.

Mayer, H. M. (1982). Geography in city and regional planning. In. J. W. Frazier (ed.). *Applied geography. Selected perspectives*, 25–57.

McKenzie, J. L. (ed.) (1983). *Webster's New Universal Unabridged Dictionary*. New York: Simon and Schuster, p. 90.

Morrill, R. L. (1969 and 1970). Geography and the transformation of society. *Antipode* 1, 6–9, and 2, 4–10.

Oak Ridge National Laboratory. (2002). Analysis of defense distribution center (DDC) supply chain. Awarded AAG applied geography project award, 2002.

Ogilvie, A. G. (1950). Isaiah Bowman, an appreciation. *Geographical Journal* 115, 229.

Peet, R. (1998). *Modern geographical thought*. Oxford: Black well Publishers.

Russell, J. A. (1954). Military geography. In P. E. James and C. F. Jones (eds.). *American Geography. Inventory and Prospect*. Syracuse : Syracuse University Press, 484–95.

Sauer, C. O. (1919). Mapping the utilization of the land. *Geographical Review* 8, 47–54.

Sauer, C. O. (1921). The problem of land classification. *Annals of the Association of American Geographers* 11, 3–16.

Schaefer, F. K. (1953). Exceptionalism in geography. *Annals of the Association of American Geographers* 43, 226–49.

Smith, N. (2003). *American Empire: Roosevelt's geographer and the prelude to globalization.* Los Angeles: University of California Press.

Sorenson, J. H. (1986). Elimination of transportation alternatives in the M-55 Rocket disposal program. *Papers and Proceedings of the Applied Geography Conferences* 9, 207–216.

Southworth, F., Peterson, B. E., Davis, R. M., Chin, S., and Scott. R. G. (1986). Application of the ORNL national highway network database to military and civilian transportation operations. *Papers and Proceedings of the Applied Geography Conferences* 9, 217–27.

Taaffe, E. J. (1974). The spatial view in context. *Annals of the American Association of American Geographers* 64 (1), 1–16.

Thomas, F. H. (1986). Personal communication, February 25, 1986.

Torrieri, N. K. and Ratcliffe, M. R. (forth coming). G. L. Gaile and C. J. Willmott (eds.). *Geography in America at the Dawn of the 21st Century.* New York: Oxford University Press.

Tuan, Y. (1971). Geography, phenomenology and the study of human nature. *Canadian Geographer* 15, 181–92.

Ullman, E. J. (1953). Human geography and area research. *Annals of the Association of American Geographers* 43(1), 54–66.

Wellar, B. (1998). Combining client-driven and curiosity-driven research in graduate programs in geography: some lessons learned and suggestions for making connections. *Papers and Proceedings of Applied Geography Conferences* 21, 213–20.

Wellar, B. (1998). *Walking security index.* Ottawa: DOT Regional Municipality of Ottawa:- Carleton, Canada.

Wellar, B. (2003). Personal communication, August 25, 2003.

Winters, H. A. and Winters, M. K. (1977). *Applications of Geographic Research.* East Lansing: Michigan State Board of Trustees.

Wright, J. K. (1952). *Geography in the making. The American Geographical Society, 1851–1951.* New York: American Geographical Society.

Zelinsky, W. (1970). Beyond the exponentials: The role of geography in the great transition. *Economic Geography* 46, 498–535.

PART III

CASE STUDIES

REGINALD G. GOLLEDGE

CHAPTER 11
DISABILITY, DISADVANTAGE, AND DISCRIMINATION: AN OVERVIEW WITH SPECIAL EMPHASIS ON BLINDNESS IN THE USA

ABSTRACT

In this chapter, I focus on an area that has, in the past, had only a limited attraction for applied geographers. Much of this may be attributed to not being aware of the nature and magnitude of the spatial problems facing disabled people. In the first part of the chapter, I discuss alternate definitions of disability, select one based on lack of functional performance, and survey the nature of disability in the US and elsewhere. Thereafter, I examine enabling legislation that provides an umbrella for disability research, examine problems of barriers, and discuss avenues of future research for which the applied geographer is eminently suited to pursue.

THE DISABLED: WHO ARE THEY?

There are several ways to interpret the term "disability" and the other concepts associated with it – including *impairment, handicapped*, and *disadvantaged*. Let us first examine some of these alternatives and justify the particular definition adopted herein.

The difficulty in obtaining data on disability is due to a lack of uniform terminology, inconsistent reporting worldwide, and a lack of institutional screening of reported statistics. Few countries have a national registry specifically devoted to collecting statistics on disability. All too often, "disability" is equated only with the problem of physical mobility. This is but *one* of the areas of disability; a list of disabilities should include:

1. Lack of physical mobility (e.g., requiring aids such as wheelchairs, crutches, walkers).
2. Vision impairment and blindness.
3. Speech impaired (or language deficient).
4. Hearing impaired.
5. Haptic impairment (e.g., touch insensitive, crippled by disease).
6. Cognitive impairment/Brain damaged (e.g., difficulty in ambulation, problems in recall and retrace, and high stress).
7. Reading impaired (e.g., dyslexia).

A. Bailly and L. J. Gibson (eds.), Applied Geography, 213–232.

8. Phobics.
9. Mental retardation (mentally challenged).
10. Neurological impairment.
11. Mental illness.
12. Physical or mental disability because of an internal medical condition (e.g., cancer, HIV/AIDS, diabetes).
13. Disability resulting from mental illness (e.g., multiple personalities).

In addition to the above, serious and profound disability is caused by health impairments (e.g., vascular, respiratory, cancer, heart disease, diabetes, etc.). Year 2000 Census estimates of the number of disabled persons in the USA reach as high as 54 million!

WORLD HEALTH ORGANIZATION (WHO) DEFINITIONS

The World Health Organization has developed a set of definitions as part of the International Classification of Impairments, Disabilities, and Handicaps (ICIDH). The ICIDH itself was developed as an extension of the International Classification of Diseases (ICD). Within this framework, the following definitions are offered:

* *Impairments* are concerned with abnormalities of body structure, organ or system function and appearance;
* *Disabilities* reflect the consequences of the impairment in terms of functional performance;
* *Handicaps* are concerned with the disadvantages experienced by an individual as a result of impairment and disabilities and the interaction of the individual with his or her surroundings (McNeil, 1992).

The emphasis here on disability concerns functional performance. Throughout the rest of this paper, disability will be used in this definitional context.

There are, however, other definitions of the above terms. For example, the Social Security Disability Insurance Program (SSDI) considers persons to be disabled if they are unable to engage in substantial gainful activity. This is a broader definition, for it takes into consideration age, education, and work history, as well as physical status and medical conditions. The landmark "Americans with Disabilities Act" (1990) has a tripartite definition including: (a) a physical or mental impairment that substantially limits one or more of the major life activities; (b) a record of such an impairment; or (c) being regarded as having such an impairment.

A definition from a different point of view is that proposed by Nagy (1991). Nagy's definition includes four components consisting of an active pathology, impairment, functional limitation, and disability. Active pathology involves some type of interference with fundamental life engaging processes. Impairment involves a deficit which is anatomical, physiological, mental, or relational. Functional limitations involve restriction on many of the senses or other activities such as hearing, seeing, walking, grasping, or performing mental tasks. Disability refers to a limitation in performing socially defined roles and tasks in interpersonal relations, family life, education, or work because of functional limitations.

In general, therefore, there are two different ways of looking at disability: First, as a physical limitation (referred to derogatively by able bodied social theorists as "the medical model") and, second, as a limitation on performing social roles (referred to as the "social construct" theory of disability). From a purely personal point of view, I prefer the former definition, as this will allow me to focus on the problems encountered in attempting to overcome physical disablement – this is of immediate concern and relevance (i.e., is an applied approach). Considering disability as relating to social roles has captured the interest of social theorists and activists concerned with societal change to accommodate disability. This is a theoretical, ideological, and long-term approach and is more academic than applied.

IMPLICIT AND EXPLICIT SPATIAL DISCRIMINATION

When the Americans With Disabilities Act was passed in July 1990, it was estimated that there were approximately 43 million disabled people in the USA (i.e., 15% total U.S. population). As the graying of America increases, so has this number and proportion, and, in the 2000 Census (where disability statistics have been deliberately collected for the second time), the figures are 54 million and 19%, respectively.

Disadvantage, disenfranchisement, and discrimination occur implicitly and explicitly with respect to disabled people in all societies. This is inevitable, considering a basic dictum of society that "the good of the many should outweigh the good of the few." Thus, at a simple level, doors have been developed with round handles that require turning, and this simple fact discriminates against the person with crippling arthritis or others with similar disabilities. This has not, however, prevented us from continuing to build doors with round doorknobs. There is no *intent* to disadvantage people by continuing this design characteristic; rather, there is an *unfortunate consequence* of using this design feature. Similar consequences exist throughout many cultures, and the result is the same in each—people who are disabled are prevented from equally participating in everyday activities by a variety of environmental and societally determined barriers.

However, many other disenfranchising and discriminatory activities take place that are more obvious than these. For many years, the blind traveler was unable to take a guide dog on transportation systems or into shops or restaurants. In the development of today's information super highways, the software is tied to visual point-and-click procedures that inhibits participation by the blind and vision-impaired or those with haptic disability. Wheelchair persons are often parked at the back of concert halls or movie theaters, in designated handicapped areas, where viewing is often impeded. Recreational and outdoor leisure time facilities (such as exploring natural environments) usually provide little access or support for the physically disabled. And many disabled people have been relegated to the lowest levels of the job market and given little chance for improvement of their economic and social status because of barrier filled and unsympathetic workplace environments. For example, considering only the working age groups (18–65), about 50% of hearing impaired people are employed, while only about 30% of blind people are employed. The blind individual working in an office environment is often "chained" to a desk without the freedom to explore facilities available in the neighborhood of the employment location. Until recently, social activities such as watching movies and television were denied to the hearing impaired. Efforts at remedying this situation (called "closed captioning") now exist.

In many countries, disabled people are educated in special classes. Their schoolmates, with whom they may only interact at recreational or lunch breaks, often deride them and label them as freaks. Mainstreaming is now taking place in some school systems, but one of the problems of mainstreaming is that the assistive facilities developed in the former special education classes (e.g., personalized instruction and access to assistive technology) may no longer be available in the general integrated classroom.

Upon release from institutions, some retarded individuals become recidivist criminals because they cannot understand or have not been taught laws of property ownership, monetary exchange systems, or civil decency. Others become victims – easy targets for a wide variety of criminal acts ranging from robbery and assault to rape and murder. Disabled people are frequently perceived to be easy targets, particularly in the inner city areas of large cities where they are forced to live in order to get ready access to public transportation. The increasing violence in many urban areas, particularly in downtown areas and on urban mass transit vehicles, makes it unsafe for disabled people to travel independently. Too easily, they become the brutalized victims of social deviants who see a disability only as an easy target with a reduced chance of retaliative response.

This brief overview of implicit and explicit discrimination could swell to large proportions. I stop here because I think I have made the appropriate point of its existence, but it is an area that should attract the research interests of Applied Geographers.

THE DISTRIBUTION OF DISABILITIES: THE CASE OF BLINDNESS

Many countries do not collect information on the distribution or numbers of disabled people, and, thus, producing a worldwide picture is very difficult. But many countries are starting to (e.g., Database on Disability in India, Nepal, Bangladesh, Pakistan, Bhutan, Maldives, and Sri Lanka 2001). For individual disabilities such as blindness, some reasonable records do exist. Although institutions such as the World Health Organization (WHO), the United Nations, and the World Blind Union try to collect worldwide information on blindness, its causes, its distribution across age groups, its occurrence in the sexes, and its occurrence by country and region, it is widely accepted that it is impossible to say how many blind or visually impaired people there are in the world today. Reasoned estimates rather than exact counts are used, and most of these estimates are extremely conservative.

A critical factor is the definition of blindness; conservative criterion would be 3/60th vision or less in the better eye. Given such criterion, WHO estimates that there were over 30 million blind persons in the world in 1984 (Watson, 1990, pp. 245), and, using simple extrapolations, they estimate there would be close to 50 million by the year 2000. It is further estimated that approximately 9 out of every 10 blind persons live in a developing country, and as many as 7 out of every 10 of those persons had a blindness that could have been prevented or cured by treatment.

According to the World Health Organization Program for Prevention of Blindness, 27 to 35 million persons have binocular vision of 0.05 or below, and, among the world population of 5.3 billion, 41 to 52 million have vision of 0.1.

Blindness rates increase greatly with increased age. Blindness caused by cataract is the most frequent reason. Insufficient numbers and spatial distributions of eye care services in developing countries results in an inability to deal with the cataract problem. In West Africa and Central America, it is estimated that 85 million people are at risk and face potential blindness.

AGE SPECIFIC PATTERNS AND CAUSES OF BLINDNESS

There are approximately 1.8 billion children below 15 years of age in the world, of which 1.5 million are blind. A major cause is Vitamin A deficiency. This is particularly so in developing countries. Blindness is more prone to occur in premature or low birth-weight babies. In developed countries, diabetic retinopathy is the most important cause of adult blindness. Other significant causes of blindness are glaucoma and macular degeneration (Nakajima, 1991).

If we modestly assume that there are at least 50 million blind people in the world, then 27 million cases would be due to cataract, 10 million to trachoma, 5 million to glaucoma, 1.0 million to xerophtalmia, and 7.0 million due to other causes. Blindness does not appear to affect one sex more so than another across all age groups (Adamsons & Taylor, 1990).

United States patterns

Some statistics for the United States, provided by the Braille Institute for 2002, are as follows:

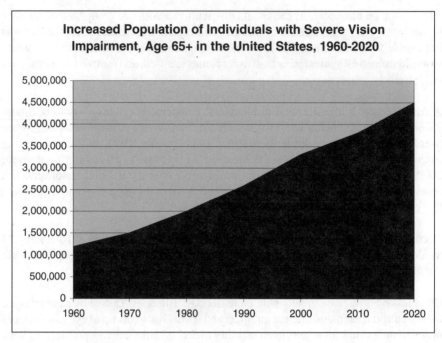

Figure 1.

- 15 million blind or vision impaired people in the USA.
- 90 percent of legally blind people have some residual vision (legal blindness is defined as having vision acuity of 20/200 or less in the better eye with corrective lenses or central vision acuity of more than 20/200 if the peripheral field is restricted to a diameter of 20 degrees or less). Severe vision impairment is defined as being unable to read an ordinary newspaper, even with glasses or other corrective lenses.
- 70 percent of severely vision-impaired persons are aged 65 years of age or older (Figure 1); 50 percent of these people can be classified as legally blind, and the current rate of blindness in the population aged 65+ is 135 per 1000 people.
- In the United States, blindness is third only to cancer and AIDS as the biggest health fear by the general public.
- In the US, vision problems affect nearly one in five (20 million) preschool age children (ages 3–5) and 25% (12.1 million) school age children (ages 6–17).

SPATIAL PROBLEMS OF DISABILITY

Within the United States, approximately 15 million people have considerable vision impairment. Of these, approximately 5 million are in a severe state (Braille Institute Newsletter, 2002). There appear to be about 2% blind people who use a blind dog in the country and around 35% are users of white canes for mobility purposes. Even given the most conservative estimates, this leaves millions of legally blind individuals and, probably, around 6–9 million blind or severely vision-impaired individuals, who either travel without assistive devices or who rely on sighted guides such as friends, relatives, fellow employees, and family for their mobility. One could also surmise that there is probably a large proportion of this group (especially those of the oldest age groups) who do not travel, even in their local neighborhoods. Lack of vision, therefore, has a considerable impact on their quality of life and their psychological and social well-being.

THE PROBLEMS OF DISABILITY IN THE WORLD TODAY: WITH SPECIAL REFERENCE TO BLINDNESS

Social and economic problems

Given the relatively significant number of blind and vision-impaired people throughout the world and their concentration in developing economies, it is obvious that an ethical problem of some magnitude exists. In developed economies, the need for financial, personal, and societal expenditure to deal with disabled populations is substantial. It is increasing as populations age and live longer, and as people are more subject to infirmity or other disability. If we are at all concerned with the quality of life for such populations, we must be aware of the levels of individual and group welfare and well-being of these partly disenfranchised and discriminated groups and be prepared to assist in some way in alleviating their problems.

Some environmental modifications have been made for the mobility disabled, and technological advances have been made to help some disabled people participate in independent travel. These include specially modified autos and dedicated para-transit, as well as terminally and route elastic group transportation devices. However, some movement patterns are still quite dependent on the assistance of others. For example, choice of outdoor recreation usually requires the assistance of a sighted guide, although some recent changes have been undertaken in national parks in some countries where talking signs, sounds, smells, and specially prepared surfaces give to the disabled person a freedom to interact with nature and travel in the natural environment to a degree heretofore impossible (Pigram, 1993).

One of the major limitations on the ability of the vision impaired to undertake spatial mobility successfully is the constraint on previewing and preprocessing information needed for travel (i.e., having information available to develop travel plans). Previewing allows a potential traveler to develop a heuristic for selecting a destination and following a path to it. While a sighted person can use a map along with other ways of obtaining information prior to travel, the sight-disabled person must rely either on a sighted guide and be a passive traveler or must be able to interpret verbal directions. Verbal directions are extremely difficult to interpret without proper attention being paid to the nuances of the spatial terms used (e.g., "the entrance is to the front and left of you" leaves a wide spectrum of travel paths open). Orientation, direction, and distance are all difficult processes to understand when tied to either an egocentric or a local relational system, as opposed to a widely accepted globally georeferenced system (e.g., "on my left" is actually "on your right," so that an instruction to turn left may have to be mentally rotated 180 degrees before it can be successfully implemented). On the other hand, tactual devices such as NOMAD (a computer-based auditory-tactile information system [Parkes & Dear, 1989]) or available tactual maps (Tatham & Dodds, 1988) can help the blind traveler both plan a route and anticipate changes of transportation mode and plan for such transfers. For the blind person, sound and touch must become substitutes for vision. But the fact remains that, without some type of prior knowledge or preprocessing of information about the environment to be traversed, successful navigation by the blind or vision-impaired individual may not be possible.

SPECIAL PROBLEMS

Perhaps the most pressing problem facing all disabled groups is that of *inaccessibility*. In spite of attempts to undertake environmental design modifications (e.g., Imrie, 1996), it is still difficult to project rehabilitation clients into normal situations of education, recreation, socialization, employment, land use, and transportation systems because of architectural design or other human-built barriers.

A second major problem is *acceptance*. Many disabled groups wish to be accepted as part of the general economic and social environment. Many activist groups argue vigorously that their disability in no way differentiates them from people without disability. They refuse to accept concessional advantages (e.g., tax breaks) and teach an ethic which says that despite their disability they are able to do anything that an able-bodied person can do. With this position in mind, they expect to be treated exactly the same as their non-disabled counterparts. However, there are a majority of disabled groups who are willing to accept the fact that their disability puts them at a social, economic, and physical disadvantage. These groups actively push for concessions, such as environmental modifications. They are particularly active today in terms of obtaining unimpeded access to buildings and to different forms of public mass transportation. For example, in the early 1990s, a disabled activist group in Australia held up the delivery of 250 new buses to state systems in South Australia and New South Wales, because the buses were not equipped to handle wheelchairs. The disabled population thus faces a dilemma: does it push for parity and equality in terms of acceptance, or does it recognize that disabled persons battle some limitations and actively lobby for environmental modifications designed to improve their quality of life? And what can be done to alter social attitudes towards disabled people that may have serious economic implications (e.g., employability)?

EXTERNALITY EFFECTS IN PROBLEM SOLVING FOR DISABLED GROUPS

Often, developments that benefit or favor segments of society provide disbenefits to others (the question of negative externality). This happens even in the limited domain of disability. For example, modifying the physical environment by putting curb cuts at intersections to allow mobility disabled (wheelchair) persons to move smoothly between sidewalk and the street provides a potential hazard for the blind or vision-impaired individual who, relying on a long cane to pick up changes in gradient and to identify a curb as a stopping point, may not be able to do so. The consequences of this error would be to continue walking out into a street in the path of oncoming traffic. In another domain, public telephone booths that have the bottom sections removed so that wheelchair persons can access the phone by moving their chair under the instrument provide no warnings for the blind person whose cane can glide smoothly under the obstruction. The result is a head and body collision between the upper projecting part of the phone booth and the traveler. Literally dozens of examples of devices ranging from signs to benches to water fountains that are placed to the advantage of one group disbenefit other groups. The whole question of locating such features in three-dimensional space is rarely given much consideration when evaluating the potential impact of a facility siting decision on other disabled groups. Certainly, Applied Geographers interested in externalities of environmental modifications have an open research landscape in front of them. Likewise, those interested in the social, economic, and environmental modifications proposed to help the disabled have virgin research territory before them.

ENVIRONMENTAL PROBLEMS

Perhaps the most widely supported movement designed to aid people with a disability is the "barrier-free" movement. This consists of legislative controls on access to buildings, transport systems, and terminals (e.g., American Standards Act, 1961). One of the major problems to date, with respect to the barrier-free movement, is the overwhelming tendency to think about environmental design changes only in terms of a single population – predominantly, the wheelchair users. Even the standard sign for modifications designed to help disabled people has the wheelchair logo (e.g., special parking places, special toilet facilities, and so on). In the USA, this is in part due to the significant efforts of the Veterans Administration which has been the most significant lobby for improving conditions for disabled veterans and disabled groups generally. Many of the design modifications undertaken to assist one disabled group have benefited many other groups. The invention of automatic opening doors is a classic example of help for wheelchair users that benefits aged and infirm groups. Similarly, the addition of auditory signals at traffic lights to inform on safe pedestrian crossing was designed primarily to help blind and vision-impaired individuals, but it appears to be widely accepted as a benefit to all in terms of focusing their attention on light changes.

The wheelchair individual faces three types of problems: (i) the problem of having to operate at a lower height than the usual standing height; (ii) having to use a cumbersome and space consuming vehicle for mobility; and (iii) the nature of the disability itself (i.e., lack of freedom of movement). Because of these disabilities, there is frequent exposure to environmental situations that provide no barriers to normal use but provide substantial barriers to a person with this form of disability (e.g., steps in front of buildings, narrow doors, narrow elevators, door handles awkwardly placed, water fountains that are impossible to reach, telephone booths that will not allow entrance by a wheelchair, emergency telephones that are too high to reach, crowded furniture, lack of ramps, lack of facilities to help use public transportation (including special hooks or locking grids for wheelchairs), curbs without curb cuts, steep gradients on sidewalks, uneven terrain, and so on). Because most of these barriers are components of the physical world, they represent a tangible and addressable problem for designers, architects, planners, policy makers, and Applied Geographers. This is not necessarily the case for other disabled groups whose barriers are often found more at the micro than the macro level or, in terms of narrow societal acceptance, of their abilities to perform everyday activities.

ETHICS AND LEGISLATION FOR DEALING WITH DISABILITY

Enabling legislation

Most of the legislation of historical significance detailed in Table 1 concerns therapeutic attacks on the macro environment rather than the microenvironment. They focus more on physical than on social remedies.

They also display two important moral axioms: (i) the goal of normalization and nondiscrimination and (ii) the value of independence. These two axioms provide dilemmas for designers interested in practical applications of the legislation. This is nowhere more evident than in the perceived paramount nature of the rule of wheelchair persons among all disabled people.

THE BARRIER-FREE ENVIRONMENT

The barrier-free movement manifested itself in the American Standards Act (1961) and consequent acts in many other countries (see Table 1). The underlying ethic focused on making macro environments amenable for disabled users rather than focusing on microenvironments. Traditionally, micro environmental concern focused on orthodox therapy. This included rehabilitation and remedial actions using medication, or via welfare, employment, education, transportation, and other systems. In today's technocratic society, it has become important to continue this focus on the microenvironment particularly with respect to things like computer equipment and access to the information superhighways of today.

Table 1. Selected legislative acts—1961–2002

Universal Declaration of Human Rights, U.N. 1948
American Standard Specifications for Making Buildings and Facilities Accessible to and
   used by the Physically Handicapped – U.S. Congress, 1961 (Revised 1980)
Canadian Building Standards for the Handicapped, 1965
Access for the Disabled to Buildings, United Kingdom, 1967
Design for Access for Handicapped Persons, Australia, 1968
(Policy Document: National Commission on Architectural Barriers to Rehabilitation of the
   Handicapped, Designed for all Americans. United States Rehabilitation Services
   Administration, 1967.)
Regulations for Access for the Disabled to Buildings, Sweden, 1969
The Chronically Sick and Disabled Persons Act, United Kingdom, 1970
Amendment to the Urban Mass Transportation Act, United States, 1970
Law 7600, Equal Opportunities Law for Persons with Disability, Costa Rica, 1976
United Nations Year of the Disabled, 1981
(Architectural and Transportation Barriers Compliance Board, United States, 1981)
The Disabled Persons Act, 1981
Committee on Restrictions Against Disabled People, UK, 1982
Building Regulations in England, Wales & Scotland, 1987
Americans with Disability Act, 1990
Building Regulations in England, Wales & Scotland, 1992
Disability Discrimination Act, UK, 1995
Disability Rights Task Force, UK, 1997
Workforce Investment Act (Section 508), 1998
Human Rights Act, UK, 1998
Special Educational Needs and Disability Act, UK, 2001
Japan Helper Dog Law, 2002

The American Standards Act of 1961, in effect, changed the traditional micro environmental emphasis for dealing with *individual* disabled people to a macro environmental concern dealing with *large groups* of disabled people. This was a remarkable and significant change in the scale of operations. Goldsmith (1983) argues that this was a form of idealism that "could only have been conceived in the self-help culture of the United States and a doctrinal leap which the social welfare culture of Britain with its caution and pragmatism would not have embarked upon of its own accord." (p. 203).

A second facet of the barrier-free movement was the reliance on the physical rather than social interventions. Implementing a macro environmental approach intimated there was no patronage, condescension, charity or philanthropy in the design ethic. The development of automatic opening doors helped not only the wheelchair-bound but many others as well (e.g., the aged). The development of telecommunication devices for the deaf and sonar or electronic guidance systems for the blind have expanded the markets for common products such as television. The NAVSTAR satellite system (which was established largely for the locational accuracy desired for military maneuvers and rocketry), has been extended to areas of location-based services and of vehicular and ocean navigation. More recently, it has become the basis for experimentation for personal guidance systems for blind and vision-impaired travelers (Loomis, Golledge, Klatzky, Speigle, & Tietz, 1994; Loomis, Golledge, & Klatzky, 1998).

The initial goal of the barrier-free movement, that of "normalization," is basically an anti-discrimination ethic. Given that barriers force inaccessibility on disabled people, they are consequently discriminated against with respect to opportunities of movement, socialization, employment, recreation, education, and so on. One tends to think of barrier-free environments primarily in terms of physical barriers (e.g., curbs, narrow doors, stairs, etc.). But, barriers exist in many facets of life. Developments in speech synthesis and digitized speech have allowed blind and vision impaired people access to computer output that was previously focused primarily in the visual domain. The development of educational devices such as NOMAD (an auditory tactile information system) now provides blind and vision impaired people access to maps and graphics, pictures and photos, an information source previously denied them. TDD systems allow deaf people to use telephones and to hear lectures, while closed captioning has given them access to TV and video soundtracks. In other words, as macro environmental changes have taken place, the hold of discrimination on disabled populations has begun to be alleviated. As this alleviation continues, "normalization" takes place. This is the primary ethic of the barrier-free movement. Crack the barriers, address them, possibly modify, adapt or remove them, and the end product will be a better integration of disabled people into the mainstream of society.

The second ethical principle built into the barrier-free movement is that of independence. Lack of independence is probably the most keenly felt disadvantage of disabled people. It is frustrating and demeaning for a wheelchair person not to be able to use a public toilet on demand because the wheelchair cannot fit through the door or there are no side rails to assist in transferring from the wheelchair to the toilet seat. It is frustrating for the blind or vision-impaired individual not to be able to take off on a whim and explore a complex environment – even for as simple a task as going to the closest place at which one could obtain a coke or a milkshake. It is difficult for the hearing impaired person to sit in a college classroom and be unable to determine what is happening because the professor turns his head to a chalkboard or addresses remarks to a side wall so that lip reading cannot take place.

All the above imply that many disabled persons cannot operate independently in the general environment. They must have guides, assistants, note takers, and so on in order to obtain some type of parity with their non-disabled peers. The barrier-free environment stretched to its full extent (i.e., not confined just to addressing the problem of physical design) aims at removing barriers and increasing independence. Achievement of this state of affairs can only be obtained with strong federal legislative support for macro environmental modifications. Such has been embodied in the Americans with Disability Act of 1990. This is basically a civil rights act for disabled people. It is not characterized as an anti-discrimination act. In effect, it established the right of disabled people to have the same freedom of access to all facets of the environment that their non-disabled peers enjoy. The ADA stated emphatically that, because there was no legal recourse to protect people with disability, there was rampant discrimination in areas such as employment, housing, education, public accommodation, public (mass) transportation, recreation, institutionalization, health services, voting, access to public services, and communication. Discrimination denies people with disability the right to compete equally for jobs, services, and health-rights for which Americans have fought since the time of the country's inception. The legislation also requires sectors of the economy that obtain or rely on federal financing to conform to the provisions of the act under penalty of withdrawal of federal funds. But, the act is not excessively punitive. It does indicate that, while every effort must be made to accommodate to disabled patrons, users, or employees, if such accommodation requires extraordinary expense then it need not be enforced. But, because of this act, many institutions that obtain federal funding have developed very minimal plans for compliance to ADA requirements. There is no doubt that considerable and constant funding is required to satisfy ADA requirements, and that achieving appropriate modifications in all manners of environments will take a considerable period of time. However, if independence and mainstreaming are the end products of these activities, then it appears to be accepted that they will be worthwhile efforts.

Transportation problems

The 1970 amendment to the Urban Mass Transportation Act in the United States stated clearly that it was national policy to provide the same rights to elderly and handicapped people as to their peers in terms of using mass transportation facilities and services. Examples where this principle was displayed include the Washington Metro and the Bay Area Rapid Transit (BART) systems.

An increasing amount of attention is being given to destination choice and transportation mode selection for disabled individuals. As mentioned earlier, in the United States, there are approximately one hundred and ten thousand white cane users and ten thousand who use a guide dog. That leaves at least one million legally blind people who have to rely on means of transportation other than personal locomotion. More than fifty percent of the disabled non-driving population of the United States do not use mass transportation. This means that approximately 500,000 blind people distributed over the United States are essentially confined to their homes or must rely on friends, neighbors, or family to take them out of their local environment. Questions immediately arise as to what are the particular features of mass transit that produce such a widespread lack of acceptance. Is it the design of the different modes? The timing? The routes? The human-mode interface? Safety or fear? Difficulty in deciding where one is enroute?

Difficulty in interfacing modes? There are, in fact, dozens of questions that need to be asked in order to examine this problem. Few of them have been pursued very far, and they represent interesting research for the Applied Geographer.

As an example of the problems that disabled people have in trying to move freely, independently, and safely about the environments in which they live, let us look at some of the actions that disabled people face when having to use mass transit - the principal form of movement for many disabled persons. These difficulties will be summarized for three categories of disability: those with hearing impairments, those blind or visually impaired, and those with cognitive impairments.

The skills needed for a disabled person to understand and use a mass transit system have been detailed by Hunter-Zaworski and Hron (1994) and include the following:

a) Understanding and performing the tasks needed to ride the system such as accessing the correct vehicle, entering the vehicle, traveling on the vehicle, departing it, and exiting the stop or terminal.
b) Identifying the potential origin and destination of the transit trip.
c) Identifying special services that might be required to allow the trip to take place such as wheelchair lifts, kneeling buses, low floor buses, ramps, special routes, door-to-door pickup and delivery, or assistance in transferring between modes.
d) Acquiring information necessary to allow use of the transit vehicle, which may be via a telephone information service, or through printed schedules.
e) Determining which part of the system will be used, including definition of specific routes, when to request for and how to use transfers, how to resolve fare payment problems, understanding fare media (e.g., tickets versus tokens, cards versus cash).
f) To initiate the trip the correct pickup point must be identified, and if more than one class of vehicle uses that stop, the potential rider must be able to differentiate between those alternate users (e.g., buses which eventually diverge to different routes).The traveler must be able to enter a vehicle in the time allocated (e.g., light rail), or be able to use the lift or stairs and negotiate gaps between cars or between curb and vehicle or platform and vehicle.
g) The potential rider must be able to adjust to the start, stops, motion, or noise of the vehicle and be able to comprehend announcements relating to stops or emergency actions.
h) The potential rider must have the skills needed to depart the vehicle.

Now, let us examine how Hunter-Zaworski and Hron suggest how different disabled groups can handle these tasks.

For people who are deaf or hard of hearing a major problem in using a transit system is the inability to receive information from driver announcements or from announcements at a terminal that come over a PA system. First, the disability impedes their ability to access normal mass communication methods. It also limits their ability to request information on a person-to-person basis, or to use standard devices such a telephone to inquire about scheduling information.

Persons with visual impairment or blindness experience a different range of problems. The critical areas are in terms of receiving system information, locating and using devices associated with the trip, and physical movement throughout the system. System information is, for the most part, available in the form of printed matter such as maps and schedules. Obviously, blind and vision-impaired people have difficulty with such media. Locating and using devices such as fare boxes or token dispensers is extremely difficult unless some assisted technology is available (e.g., some type of tactile map or spatial representation [Andrews, 1983]). To move through a transit terminal, to change platforms, and so on, the blind or vision impaired person must have an accurate mental model or cognitive map of the terminal and be able to recover orientation and direction information if they are randomly disturbed such as by being rotated or disoriented by a jostling crowd. Whether finding one's way through a terminal, entering or exiting a vehicle, or trying to determine where one is on route, orientation and access to information are the two major problems faced by the vision impaired transit user.

Persons with cognitive or emotional impairments face yet another set of problems in trying to cope with mass transit. Some of these problems may result from a system failure (e.g., a vehicle running later or leaving early) which may require a change in the routine usually followed, or require a change in the route itself, with consequent differences in the degree of crowding, stairs, heights, or noise, any of which may have a negative impact. For such individuals increased stress may result in confusion, shyness or withdrawal, hesitancy, or at the other extreme, raucous or boisterous behavior. Those with learning disabilities may suffer most from confusion resulting in an inability to follow directions, or an inability to make his or her needs known to a driver, conductor, or other assistive individual.

Those with brain injury or with mental retardation may be unable to comprehend written schedules or verbal announcements over the PA systems. They may have difficulty comprehending scheduling, particularly where intramodal transfer is required.

What is obvious from this brief overview is that there is no single set of solutions that will make mass transit more accessible to all groups of disabled people. Solutions must be multi-media and multi-dimensional in order to compensate for the problems of transit use by disabled people. Since most disabled people are non-drivers, mass transit is the primary form of travel for those who wish to develop a feeling of independence – the same privilege that is enjoyed by the non-disabled majority. If such privileges are denied to disabled subgroups, society is discriminating against them, negatively impacting their freedoms and their quality of life.

SOCIAL PROBLEMS AND DISABILITY

Social problems relating to blindness

Although orientation and mobility training tries to instill some corrective strategies in individuals who are taught to travel independently in complex environments (e.g., retracing routes, asking passers-by for directions, shorelining, etc.), feelings of incompetence, frustration, anger, and despair frequently arise. Emergence of such feelings is inevitable given there is considerable variation in the ability of people to solve spatial problems, to make spatial judgments, just as there is variation in their willingness to take risks, to search and explore, or to approach strangers for help. For many blind or severely vision-impaired people, it is difficult or impossible to free themselves from a feeling of dependence. The vision impaired young adult who has to lie about his car being "serviced" when he and his date must use public mass transit to get to an evening's entertainment; the independent worker who stands forlornly waiting for the ride to work that never turned up; the independent blind traveler who could not see a construction sign saying "please cross the street here" and is instead turned into a blind alley where he/she is assaulted and robbed, all face problems in mobility and travel above and beyond those faced by the sighted traveler. It is harrowing enough to face the obstacles found in a complex physical environment without also having to bear the social and psychological stigma of disability. Ideas, suggestions, and inventions that can help lower the degree of disability and improve the quality of life for disabled people make a contribution that is beyond evaluation.

NIMBY

One ongoing problem that has faced both public and private sectors for some considerable time is the question of locating group homes or halfway houses in various neighborhoods within a city and, particularly, getting the location decision accepted both within a neighborhood and within the general community. This activity is a classic example of the NIMBY attitude (i.e., Not in My Back Yard!). Most people will agree that integrating de-institutionalized disabled people into the community at large is a good thing. However, equally as many then draw a line with respect to the precise location of group homes designed to help this integration. It is okay to place a group home for the homeless, the mentally disturbed, those in drug rehabilitation, the physically disabled, or the mentally retarded in someone else's neighborhood, but it becomes less acceptable when placed in one's own. An immediate consequence of the location of such facilities, which are usually located with the best of ethical and moral intentions, is to produce change in the behavioral patterns of neighborhood residents and to produce changes in the administration and governance and protection of such neighborhoods. For example, a group home for de-institutionalized residents of some mental hospitals could include some mildly retarded individuals as well as others who may have had a history of criminal offense or violence. Often, the reaction of local residents is to change movement patterns to paths away from such a group habitat. There is also sometimes additional protective actions and commands given with respect to the play areas of pre-teenage children. And, finally, the reaction of many local authorities is to increase the frequency of police patrols within the area. Thus, an act that is driven by social conscience can have both physical effects (e.g., increased police patrols, lowering of rental and owned housing values in the immediate area, thus producing potential tax loss); changed activity and behavior patterns by individuals in the neighborhood, often to the detriment of existing neighboring ties; an increase in the potential number of homes for sale; and potentially increased out-migration.

This type of responsibility and action gives some clear indication that non-economic variables will have to be considered and complemented by spatial and affective variables including the strength of emotional reactions to different decisions, attitudes, and preferences. Such variables often turn out to be equally as important as more conventional environmental or economic variables in helping to understand activities and processes occurring within specific areas of built environments.

Concerns for the future

Scientists and decision makers in private, public, or institutional settings can make significant inroads into the state of disenfranchisement of disabled people. Removing or cracking the barriers that restrain disabled people, whether they are physical, human, or social, is one of the most challenging tasks for the future. As the world becomes more electronically organized, the barriers become even more subtle. Is a physically disabled person with arthritic hands or Carpel Tunnel Syndrome denied forever access to a computer because access is tied to keyboards? Is a blind or vision impaired person denied access to the information super highway because it is tied to the vision-based point and click technology of graphic user interfaces? Can blind people and wheelchair-bound people travel the same streets with the same levels of safety and confidence as non-disabled people? Just how much does a small change in the physical or socio-cultural environment contribute to a change in the quality of life for a disabled subgroup? The questions to be pursued in this context are of considerable importance in both an ethical and moral domain and in a practical economic and welfare domain.

Design and myths

Designing for human environments, particularly for those with vision, is a completely different practice than designing for human environments for disabled people. For those with vision, exotic layouts with carefully placed vegetation, rest places, and exciting visual experiences, may dominate a particular design layout. For the vision-impaired individual, all those things that make the visualized layout attractive and preferable may turn out to be obstacles and barriers of the worst kind. Signs, one of the most significant portrayers of information to the sighted for example, are often just hanging obstacles to the blind and other disabled people. Except for simple signs, problems of interpretation may arise for the de-institutionalized borderline retarded or poorly educated disabled person. Sidewalks and stairs are everyday features for the non-disabled but can present impenetrable barriers to the wheelchair-bound.

There are some common misconceptions among sighted people. Many assume, for example, that immediately a person loses sight, their other senses improve dramatically to compensate for the loss of vision. This is not so. Of course, one has to use the other senses more, and, with practice, one develops a greater acuity in using them. But, if you have a hearing impairment when you lose your sight, the hearing impairment stays. You still have to make do with it.

A second myth that appears to be commonly accepted is that, when a person loses their sight, they immediately (perhaps through osmosis?) develop the ability to read Braille! Based on this assumption, designers, architects, and planners for years have been putting Braille lettering in every elevator, use it to indicate bathrooms, indicate directions, and so on. In the United States, a well developed country, only about twelve percent (at the very most) of blind people read Braille! In developing economies, this figure is considerably less. One must, then, consider what are the alternatives? Obviously, audition (i.e., verbal messages) may provide part of the answer. Digitizing natural sounds will also help. For example, elevator buttons labeled with Braille may not help a person who cannot read Braille; but a button that speaks its number when touched certainly would. Sound signals that indicate the passing of each floor are also of considerable help to the blind, but are no assistance to the deaf. In the latter case, flashing lights as each floor is passed are much more useful. This, of course, is of no use to the blind. In airplanes, initial instructions are given to "follow the blinking lights" if there is some type of internal problem and evacuation is called for. This is quite useless for the blind traveler, who, for the most part, is advised to sit and be the last person helped from the plane. I hardly find this type of treatment equitable!

For each disabled population, therefore, there is a different world full of unique barriers and obstacles that need to be charted, explored, and for which methods of representing the environment with all its hazards need to be developed. In transportation planning, we have dozens of algorithms for working out optimal routes between origins and destinations for people in vehicles. But we have few, if any, specifically designed to work out optimal routes for disabled travelers requiring obstacle-free routes. We do not even know how existing path selecting and network models have to be modified for use in solving such problems, or whether something completely new has to be developed.

Disability research in the USA benefits from a variety of funding sources. At the national level, the National Science Foundation has a "Program for People with Disabilities" that has many facets covering ethnic and cultural minorities, minority female representation (e.g., Women in Science and Engineering), and programs for the blind, deaf, speech defective, and mobility impaired (NSF/PPD). The National Institute for Health has similar programs (for example, NIH/National Eye Institute [NEI]). The Department of Transportation (DOT) has a variety of Mass Transit and private vehicle disability programs, while state DOTs (e.g., Ohio DOT, Cal Trans) fund more local or regional level projects. While the 1990 Americans with Disability Act (ADA) is pervasive and has stimulated concern for informed policy and practice with respect to disability access, there is a very uneven spread of funding generally. The NIH/NEI, for example, emphasizes funding for blind, vision-impaired, and low vision related projects. About 90% of DOT funding has been oriented towards wheelchair access, while attention to the equally significant accessibility problems of the blind have received little support. As more research across the board on the accessibility needs of all disabled groups is undertaken, some slight rearrangement of funding is taking place. But, it has usually taken a specific Act (e.g., Senate Bill 508 requiring government departments to make data and publications equally available to all) before major changes are made. But Acts like ADA and Bill 508 have served to make state and local authorities aware of the various public needs of disabled people, and both larger budgetary allocations for applications that modify environments and increased local funding for applied research relating to accessibility are becoming an important practice.

I sincerely hope that one of the dominant outcomes of this chapter will be to direct attention of Applied Geographers to the solution of problems such as those I have raised in this paper. This is a new and challenging arena for the Applied Geographer – far from the traditional emphasis on regional, economic, and developmental concerns. But, in a world where attention is focusing more and more on human resources and human-environment relations, investigating the geographic problems relating to groups who are disadvantaged, discriminated against, and disabled, can represent a new direction for research and the widening of research horizons for the Applied Geographer.

ACKNOWLEDGMENTS

I wish to acknowledge the assistance of Tony Richardson and Dave Lemberg, graduate students in geography at UCSB, for helping to assemble tables of barriers and assistive devices. Partial financial support was provided by the California Department of Transportation PATH Project, Grant #MOU167, and NSF Grant #HRD-0099261.

REGINALD G. GOLLEDGE
Department of Geography
and
Research Unit in Spatial Cognition and Choice
University of California Santa Barbara
Santa Barbara, CA 93106

Prepared for book on Applied Geography
edited by A. Bailly and Lay Gibson
March 2003

REFERENCES

Adamsons, I., and Taylor, H.R. (1990) Major causes of world blindness: Their treatment and prevention. Current Opinion in Ophthalmology, 1, 635–642.

Andrews, S.K. (1983) Spatial cognition through tactual maps. In J. Wiedel (Ed.), Proceedings of the 1st International Symposium on Maps and Graphics for the Visually Handicapped. Washington, DC: Association of American Geographers, pp. 30–40.

Braille Institute Newsletter, September 2002, **http://www.brailleinstitute.org/Education-Statistics.html.**

Goldsmith, S. (1983) The ideology of designing for the disabled. In: Proceedings of the Fourteenth International Conference of the Environmental Design Research Association (EDRA 14). Lincoln, Nebraska, pp. 198–214.

Hunter-Zaworski, K., and Hron, M. (1994) Improving bus accessability by people with sensory and cognitive impairments. Federal Transit Administration, Washington, DC.

Imrie, R. (1996). Disability and the City: International Perspectives. London: Paul Chapman Publishing.

Loomis, J.M., Golledge, R.G., Klatzky, R.L., Speigle, J.M., and Tietz, J. (1994) Personal guidance system for the visually impaired. In Proceedings of the First Annual International ACMISIGCAPH Conference on Assistive Technologies, Marina Del Rey, California, October 31-November 1.

Loomis, J.M., Golledge, R.G., & Klatzky, R.L. (1998). Navigation system for the blind: Auditory display modes and guidance. Presence: Teleoperators and Virtual Environments, 7(2), 193–203.

McNeil, J.M. (1992) Americans with disabilities. U.S. Department of Commerce, Economics and Statistics, Bureau of the Census, Washington, D.C.

Nagy, S. (1991) Disability concepts revisited: Implications for prevention. In A. Pope and A. Tarlov (Eds.), Disability in America: Towards a national agenda for prevention. Washington, DC: National Academy Press.

Nakajima, A. (1991) Epidemiology of visual impairmnent and blindness. Current Opinion in Ophthalmology, 2, 733–738.

Parkes, D., & Dear, R. (1989). NOMAD: An Audio-Tactile Interactive Graphics Processor for the Visually Impaired and Blind. Reference Manual (NOMAD Manual Version 1). Sydney, Australia: Quantum Technology Pty. Ltd.

Pigram, J.J. (1993). Human-nature relationships: Leisure environments and natural settings. In T. Gärling & R.G. Gollege (Eds.), Behavior and Environment: Psychological and Geographical Approaches (pp. 400–426). North-Holland: Elsevier Publishers.

Tatham, A.F., & Dodds, A.G. (1988). Proceedings of the Second International Symposium on Maps and Graphics for Visually Handicapped People. Nottingham, England: University of Nottingham.

Watson, D.A. (1990) A message from the World Blind Union. Journal of Visual Impairment and Blindness, 84, 6: 245.

REGINALD G. GOLLEDGE

CHAPTER 12
HUMAN WAYFINDING

ABSTRACT

While much attention has been paid to examining spatial movements as behavioral traces that are measured by distances, directions, and volumes of movement—as in migration and mobility research—less attention has focused on the cognitive and behavioral components of the wayfinding act. This lack is addressed in this chapter. Wayfinding processes, such as path integration or dead reckoning, landmark navigation or piloting, spatial search, and spatial updating and layout recognition are emphasized. In other words, wayfinding processes rather than the spatial manifestations of wayfinding acts are emphasized. A role for Applied Geography is suggested in terms of designing and testing technologies to help wayfinding decision-making and creating Smart Environments that tie to Intelligent Transport Systems and to the Location Based Services Industries.

INTRODUCTION

The movement of humans from place to place and the spatial interactions generated by human wants and needs are a natural domain for the applied geographer. Generally subsumed under titles such as "migration and mobility" or "transportation geography" the patterns of flows and the reasons why they occur provide a natural forum for the application of geographic thinking, reasoning, and analysis. But much of this research has been declarative—i.e., it details the facts of network flows of goods, services, people, and even ideas, at times summarizing these in vivid visualizations or in mathematical models that predict optimum network flows or other types of spatial interactions (Church, et al., 1991; McFadden, 2002; Fotheringham, et al., 2000). Many times the explanation for such movements and interactions are couched in physical structural terms, such as place to place differences in the existence or production of resources, place to place differences in levels of economic well being or cultural and ethnic concentration, or by illustrating what would be the case if everyone was located in a perfect world with complete information. Sadly, these models do not always relate well to the real world where incomplete information, lack of awareness, variations in personal tastes and preferences, political, social and economic alliances, and a host of other factors produce uncertainty and incomplete information. In this chapter, I depart substantially from the traditional emphasis in this area and focus more explicitly on a search for an understanding of the decision making processes that lay behind revealed behaviors

A. Bailly and L. J. Gibson (eds.), Applied Geography, 233–252.

and observable flows. In the latter stages of the chapter, I focus on recent technical innovations that are aimed at supplying missing information for wayfinders when they travel.

TRAVEL BEHAVIOR

Travel behavior has been described by Golledge and Gärling (2002) as a movement through space using a particular mode of travel. Travel behavior is a revealed phenomena, and it can be represented as a trace through an environment. This trace may be called a path, a route, or a course depending on whether the guiding process is wayfinding or navigation. Paths or routes are defined by selecting a set of linked segments from a network of connected nodes and links. The nodes are places where links can be joined or intersect and may range from mere road intersections to places of significant human interest (e.g., cities as nodes in a national highway network), depending on the scale at which they are examined. A course is regarded to be different from a route in that it is usually carefully plotted and strictly adhered to, with travel often being monitored by agencies such as air traffic controllers. Travel behavior takes place over courses, paths or routes in manners that are either strictly controlled (e.g., Just-in-Time delivery systems) or as a result of spontaneous en-route human decision making and choice behavior. The latter is said to be most typical of wayfinding.

There is a tendency today to distinguish between wayfinding and navigation (Golledge, 1999). Wayfinding refers to the ability to find a route, learn it, and retrace it or reverse it from memory (Golledge, 1999). It also refers to a person's ability to modify the route for any specific reason but still retain the ability to eventually reach the same à priori given destination. It is universal to all cultures and is the dominant form of movement in primitive or less-developed societies where comprehensive transportation networks have not yet been constructed. In developed economies, wayfinding is involved in daily and longer term episodic activities. These range from food shopping to holidaying or other types of variety seeking behavior. Wayfinding often involves search and exploration, and may not necessarily require retaining knowledge of how to get back to a home base (a move after choosing a new residence). Navigation, on the other hand, often refers to the deliberate act of using mathematical or technical aids to plot a course that should be adhered to when traveling from an origin to a destination (Loomis, et al., 1999). Unless there are emergency problems, such as equipment failure, or other factors that require deviation from the course, it is followed religiously and movement can usually be controlled by an electromechanical device (e.g., an autopilot). Individual travelers mostly use wayfinding rather than navigation practices, while commercial travel usually is based on navigation procedures. There are exceptions, however, and later in this chapter we will discuss some of these.

Humans are generally effective, though frequently inefficient, wayfinders. Consequently, human history is replete with examples of a search for assistive devices that help to turn wayfinding into navigating. The devices aim to reduce fluctuations in travel behavior at various scales and include such things as chronometers, sextants, odometers, compass, charts and maps, laser measuring devices, and Global Positioning Systems (GPS). Their use is designed to improve effectiveness, efficiency, and accuracy of movement by conforming to principles such as shortest paths, least effort, optimally covering or connecting places in trip chains, and so on.

But most human travel does not involve the use of assistive devices like these. People tend to rely on the information stored in *cognitive maps* for developing travel plans and implementing them, particularly within the context of local travel. For long trips (e.g., cross regional, cross national, or international) freedom of movement is usually restricted by the availability of commercial carriers, and consequently travel behavior may become closer to optimal movement given the available choices of a restricted transportation system. The freedom associated with vehicular movement, or movement by cycle or on foot, however is not subject to these same constraints.

Most humans do not search for optimal routes through the networks with which they interact on a regular basis (Gärling and Gärling, 1988; Golledge, 1997). They're often more properly described as satisficers or boundedly rational travelers in that they're content with routes that are good enough, seem simple, or allow a person to get there on time, rather than trying to find optimal paths through network systems depending on modes that are constrained by specific limiting criteria (Mahmassani and Jou, 1998).

Most regular human travel is purpose-dependent. Some purposes, such as journey to work, lend themselves to optimal path selection strategies or routines. In this way, travelers try to minimize en-route decision making and develop confidence in the fact that they can schedule certain time periods for activity-related travel and almost invariably achieve their goals within predetermined time or other travel constraints (Hägerstrand, 1970). But many trip purposes are not so easily constrained. Trips for recreation, socializing, dining out, visiting friends, and others that can make up to 30–40% of the daily activity pattern, tend to rely more on wayfinding rather than navigation practices. They also tend to rely more on stored knowledge in long term memory and the consequent use of this information to develop travel plans. Consequently, path selection strategies may be considered purpose dependent. Thus two neighbors going to the same destination for different purposes may select substantially different routes. This purpose dependent variation in travel behavior makes it hard to model.

In general, however, travel behavior requires the following:

- the ability to recognize a destination when in its vicinity
- the ability to select a sequence of path segments and turn angles that will allow arrival at or near possible destinations

- the ability to store information about a route traveled so that it can be retraced at some time in the future
- the ability to construct a route of segments and turn angles in the reverse order to that originally experienced so that return trips can be undertaken
- the ability to identify on-route and off-route landmarks for location, direction, and orientation purposes
- the ability to examine experienced and cognitively stored information to enable shortcutting or alternate segment use when desired or needed (e.g., when barriers impede following the original route)

All these abilities involve deliberate mental activities. They can be supplemented by reference to external aids. These external aids include:

- topographic or cartographic maps of a network through which travel takes place
- strip maps (e.g., Automobile Association "Trip-Tiks")
- hand-drawn sketches of routes or segments of routes
- written or verbal instructions based on the use of spatial language
- photos, videos, or other images of the route or section of it
- computer representations in various formats (including those listed above) or as In Vehicle Guidance Systems (IVGS)

Basic research on wayfinding behavior has tended to split along the lines articulated by Sholl (1987). She suggested that there are two broad types of environmental knowledge relevant for travel: *egocentric*, in which the person is the focal point of a human-environment interaction and environmental knowledge is compiled in terms of person-to-object sensing; and *allocentric* information, consisting of object-to-object relations that are independent of a person's location or movement. Scholl's experiments indicated that humans can point successfully to unseen but familiar locations – a gesture-based "egocentric" representational mode for indicating knowledge of distant places based on spatially updating a person's current location in a cognitive map. They also suggest that humans have the ability to make interpoint distance and direction judgments that are constant regardless of the positional change or change in the perspective taken by a moving traveler. In other words, humans create layout knowledge consisting of the geometry of environmental settings and structures. These do not change as a person changes location (i.e., the distance from Chicago to St. Louis is the same whether viewed from a location in Iowa or Florida. Travel planning (according to Gärling and Golledge, 1989) involves an ability to integrate these two approaches. In other words, some recognition of the elemental locational, connective, and physical structure of places in an environment must be comprehended at the same time as the traveling individual must be able to simultaneously move and spatially update their position in relation to stable environmental features. These acts, environmental comprehension and spatial updating, involve understanding essential spatial relations such as location, distance, direction, reference frame, connectivity, and so on. To produce a travel plan (i.e., a daily schedule of potential activities), an individual must be able to coordinate the proposed movement patterns with desirable or feasible places at which a purpose can be achieved.

THE ROLE OF SIGNS

To facilitate wayfinding, through its history humankind have used signage systems (Golledge, 2002b). These have ranged from simple rock cairns that acted as place markers or choice point signifiers to selecting or marking specific vegetative or landform objects to indicate location, orientation, or direction, to creating maps and charts of sticks and stones or shells (Gladwin, 1970), to recording information in rock paintings or carvings (Uttal, 2000), to developing Portolan charts for nearshore ocean navigation to inventing naming and numbering systems for Gazeteers (from Cartesian coordinates to numbered residential blocks) (Hill and Goodchild, 2000) to using the configuration of celestial bodies (e.g., star charts) or sun angle or wind direction or water flow to provide signs for interpreting where one is and where one should go (Wiltkschko and Wiltschko, 1999).

Signage is a way of minimizing the chances of becoming lost. A critical feature of being lost concerns not being able to recognize signage in one's vicinity or being unable to interpret signage as used by local residents. Being lost involves: i) not knowing where you are; ii) not knowing where to go next; iii) possibly knowing where to go but not knowing how to get there; or iv) knowing where you are relative to a specific place but not knowing where that place is in the general scheme of things, (e.g., knowing you're outside Anna's Bakery but not knowing where Anna's Bakery lies within a city). One also can become lost by erroneously believing you know where you are or erroneously believing you know how to get to your next destination.

Signage, in one form or another, is designed to alleviate these concerns and views and to restore a sense of knowing and purpose to a traveler. Signage is particularly important for newcomers in an environment as signs provide the markers and landmarks to help wayfinding activities. Signage carries with it assumptions that a potential user can understand the signage system (e.g., comprehend advisory symbols), can read the written information on a sign, or be able to interpret multimodal information on the sign. One such widespread signage form is the "You Are Here" map. This generally consists of a vertically mounted map of the local area with the mapstand's location clearly marked. However, such maps are not always aligned properly with the real world. Problems may arise when people try to relate direction and orientation of misaligned You Are Here map information with the real world when their orientations do not coincide. Interpreting a sign such as this may require mental rotation, mental transformation, or mental translations between the map and the environment: cognitive acts which are not easy to perform. These are in fact complex cognitive tasks and can result in substantial errors of interpretation (Levine, 1982; Levine, et al. 1984).

Like disoriented "You Are Here" maps, a person's cognitive maps are often incomplete, distorted with respect to the real world, and require mental rotation, alignment, transformation, translation, and matching when being used in travel planning or natural wayfinding (Kitchin, 1994; Portugali, 1996). But somehow or other, humans are still usually able to complete wayfinding tasks even if they do them inefficiently and ineffectively.

ENVIRONMENTAL LEGIBILITY

One avenue of research related to wayfinding and travel behavior that can profit from examination by applied geographers is that of measuring and assessing environmental legibility. The legibility of an environment is dominated by two dimensions. The first, initially identified by Lynch (1960), concerns the spatial representation of ones surroundings (including the number of environmental features such as landmarks). It focuses on both physical characteristics and spatial relations of objects and features in that environment (such as connectivities and spatial associations). From this perspective, one's representation of an external environment is structured to be as isomorphic as possible vis-à-vis the surrounding physical structure. This approach argues that environments can be examined as coherent structures. According to this view, a legible environment is one where the spatial structure is relatively obvious. In these cases, legibility depends on the ability to organize the complexity of the surrounding environment, the ease of differentiating particular components in the environment, and the nature of its visual perceptual form. Thus, a legible environment allows for object clustering and feature characterization as well as hierarchical ordering of phenomena. Legibility becomes the degree of distinctiveness that enables viewers to comprehend their surroundings.

A second interpretation of legibility focuses on behavior, particularly on travel. Weisman (1981) suggested that legibility is essentially the facility with which travelers can find their way through an environment. While again this concept is based on the quality and complexity of the surrounding spatial structure, this interpretation focuses more on the ease with which transfer takes place between particular origins and destinations. In this situation, a legible environment is one where destinations can be directly observed or estimated and where travel can be guided by directly viewing elements of the surrounding space or by matching the surrounding space with a travel aid such as a map. Thus two environments that are approximately the same size can have different levels of legibility if one is complex and difficult to move through while the other facilitates travel in a relatively unimpeded way.

In recent years a third dimension of legibility has been suggested that combines both spatial and functional characteristics of the setting. This emphasizes the socio-cultural meanings of objects and features in the surroundings, incorporating qualitative and emotional characteristics along with physical and spatial (e.g., a sense of apprehension or fear) (Amedeo and York, 1990). In this view, emotional, spiritual, or religious facets of particular surroundings have significance to various socio-cultural groups which, in turn, makes the environment more or less illegible to other groups. For example, symbols which are not directly perceivable may replace physical form as the major differentiating characteristic that facilitates environmental knowing. This latter dimension allows for elements of an environment to be imbued with significance and elevated to the state of landmarks even where they are not distinctly different in physical form or appearance from surrounding areas (e.g., the particular house in a row of terrace houses in which Ulysses S. Grant slept). This consideration underlies the definition of idiosyncratic landmarks, which often prove to be significant for a single individual from a particular socio-economic or cultural group while simultaneously having little meaning to others. The symbolic landmarks facilitate intra-group communication, including information relative to travel, but for non-in-group members, directions based on such idiosyncratic information may be incoherent. Landmarks vary from the parochial (e.g., a local church steeple) to internationally well known physical or built features (Golledge, 2002d).

Legible environments, therefore, include the following characteristics:

* objects and features with strong symbolic meanings
* landmarks with distinct and highly visible form
* excellent place based and direction giving signage
* a number of clearly defined anchor points in both the surroundings and on a transport network
* a clearly defined hierarchy in travel related systems (e.g., interstate highways, state highways, major arterials, arterials, suburban streets, roads, lanes and alleys) (Golledge, 2002c).

Environmental legibility thus can be said to consist of three forms: a) physical or spatial relationships made obvious by characteristics such as shape, size, dominance of physical form, proximity, or hierarchical order; b) socio-cultural characteristics such as religious, spiritual, aesthetic, functional, or historical factors; and c) behavioral legibility which is tied to use for travel or ease of imparting communicable information about how to move within a given environment. Legibility thus has physical, spatial, social, or cultural markers as well as behavioral dimensions, and in any given setting, one or more of these can dominate.

According to Gärling, Böök and Lindberg (1984), environmental legibility develops through three stages: exploratory, adaptive, and abstract. At the first stage, a person's understanding of an environment is dominated by a visual experience based on travel and is essentially concrete, i.e., spatial-geographic. Cognitive mapping research (Golledge, 1974; Kitchin, 1994; Golledge, 2002a) corroborates this hypothesis. Evidence can be obtained by having people produce spatial products, (i.e., external representations of stored environmental knowledge). These tend to be dominated by route map structures rather than be represented as configuration or layout structures. Environmental knowing takes place primarily to facilitate travel and to allow individuals and families to establish daily activity patterns. This Gärling, et al. called an exploratory stage.

As time progresses, social and cultural meanings and symbolic interpretations of place become more prominent. These cultural codings and physical signs increase the legibility of an environment by focusing attention on specific features and relationships. This period is called the "adaptive" stage by Gärling, et al. At this next stage, individuals move more easily around an environment, explore with greater confidence and do not hesitate to investigate unknown places.

The third level is termed an "abstract" phase. Here, a survey or configurational type knowledge structure is developed that includes understanding the abstract geometry of objects and features in the environment itself. It is the stage that provides the greatest variety of input for developing travel plans for different trip purposes. It also contains the most information that may facilitate shortcutting rather than trip repetition or trip retrace, and as such becomes a critical developmental feature concerned with adapting to and using environmental settings. This final stage combines a metric understanding of points (e.g., landmarks and other places), lines (e.g., connecting routes between environmental features), and layouts (including the incorporation of points, lines and areas into a comprehensive configuration). Landmarks act as identifiers or choice points on routes and can be linked in sequence to facilitate the wayfinding or navigation task. This procedure is generally called "piloting" (Loomis, et al., 1999). Distant landmarks provide orientation and frame of reference information and sometimes provide a heading vector to help confine travel to a specific sector or corridor of space.

THE ROLE OF LANDMARKS

When traveling, landmarks also act as primers, such as knowing that one has to "turn right one block after the cathedral." Landmarks may differ from place to place and may consist of natural or built features (e.g., Niagara Falls or the Empire State Building). Landmark definition is both culture and region dependent. In the travel domain landmarks are regarded as strategic foci towards or away from which travel takes place. They may also occur as intermediate foci on routes that assist in spatial decision making by priming "next move" choices. Landmarks must perform two primary functions in wayfinding: the first because they are recognized as outstanding features in the environment; the second is because they accrue specific significance and are easily identified when communicating by common speech. Landmarks also act as organizing features (anchor points) which dominate local knowledge structures.

Most landmarks are the dominant point factors involved in determining environmental legibility. Other travel related factors such as paths and networks are more difficult to assess as legibility components. Points are connected by lines of travel. In the distant past these were called tracks or paths and were often imprecisely defined. They wandered depending on environmental obstacles or who made the paths or tracks (e.g., animals vs. humans). Today, travel via a transportation system laid down in accordance with specific design characteristics and rules of operation is the most universal way of moving through an environment.

KNOWING TRANSPORTATION NETWORKS

As much of our environmental knowledge derives from travel experience (Hayes-Roth and Hayes-Roth, 1979) and since such travel is generally confined to a limited number of routes, it stands to reason that one's knowledge structure of an environment will be partial, incomplete and may be disconnected, and distorted. Given the complexity of even simple networks, throughout history humankind has tried to make environments legible and travel feasible by locating signs, developing strip maps, planar maps, topographic maps, visualizations, verbal descriptions, written descriptions, and immersive virtual settings as ways of experiencing and learning routes through networks. Just as collections of locations or places can be combined to form configural layouts, so can individual routes be combined to form networks.

Networks, cognitive and spatial, are developed to link places and to allow connections between multiple places to be formalized. Networks provide for both local and global movement systems. They often are based on hierarchies of importance with that characterization being defined in terms of flow capacity (e.g., freeways vs. local roads). Wayfinding skills are used to define routes through networks. Specific routes depend on mode choice (e.g., pedestrian, cycle, car, bus, rail, water transport, or air transport). In today's societies, "public" transportation removes the need for wayfinding by some individuals, but these mass transit trips rarely comprise more than 10–15% of total daily trips. In general, being able to clearly and quickly define routes through networks make environments legible in the sense of organizing and facilitating travel through them.

In developing economies where transportation systems are often in early stages of growth, networks may be simple, often highly linear with few connecting branches. They can be learned quickly and used efficiently. As economic and social complexity of a society increase, network complexity also increases to a level where it is virtually impossible for a single person to know the entire network of possible paths that they could follow to any given destination.

Research has shown that when communicating about potential travel, routes are best described using only ordinal or sequential information. No attempt should be made to measure or describe the entire network or to consider all possible paths through it.

ACCESSIBILITY

In a travel sense, I've argued that a network is legible if travel can take place effectively, efficiently, and with minimal stress. While legibility is an important cognitive and environmental characteristic that influences wayfinding and travel generally, an equally important concept is that of *accessibility*. Even when an environment is legible, for certain trip purposes, it may not prove accessible. For example, a network may be developed as a tree-like structure which is quite legible to the user, but this configuration may make reaching adjacent places (e.g., one "branch" from any other "branch") a long and complex process. Accessibility may also defy scale for a large and seemingly complex environment may be made accessible by human intervention (e.g., restriction of the direction of traffic flow or excellent signage), while a network in a small simple environment, lacking signage or flow restrictions, may prove extremely difficult to negotiate.

Accessibility generally is considered a major goal pursued by transport planners and policy makers (Talen and Anselin, 1996). It promotes effective and efficient use of a system, reduces stress and fear in travelers, and may help to routinize behavior. But, like legibility, accessibility does not have a widely accepted form of measurement. For example, consider two places that are proximate in space but inaccessible (e.g., two places on opposite sides of a river with no connecting bridge or ferry). One may have to travel to the extremity of a network and back in order to interact! This raises the question as to whether accessibility can be regarded as an efficiency concept or as a value or quality concept? Is a place accessible by a single route equivalent in its accessibility to a place that can be reached via a thousand different routes? While researchers such as Talen and Anselin (1996), Southworth (1981), Chen, Recker, and McNally (1997), Marston and Golledge (2001a; 2001b), Miller (1991; 1999), Ingram (1971), Kwan (1998) have investigated accessibility in a variety of contexts and settings among different economic, demographic, social, cultural, and gender groups, there still is no widely accepted measure of accessibility nor any one distinct model to define and measure it. Completing this task will help clarify and model wayfinding activities.

One area of increasing importance which has attracted very little attention and provides a rich research opportunity for applied geographers is accessibility in the context of the movement of disabled people. Mandated in the United States by the Americans with Disabilities Act (Public Law 101-336, 1990), accessibility was seen as an important part of the process of providing equal opportunity and civil rights to disabled people who wanted to travel independently in large scale environments. Initially interpreted in a limited way in the transport domain (e.g., in terms of providing space for wheelchairs in buses, trains, and other forms of mass transit), ensuring accessibility soon became a problem of significant magnitude that extended well beyond its initial application. For example, wheelchair travelers were often thwarted in attempts to move independently around urban environments because of the barriers provided by curbs. Installing curb cuts was touted as the solution for this particular problem and has achieved a certain degree of success. However, there are still no adequate standards as to where curb cuts should be located and how many need to be implemented in an environment to facilitate accessible travel in many directions. Like many innovations, this one had unexpected impacts (externalities) on a different disabled group—the blind traveler. Taught by orientation and mobility experts to walk along sidewalks until a device (such as a long cane being used as an obstacle avoider) found the curb, the blind pedestrian was trained to stop there and listen to traffic sounds before crossing. Unfortunately, some curb cuts were of gradual gradient and were located in such a way that a blind traveler would easily walk down the curb cut and into the traffic lanes without recognizing a curb as such. So what became accessible for one disabled traveler became a peril for another disabled traveler. In addition, modifications to buses to make them accessible to wheelchair users did nothing to alleviate the problems of the blind traveler trying to identify which bus to signal and which bus to board.

Church and Marston (Submitted) have addressed this difficult applied problem of measuring accessibility. They first use a definition offered by Handy (1993) that suggests there are three components of accessibility: i) the spatial distribution of potential destinations; ii) the ease of reaching each destination; and iii) the magnitude and the quality and character of the activities found there. In this way it is assumed that the more destinations that are available within some discrete time or distance range, then the greater is accessibility. Then in their review of modern literature, Church and Marston suggest there have been seven types of measurement explored: a) counting; b) total sums and distances; c) closest available; d) close interaction potential; e) probabilistic choice; f) net and maximum benefits; and g) absolute accessibility. Accessibility measures, therefore, range from the simplest (i.e., a count of the number of places within some specified distance at which a particular activity could take place), to more complicated definitions based on the number and nature of spatial separations from other possible or feasible places to a new origin.

But perhaps the simplest measure of all is a binary designation as to whether or not a destination can be "accessed" from another point somewhere in space. This latter definition has been most commonly used in determining accessibility in conformance with ADA requirements. For example, for a wheelchair person, a building either has accessible ramps in addition to steps, and/or elevators in addition to stairs, or it is deemed inaccessible to wheelchair users. But Church and Marston argue that even when an absolute accessibility measure is satisfied, accessibility may still be problematic. They then go on to develop a measure of *relative accessibility* in which they compare the distance, cost, or time taken by a disabled traveler to get between an origin and a destination to the same measure compiled for an able bodied person traveling between the two places. This relative measure can be determined for ground level movement in urban areas where direct travel may be impossible for the disabled person because of say, the absence of curb cuts, requiring a longer or more time-consuming roundabout route if it proves ultimately possible to travel to a desired place. The authors point out that the calculation of relative accessibility measures in various domains of travel for activity purposes by different disabled groups (particularly the blind or the wheelchair users) is a problem amenable to solutions using the skills and methods of applied geography.

CREATING SMART ENVIRONMENTS

There are essentially three ways to improve travel through complex environments. The first is to modify the environment; the second is to carry information while traveling; the third is to create a Smart Environment.

Environments can be modified by removing barriers, installing signs, constructing missing network links, facilitating shortcutting, and making the environment more legible. The second way is for the traveler to carry the information needed to help make en-route choices. This can be accomplished by providing travel aids such as cartographic maps, written or verbal descriptions, or personal guidance systems that use a wearable computer or Personal Digital Assistant (PDA), that include a digitized base map of the area of proposed travel and a software query system for finding out where you are and what's in the vicinity. The development of such a system has intrigued a variety of researchers in geography, planning, psychology, engineering, and disability studies for more than a decade (e.g., Makino, et al., 1996; Helal, Moore, and Ramachandran, 2001; Fruchterman, 1995). For example, a Personal Guidance System (PGS) was conceptualized in 1985 (Loomis, 1985) developed in the early 1990's by a group of researchers from psychology and geography at the University of California, Santa Barbara (Loomis, Hebert, and Cincinelli, 1988; Golledge, et al., 1991; Loomis, et al., 1998). (For a complete description of this device, see Loomis, et al., 1998 and Golledge, et al., 1998). In this voice-activated system, a GPS tracks an individual as he moves through an environment, downloading position to a digital database (geocoded map of a local environment), upon which a number of functionalities of a Geographic Information System (GIS) operate. These include developing a shortest or obstacle-free path between the traveler's current location and a stated destination, identifying when the traveler takes a wrong turn and goes off route, specification of key environmental features in the vicinity of a traveler (using a dynamic buffering procedure), and presenting layout and route information to the traveler in the form of a virtual auditory reality (Loomis, Hebert, and Cicinelli, 1988).

The third alternative is to create what I call a "Smart Environment". A Smart Environment is one that provides a traveler with relevant necessary real-time information for making decisions and choices (such as identifying location of sites for shopping, recreating, etc). It also can consist of informative messages that are accessed individually on demand using infrared signal detectors, laser beam activated signals, Talking Signs, smart clothing, and other devices.

For example, Remote Infrared Auditory Signage (RIAS) is a technique that is rapidly growing in importance and which is becoming essential for the development of many location based services. This technology requires the installation of infrared transmitters at specific places in an environment, (e.g., street intersections, entrances to buildings, in conjunction with building directories, for locating elevators and other convenience or service devices, etc). The transmitter contains an auditory message on an inductive loop and a set of diodes which control the distance and angle over which a message is distributed (e.g., 15 feet, 51 degrees). The message is usually just a label (e.g., "elevator", "ticket booth", "corner of Spring and High Streets") but can contain more related information (e.g., the menu of a restaurant). The message is intercepted by a handheld receiver that translates the talking sign information and reports it to a traveler. This translation and reporting can potentially be done in a variety of languages such that even if a person did not read or speak a specific language they could still obtain a translation of what the message says (translation opportunities may be limited to, say English, Spanish, Japanese, or Chinese). Using a talking sign, a visitor in an unfamiliar environment can decipher where they are and what's around them by listening to sets of identification messages (Crandall, et al., 1998; 1999; Marston and Golledge, 2001a).

The RIAS are an elaboration of a simple verbal landmark which provides a message when activated, (such as by a hand pressure on a touch sensitive plate as is often found in museums). These verbal landmarks are also sometimes experienced in the form of "You Are Here" maps (Levine, et al., 1984), which provide the traveler with information about their current location.

Recent advances in information technology and biotechnology have provided new and innovative ways for creating elements of Smart Environments. For example, Global Positioning Systems (GPS) have been used to locate and track people through natural and built environments. A particular benefit has been the GPS/cell phone combination (Makino, et al., 1996) which have been used to locate some lost or disoriented Alzheimer's sufferers in Japan. The individual, or anyone else for that matter, simply has to press a button on a cell phone carried by the afflicted person, which activates a GPS and determines a location that is sent to a central server. The server then informs who's responsible for the Alzheimer's patient as to their location and allows prompt recovery.

Even more recently, biotechnologists have produced "smart fabrics" woven from electronic fibers. The fibers are tied to a variety of sensors in a wearable computer. As one moves around an environment, the sensors, woven into clothing, check temperature, wind direction, humidity, pollution levels, and other environmental indicators and relay them via an ear jack to a traveler for evaluation of the feasibility of undertaking specific trips given real-time environmental constraints. Other researchers have looked at the possibility of tapping the unused portions of energy in fluorescent lighting systems to encode local messages that can act as wayfinding aids. It is suggested that these could be located in environments such as airports, shopping malls, institutional buildings, or other enclosed settings, where they could provide location messages and information about nearby facilities. The new fabrics are called "Smart Clothing" and the proposed information based fluorescent systems are termed "Smart Lighting".

As an example of a practical use of how a Smart Environment can help people, consider a situation within a building when an emergency occurs (such as fire, or smoke and dust resulting from explosion or earthquake damage). In these cases, the environment could be filled with smoke and particulates to such a degree that vision is not useful. The frequently used visible signage that points to exits may not be accessible. A Smart Environment would automatically initiate alternate evacuation directions, including verbal directions.

Modern technology is thus paving the way for the creation of Smart Environments. In a short time, Talking Signs will begin appearing on most busses, trolleys, trams, light rail, and indeed, suburban rail. They will give notification of things such as route number and destination, probable time of arrival at destination, time of arrival of next vehicle at a pickup point, and so on. Talking buses of this type using RIAS have already been implemented in some of the larger US cities, with San Francisco being the current leader in terms of providing Talking Signs in public buildings and transportation systems (Marston and Golledge, 2001b).

Research on feasibility and practicality of RIAS has been undertaken by among others, Bentzen, et al. (1999); Crandall, et al. (1998); Marston and Golledge (2001b); and Marston (2002).

For example, Marston has shown that RIAS installed in a suburban train terminal and associated light rail terminal, and nearby bus stops and taxi stands dramatically improved the ability of blind travelers to make modal transfers – an important part of wayfinding. He also investigated whether or not people who are currently not willing to travel extensively because of lack of access to information, (in this case because of lack of sight), would expand their activity spaces if they had access to more information about an environment. Their responses showed substantial willingness to enlarge their activity space and travel behavior pattern if such information was available.

Another and more widespread way of creating a Smart Environment consists of creating Intelligent Transportation Systems (ITS). Examples include establishing Highway Advisory Signs (HAS) and evaluating the extent to which the on route advisory information produces changes in travel plans and consequent traffic flow (Mahmassani 2002). Information may be made available through Highway Advisory Radio (HAR), Advanced Traveler Information Systems (ATIS), In Vehicle Guidance Systems (IVGS), Advanced Traffic Management Systems (ATMS), Advanced Mass Transit System (AMTS), and Intelligent Highway Systems (IHS). Examples of the latter are becoming common now and include the installation of ID numbers on vehicles that can be read by an optical scanner (often laser based) as the vehicle enters and leaves a toll road or other limited occupancy roadway for which a charge is made. The automatic counter records the time and place of entry and exit and sends the data to an automated billing service.

Geographic contribution to the general area of ITS, has focused on the provision of map based information systems available to the driver while traveling (IVGS). There is still controversy over where in the vehicle the map display should be located and even more controversy over how much information should be displayed on such a map. For decades cartographers and geographers have experimented with simplification and generalization of maps to improve legibility (Monmonier, 1996; Clarke, 1999). With in-car devices, two goals have to be pursued: the first to provide a quick visual fix on the general location of the vehicle within the larger environment through which it is traveling, and the second to provide a more detailed visualization of the nearby area that is needed for on route decision making. Fundamental research is continuing into where this information should be displayed in typical visualization (map) form or whether it should be provided by spatial language. Unfortunately, spatial language is often opaque to non-geospatial experts. There are what appear to be simple problems of interpretation such as:

• determining whether a traveler understands where the cardinal directions lie (i.e., within a larger environment and with respect to the current travel heading);
• whether directions should be provided in terms of bearings from north, angles from current heading directions, clock face from current heading directions, or simply quadrant based "left, right and ahead" type of commands (Gopal and Coucleleis, 1989; Golledge and Gärling, 2002).

There is little doubt that as we move into the new age of information technology, that Smart Environments are going to play an increasingly important part in wayfinding and travel at all scales and in all types of environments. This is certainly a large potential research area for the applied geographer who can combine skills in transportation modeling, movement analysis, optimization methodology, cognitive behavioral concepts, spatial analysis, and location-based service modeling to solve future problems.

SUMMARY

Wayfinding is an activity fundamental to all humanity. As civilization has matured, more innovative ways to assist the wayfinder have been part of growth and development. Today it would be difficult to imagine a world without signs, although one must remember that populations such as the blind or severely vision impaired actually live in such a world. Without signs, people create landmarks and other markers, limit their experience with networks, try to integrate experiences into comprehensible wholes to provide an understanding of settings of where they live and travel, and now are creating Smart Environments that take much of the decision making problems away from travelers.

Applied geographers have focused on the transportation sector generally as a critical area in which to apply their expertise. Network models, flows of goods in the form of freight traffic, and determining the optimal provision of services (such as fire, police, and emergency medical) have attracted their attention. Movement of people at scales ranging from episodic daily behavior (i.e., trips to work, shop, education) to the infrequent major migratory moves (e.g., cross national or international) all involve some wayfinding. For the geographer interested in travel behavior for episodic activities such as durable shopping goods, recreation, health and medical services, and other less frequent activities, it has proven to be very difficult to explain and model travel behavior. Nevertheless, within transportation geography, research on these topics has amassed a considerable number of explanatory and predictive capabilities. As one focuses more on explaining human behavior, then model building becomes more difficult and less easily determined. However an interesting question is *how will the creation of Smart Environments improve our ability to predict different types of travel behavior?* Smart Environments will help remove uncertainty about environmental information, and consequently should reduce stress and or fear when undertaking trips. Because travel is such an important part of everyday life, the development of technologies that will influence travel should concern applied geographers interested in planning, prediction, policy making or modeling.

ACKNOWLEDGEMENTS

This research was partly funded by UCTC Grant #DTRS99-G-0009 and NSF Grant # HRD-0099261. I gratefully acknowledge the assistance of geography graduate students James Marston, Jianyu Zhou, and Psychology Professors Jack Loomis (UCSB) and Roberta Klatzky (Carnegie Mellon) for ongoing contributions to the development of the ideas in this paper.

REGINALD G. GOLLEDGE
Department of Geography and
Research Unit on Spatial Cognition and Choice
University of California, Santa Barbara
Santa Barbara, CA 93106
Email: golledge@geog.ucsb.edu

REFERENCES

Amedeo, D., & York, R. A. (1990). Indications of Environmental Schemata From Thoughts About Environments. *Journal of Environmental Psychology, 10*(3), 219–253.

Bentzen, B. L., Crandall, W. F., & Myers, L. (1999). Wayfinding system for transportation services: Remote infrared audible signage for transit stations, surface transit, and intersections. *Transportation Research Record, 1671*, 19–26.

Chen, C., Recker, W., and McNally, M. (1997). "An activity-based approach to accessibility", paper presented at the Annual Meeting of the Western Regional Science Association, Kona, Hawaii, February.

Church, R. L., & Marston, J. R. (Submitted). *Measuring accessibility for people with a disability.* Geographical Analysis.

Church, R. L., Loban, S. R., & Lombard, K. (1991). *Exploring spatial alternatives for a corridor location problem*: Department of Geography, University of California Santa Barbara.

Clarke, K. C. (1999). *Getting Started with Geographic Information Systems (2nd Edition).* Upper Saddle River, NJ: Prentice-Hall.

Crandall, W., Bentzen, B. L., & Myers, L. (1998). *Talking Signs® remote infrared audible signage for transit stations, surface transit, intersections and ATMS* (CSUN' 98 papers (http://www.dinf.org/csun_98/csun98_063.htm)): Smith-Kettlewell Eye Institute, Rehabilitation Engineering Research Center, San Francisco, CA.

Fotheringham, A. S., Brunsdon, C., & Charlton, M. (2000). *Quantitative Geography: Perspectives on Spatial Data Analysis.* Thousand Oaks, CA: SAGE Publications Inc.

Fruchterman, J. (1995). Arkenstone's orientation tools: Atlas Speaks and STRIDER, *Proceedings of the Conference on Orientation and navigation Systems for Blind Persons.* Hatfield, England.

Gärling, T., & Gärling, E. (1988). Distance minimization in downtown pedestrian shopping behavior. *Environment and Planning A, 20,* 547–554.

Gärling, T., & Golledge, R. G. (1989). Environmental perception and cognition. In E. H. Zube & G. T. Moore (Eds.), *Advances in Environment, Behavior, and Design, Volume 2* (pp. 203–236). New York: Plenum Press.

Gärling, T., Böök, A., & Lindberg, E. (1984). Cognitive mapping of large-scale environments: The interrelationship of action plans, acquisition, and orientation. *Environment and Behavior, 16,* 3–34.

Gladwin, T. (1970). *East is a Big Bird: Navigation and Logic on Pulawat Atoll.* Cambridge: Harvard University Press.

Golledge, R. G. (1974). *On Determining Cognitive Configurations of a City: Vol. 1—Problem Statement, Experimental Design and Preliminary Findings* (Final Report, NSF Grant #GS-37969).

Golledge, R. G. (1997). Defining the criteria used in path selection. In D. F. Ettema & H. J. P. Timmermans (Eds.), *Activity-Based Approaches to Travel Analysis* (pp. 151–169). New York: Elsevier.

Golledge, R. G. (1999). Human wayfinding and cognitive maps. In R. G. Golledge (Ed.), *Wayfinding Behavior: Cognitive Mapping and Other Spatial Processes* (pp. 5–45). Baltimore, MD: The Johns Hopkins University Press.

Golledge, R. G. (Ed.). (1999). *Behavior: Cognitive Mapping and Other Spatial Processes.* Baltimore, MD: Johns Hopkins University Press.

Golledge, R. G. (2002a). Cognitive Maps. In K. Kempf-Leonard (Ed.), *Encyclopedia of Social Measurement* (Vol. Submitted,). San Diego, CA: Academic Press Inc.

Golledge, R. G. (2002b). Human wayfinding behavior. In M. Rockman & J. Steele (Eds.), *Colonization of Unfamiliar Landscapes: The Archaeology of Adaptation.* London: Routledge: Submitted.

Golledge, R. G. (2002c). *What is a landmark and what really are the world's most significant ?* Directions Magazine (http://www.directionsmag.com).

Golledge, R. G., & Gärling, T. (2002). Spatial behavior in transportation modeling and planning. In K. Goulias (Ed.), *Transportation and Engineering Handbook*: Submitted.

Golledge, R. G., Loomis, J. M., Klatzky, R. L., Flury, A., & Yang, X. L. (1991). Designing a personal guidance system to aid navigation without sight: Progress on the GIS component. *International Journal of Geographical Information Systems, 5*(4), 373–396.

Golledge, R. G., Klatzky, R. L., Loomis, J. M., Speigle, J., & Tietz, J. (1998). A geographic information system for a GPS based personal guidance system. *International Journal of Geographical Information Science, 12*(7), 727–749.

Gopal, S., & Couclelis, H. (1989). *Navigational directions in a college campus* (Unpublished manuscript): Department of Geography, University of California, Santa Barbara, California.

Hägerstrand, T. (1970). What about people in regional science? *Papers of the Regional Science Association, 24*, 7–21.

Handy, S. (1993). *Regional versus local acessibility: Implications for nonwork travel* (Reprint 234): The University of California Transportation Center (UCTC), University of California, Berkeley.

Hayes-Roth, B., & Hayes-Roth, F. (1979). A cognitive model for planning. *Cognitive Science, 3*, 275–310.

Helal, A., Moore, S., and Ramachandran, B. (2001). "Drishti: An Integrated Navigation System for Visually Impaired and Disabled," Proceedings of the 5th International Symposium on Wearable Computer, October 2001, Zurich, Switzerland.

Hill, L. L. and Goodchild, M. E. (2000). *Digital Gazeteer Information Exchange (DGIE)* Workshop Report. Department of Geography, UCSB.

Ingram, D. R. (1971). "The concept of , a search for an operational form," Regional Studies 5, 101–117.

Kitchin, R. M. (1994). Cognitive maps: What are they and why study them? *Journal of Environmental Psychology, 14*(1), 1–19.

Kwan, M.-P. (1998). Space-time and integral measures of individual : A comparative analysis using a point-based framework. *Geographical Analysis, 30*(3), 191–216.

Levine, M. (1982). You-are-here maps: Psychological considerations. *Environment and Behavior, 14*(2), 221–237.

Levine, M., Marchon, I., & Hanley, G. (1984). The placement and misplacement of you-are-here maps. *Environment and Behavior, 16*(2), 139–157.

Loomis, J. M. (1985). *Digital map and navigation system for the visually impaired* (Unpublished manuscript): Department of Psychology, University of California, Santa Barbara.

Loomis, J. M., Golledge, R. G., & Klatzky, R. L. (1998). Navigation system for the blind: Auditory display modes and guidance. *Presence: Teleoperators and Virtual Environments, 7*(2), 193–203.

Loomis, J. M., Klatzky, R. L., Golledge, R. G., & Philbeck, J. W. (1999). Human navigation by path integration. In R. G. Golledge (Ed.), *Behavior: Cognitive Mapping and Other Spatial Processes* (pp. 125–151). Baltimore, MD: The John Hopkins University Press.

Loomis, J. M., Hebert, C., & Cicinelli, J. (1988). *A computer based auditory system for tracking blind navigators*.

Lynch, K. (1960). *The Image of the City*. Cambridge, MA: The MIT Press.

Mahmassani, H. S. (Ed.). (2002). *In Perpetual Motion: Travel Behavior Research Opportunities and Application Challenges*. New York: Pergamon.

Mahmassani, H., & Jou, R.-C. (1998). Bounded rationality in commuter decision dynamics: Incorporating trip chaining in departure time and route switching decisions. In T. Gärling, T. Laitila, & K. Weston (Eds.), *Theoretical Foundations of Travel Choice Modeling*. Oxford: Elsevier.

Makino, H., Ishii, I., & Nakashizuka, M. (1996). Development of navigation system for the blind using GPS and mobile phone connection, *Proceedings of the 18th Annual Meeting of the IEEE EMBS, Amsterdam*. Amsterdam, The Netherlands.

Marston, J. (2002). Towards an Accessible City: Empirical Measurement and Modeling of Access to Urban Opportunities for those with Vision Impairments Using Remote Infrared Audible Signage., UCSB, Santa Barbara.

Marston, J. R., & Golledge, R. G. (2001a). Empirical measurement of barriers to public transit for the vision-impaired and the use of remote infrared auditory for mitigation. Paper presented at the CSUN's 16 Annual International Conference, "Technology and Persons with Disabilities", Los Angeles, CA.

Marston, J. R., & Golledge, R. G. (2001b). Equal access or transit subsidies, what does the ADA mandate? Limits to transit choice and activities (unequal access) and monetary trade-offs for equal access reported by the vision impaired. Paper presented at the 97th Annual Meeting of the Association of American Geographers, New York, NY.

McFadden, D. (2002). Disaggregate behavioral travel demand's RUM side: A 30-year retrospective. In D. A. Hensher & J. King (Eds.), *The Leading Edge in Travel Behavior Research* (Vol. In Press). Oxford: Pergamon Press.

Miller, H. J. (1991). Modelling using space-time prism concepts within geographical information systems. *International Journal of Geographical Information Systems, 5*(3), 287–301.

Miller, H. J. (1999). *GIS software for measuring space-time in transportation planning and analysis*. Paper presented at the International Workshop on Geographic Information Systems for Transportation and Intelligent Transportation Systems, Hong Kong.

Monmonier, M. (1996). *How to Lie With Maps*. Chicago: University of Chicago Press.

Public Law 101–336 [S. 933], July 26, 1990, Americans With Disabilities Act Of 1990, United States Public Laws 101st Congress–Second Session

Portugali, J. (Ed.). (1996). *The Construction of Cognitive Maps*. Dordrecht: Kluwer Academic Publishers.

Sholl, M. J. (1987). Cognitive maps as orienting schemata. *Journal of Experimental Psychology: Learning, Memory, and Cognition, 13*(4), 615–628.

Southworth, F. (1981). Calibration of multinomial logit models of mode and destination choice. *Transportation Research A, 15*, 315–325.

Talen, E., & Anselin, L. (1996). Assessing spatial equity: An evaluation of measures of to public playgrounds. *Environment and Planning A, Forthcoming*.

Uttal, D. H. (2000). Seeing the big picture: Map use and the development of spatial cognition. *Developmental Science, 3*, 247–264.

Weisman, G. D. (1981). Evaluating architectural legibility: in the built environment. *Environment and Behavior, 13*(2), 189–204.

Wiltschko, R., & Wiltschko, W. (1999). Compass orientation as a basic element in avian orientation and navigation. In R. G. Golledge (Ed.), *Behavior: Cognitive Mapping and Other Spatial Processes* (pp. 259–293). Baltimore & London: The Johns Hopkins University Press.

JESSIE P.H. POON AND JAMES E. MCCONNELL

CHAPTER 13
INTERNATIONAL TRADE

1. INTRODUCTION

The history of Geography is characterized by a long-standing interest on the international and spatial dimension of commercial activities. One of the earliest and most influential geography books was *A Handbook of Commercial Geography* written by George Chisholm in 1889. Chisholm's book quickly became a handbook for British merchants, traders and manufacturers (Barnes, 2002). Hence from the start, geographers' interest in international trade was motivated by its potential for socio-economic relevancy and the relation of national production to international development and competition.

The internationalization, and more recentlyglobalization, of commercial activities has attracted geographers for three major reasons. First, some 70% of world trade is conducted by fewer than 25 countries. A clear core-periphery geographic pattern in international exchanges may therefore be detected. Research associated with uneven international development constitutes one main theme among applied geographers in the field. Second, the emergence of business geography as a sub-field in the discipline has seen the more macroeconomic analysis of international trade and development in the past being increasingly grounded in microeconomic analysis, namely, the role that firms and industries play in shaping and driving commercial transactions, and thereby their impact on the allocation of resources. Understanding the relationship between firm activities and the exports that they generate has practical implications for the economic health of a country or region such as employment growth. Third, concern for geographic sensibilities has resulted in a more multiscalar approach to the study of international trade that emphasizes both sub-regional and the trans-regional dimensions of national and global production. This is because a large share of national production may frequently be attributed to a few productive regional economies in a country. At the same time, sub-regionalism is paralleled by a trend of trans-regional pooling of national markets that has evoked different responses from policy-makers in industrialized and developing countries. In both cases, the impact on regional, national and international growth can be significant.

A. Bailly and L. J. Gibson (eds.), Applied Geography, 253–272.
© 2004 *Kluwer Academic Publishers. Printed in the Netherlands.*

The above three themes form the basis of review in this chapter. Each of the themes will be further elaborated below.

2. INTERNATIONAL TRADE AND DEVELOPMENT

Historically, the relationship between international trade and economic growth and development has been perceived as an important one, not only by trans-national organizations such as the World Bank and the International Monetary Fund, but also by government agencies at national and "local" levels. For example, international development experts deliberate whether the economic policies of national governments should be designed to promote exports or protect domestic industry from competing imports. National government leaders are regularly subjected to political and economic pressures from various industry sectors and other constituencies to develop and implement policies regarding such trade-related issues as human rights, food safety, the environment, intellectual property rights, labor standards, and income inequalities among regions. Similarly, government officials and planners at sub-national and metropolitan levels, who among others are responsible for the economic health and prosperity of their respective territories and communities, raise questions about the relative importance of exports in creating jobs, and are concerned about what support, if any, should be given to assist local companies in developing their prowess in international markets.

In each of these situations, research on the relationship between trade and development is vital in helping policymakers and economic planners make appropriate decisions. Concern for this relationship is a long-standing one, both among economists and geographers.

One of the first efforts of geographers to examine the connections between trade and development is the *Atlas of Economic Development* (Ginsburg, 1961). The author states that " . . . trade not only provides a good index to international interaction, but also is a measure of the degree to which an economy organizes itself for production and exchange and therefore has developed beyond a subsistence stage" (*Ibid.*, p. 102). A few years later, Thoman and Conkling (1967) wrote the first comprehensive text on the geography of international trade in which they examine the changing patterns and relative importance of commodity trade for countries at varying levels of economic development. Some ten years later, Johnston suggests that " . . . alteration of the trade system, or of the trading patterns of countries, will assist in the development process, thus inferring a temporal and causal relationship from the cross-sectional associations between trade parameters and 'development levels'" (1976, p. 138).

Although historically geographers have arguably paid less attention to trade and economic development issues than have economist, recent research by geographers has been increasingly focused upon this relationship. What is the nature of this research, and how might the results of these investigations be applied to ongoing issues at the global, national, and sub-national levels? Three geographic levels of decision making are set forth below, beginning with issues and policies of concern to practitioners working at the international and national levels, and concluding with matters of importance within nation-states at regional and local levels.

2.1. Multinational issues and policies

The challenge for international development experts at the World Bank or the World Trade Organization (WTO), for example, is to address issues and formulate policies that are of concern to many, if not all, nations in the world. Many of these concerns have a direct connection to international trade and development. They include the persistent income and economic development gap between rich and poor countries, and decisions about how best to assist nations that are in economic difficulties; the debate over whether a multilateral or regional approach is the better procedure to follow in order to achieve consensus among nations regarding international trade and development issues; and the impacts on the world economy of currency fluctuations, flows of investment capital, and changing patterns of cross-country trade in goods and services.

Geographers have examined many of these issues. For example, Johnston (1989) calls upon geographers to develop the theoretical underpinnings of international trade so that we have a better understanding of how the spatial system of international trade has "... come into being, why it sustained major international [economic] inequalities, and ... how manipulation of the pattern of trade could right those inequalities" (Ibid., p. 338). He describes a theory of "combined and uneven development" in which the spatial structure of the world is divided into core (developed) and periphery (developing) territories, and where "development in the core and underdevelopment in the periphery are interdependent processes, and the pattern of trade is a consequence of their interaction" (Ibid., p. 342). He also recognizes the important role of the state and the changing hegemonic power of nations in accounting for the direction and intensity of international trade flows and the cycles of economic growth and decline. This perspective, which parallels the earlier work of Myrdal (1957), supports the view that trade does not necessarily work for equality, at either the international or national level. In fact, as Myrdal notes, trade "... may, on the contrary, have strong backwash effects on the under-developed countries" (Ibid., p. 51).

Very recent work by O'Brien and Leichenko (2003) provides further challenge to the notion that international trade leads naturally to economic growth and increased living standards for all parties. In their discussion of who is winning and who is losing as a result of globalization, the authors argue that:

> "... trade is not necessarily a driver of growth, but rather that economic growth often tends to drive the growth of trade ... [Moreover] ... growing evidence suggests that globalization is not promoting convergence in incomes between rich and poor countries, but, instead, is increasing global income inequality". (Ibid., p. 95).

The authors also emphasize that government trade policies, such as trade liberalization, may have different impacts on economic development at the global level than at the sub-national level, and the results may also differ if measurements of the impacts on inequality incorporate both "national and individual-level measures" (Ibid., p. 97). Thus it is important for international trade and development specialists and policymakers to reconsider the truism from much of the traditional literature on international economics that trade is an engine of growth and that trade liberalization benefits all of the trading partners.

Similarly, the work by Gaile and Grant (1989) takes a critical view of the world-systems theory that a country's economic power, as defined by its position in the hierarchy of economic development, determines what types of goods and services are traded and with whom. The authors demonstrate that it is not just economic power that helps to explain the global patterns of international trade, but also military power, the educational attainment of a country's populace, and the relative location of a country with regard to its potential trading partners. In sum, this research on the dynamic relationships between trade and political-economic development is useful for policymakers in that it offers insights into the different impacts trade can have, depending upon a country's relative location and level of economic development, and it articulates the role of national governments in utilizing the potential benefits from international trade to foster economic growth.

Another international concern of policymakers relates to how best to deal with the forces of globalization, the resulting commercial and political conflicts that arise among trading nations, and the natural desire of nation-states to protect domestic economic interests. Inherent in these issues is how best to proceed to resolve commercial and diplomatic differences, while maintaining national economic development objectives. Some would champion a multilateral approach, whereby, under the auspices of the WTO, member countries work out their differences together, as one institutional body. In contrast, others argue that the multilateral approach is a slow, tedious process, and that resolving trade and development issues through bi-national or regional agreements is a much more timely and effective way to respond to the changing nature of the international trading system. The research by Michalak and Gibb (1997) is representative of the attention geographers are giving to this issue. They conclude that nation-states are responding to globalization by "... weakening their commitment to the principles of multilateral trade and engaging the process of regional integration" (Ibid., p. 274). Moreover, they argue that nation-states, allied into regional trade blocs, are the principal agents in reconstructing the world economy. This realization of how the global space economy is evolving is important for trade negotiators at the WTO and at the central offices of the various regional trade blocks who must reconcile preference for a more multilateral approach with these multiplying regional economic structures.

2.2. National-level policies

Geographers have also been interested in the impact of international trade patterns and trade policy on the economic well-being of nation-states. This work provides assistance to government officials and planners working at the national level who are responsible for developing trade and development policies for their respective countries. As Bhagwati (1988) suggests, trade and industry policies of nation-states are typically a mix of ideological factors (e.g., free trade versus protectionist biases), national interests (e.g., economic prosperity, political influence, and security), and institutions (e.g., legislative bodies, policies, and programs). Thus, policy decisions are molded not only by events and conditions at the global level, but also by the many political and economic pressures being generated by various domestic constituencies. Such pressures may originate, for example, from industry associations, labor unions, exporting and importing companies, environmental organizations, and advocates for farmers and consumers. It is necessary, therefore, for practitioners and business executives within nations to be cognizant of how the country's trade patterns and policies are affecting the international competitive position of their respective nations, and, hence, the economic health of their domestic businesses and local communities.

The concern of geographers for trade policy and the implications of such for national governments is represented by the work of Grant (1993). He favors an "institutional" analysis of trade policies, which recognizes the important role institutions have in mediating the interactions between specific national societal and governmental groups and the pressures arising from the international political economy. After examining trends in the national trade and industry policies and policymaking institutions of the United States and Japan, he concludes that the trade policies of these two nations have been reversed, with Japan pursuing a more liberalizing trade policy and the U.S. following a more protectionist trend. This kind of an investigation by geographers demonstrates the important role policy-making institutions have in creating a domestic environment that can have significant impacts upon the international competitive posture of domestic companies and thus upon prospects for national economic growth and development.

One of the key trade policy issues for nation-states is the extent to which countries should promote trade liberalization and export promotion or favor a more protectionist position designed to make it difficult for foreign companies to compete with domestic industries. Basic to this issue is the question of whether exports and the promotion of exports result in the economic growth of nation-states. One of the comprehensive analyses of this question is by the geographer, Poon. She concludes that "empirical support for the notion that the export-promotion strategy is universally valid does not exist" (Poon, 1994, p. 49). Basing her analysis on the average annual growth rates of exports and of GDP per capital over the 1960–70 and 1970–80 time intervals, she finds that the relationship between exports and economic growth for a sample of 49 developing countries depends upon the country's stage of economic development. More specifically, the positive impacts of exporting on a country's economic growth increase as the nation moves from a low to an intermediate stage of development, and then they level off in later development phases. Thus, the relationship between a country's trade-promotion efforts and economic growth is a conditional and contextual one, which emphasizes one of the geographer's primary themes, namely, that place matters.

That the relationship between a country's trade policies and economic growth is conditional and contextual is a central theme reflected in much of the more recent research by geographers. For example, Gwynne (1996) provides additional insights into the problems developing countries in East Asia have of utilizing international trade and industry policies to achieve sustained economic growth. He shows that the basis for sustained international competitiveness and economic growth is an "outward-oriented industrialization" effort and the development of human resources (Ibid., p. 260). However, he notes that these national policies regarding trade and human capital development must also be joined with a supportive domestic culture, appropriate political organizations, and effective macroeconomic policy. Duplicating this combination of attributes may be very difficult in other developing countries; nevertheless, the model describes the kinds of changes that are necessary for sustained economic growth to be realized by developing nations.

Similarly, Webber (1994, 2000) examines the development of trade and economic policy in Australia, and concludes that the forces of globalization take a variety of forms, each of which may require a variety of policy responses. Based upon an assessment of the economic successes of the newly developed nations in East Asia, he concludes that countries develop their trade and development policies in ways that reflect their local political and social structures, as well as their individual perceptions of global conditions and circumstances; hence, it must not be assumed that policies that work in one place can be copied to work automatically in different national environments. Thus, because places are inherently different, so, therefore, is the likelihood that their respective trade and development policies will need to be crafted to suit their specific circumstances.

Although much of the research by geographers on the relationship between trade and development is based upon the movement of commodities, increasing attention is being given to the rapidly growing international trade in services. For example, an analysis of Australia's exports of advanced producer services to its Asia-Pacific neighbors suggests that only a small part of that country's trade is in these activities. The explanation offered by O'Connor and Daniels (2001) is that, while Australia has a well-developed advanced producer services sector, economic development conditions in many of the other Asian nations are not yet sufficient to stimulate major growth of this sector. Hence, Australia's exports of services in this sector are more successful in entering the markets of the United States and the United Kingdom, where host-country conditions, institutional structures, and business and commercial practices are more similar.

For practitioners, these analyzes emphasize that for places that are very different, where home- and host-country levels of economic and institutional development are dissimilar, and where social and cultural systems are unalike, one may expect to find very different patterns of trade. Furthermore, countries that wish to increase their participation in the international exchange of goods and services must be prepared to make the necessary investments in infrastructure and changes in institutional structures that move them toward a more equal footing with more advanced trading nations. In talking about the gains from trade and the importance of trade policy to a country's economic well-being, the economist Irwin points out that:

> "Recent empirical research has uncovered an indirect link between trade and growth: the share of investment in GDP is positively correlated with growth in per capita income, and trade is positively correlated with investment. This means that while trade may not be directly correlated with growth, it may stimulate growth indirectly through investment" (Irwin, 2002, pp. 40–41).

Research by several geographers (MacPherson and McConnell, 1992; Perry and Hui, 1998; Poon and Thompson, 1998; Edgington and Hayter, 2000; Pantulu and Poon, 2003; and Yeung, 2002) examines these indirect links between investment, trade, and economic growth.

2.3. Sub-national (regional and local) concerns

Practitioners who are interested in the relationship between trade and development at any level should perhaps heed the advice of the economist Krugman: "One of the best ways to understand how the international economy works is to start by looking at what happens inside nations. If we want to understand differences in national growth rates, a good place to start is by examining differences in regional growth; if we want to understand international specialization, a good place to start is with local specialization" (Krugman, 1991, 3). For many geographers, therefore, the spatial frame of reference within which the connections between trade and development are being investigated is the sub-national level. Most of this work is focused upon the relationship between exports and the economic growth of states and regions. These empirical studies are also designed to advance the theoretical framework underlying the ties between trade and development, and to prescribe appropriate policies and programs for local and regional politicians, planners, and business executives.

Two of the earliest studies on the tie between exports and regional economic growth in the United Kingdom and the United States are by Hoare (1985) and Erickson (1989), respectively. Both studies report a modest correlation between exporting and regional economic growth. For example, Erickson's study reveals that although the growth in employment and manufacturing value added for states in the United States is positively and significantly correlated with increases in the value of export shipments, the primary determinant of a state's economic growth is the growth in the domestic manufacturing sector. Moreover, the results indicate that shifts in the sales of manufacturing goods from domestic to foreign markets are not significantly correlated with higher rates of state-level industrial growth. What is uncertain from these early studies, therefore, is the direction of the causality involved in the relationship between exporting and economic growth.

The uncertainty of these early findings creates difficulties for state and local practitioners. The traditional belief has been that an important strategy for increasing regional and local economic growth and development is to promote exporting. More recent studies by geographers, however, suggest that this relationship is less than clear-cut. For example, Glasmeier and Leichenko (1996) find that the growth of exports for a region actually occurs as a result of the growth in the area's labor and/or capital supplies. This suggests that the growth of exports in a region follows from and is the result of growth in the area's economy. This raises important planning issues for practitioners in that if the growth in exporting follows from the growth of the economy, then *initial* expenditures of development funds to advance a region's economic development might be better allocated to projects other than promoting exports.

Two recent studies have provided much more detailed assessment of the role of exports in regional and local economic growth. Leichenko and Coulson (1999) establish that, in general, a bi-directional relationship does exist in the United States between trade and industrial growth across two-digit manufacturing sectors. However, the relationship is a complex one, and it tends to be sector dependent. Thus, for some of the sectors, such as electronics and chemicals, promoting exports is a good strategy to create industrial growth. In contrast, in other sectors, manufacturing growth leads export growth, and in other sectors the causality is actually negative, implying that an increase in exports leads to a reduction in employment and production.

Perhaps the most comprehensive analysis of this relationship is the recent work by Leichenko (2000). Her study examines the causal relationship between manufacturing exports and manufacturing employment, productivity, and output for 48 states and four census regions of the United States over the period from 1980 to 1991. The results of her investigation suggest that a bi-directional relationship exists across most of the country, which implies that the growth of exports promotes increased production and higher productivity, and, in like manner, higher productivity and production result in the growth of exports. However, for regional and local policymakers, one of the most important discoveries of her study is that while the growth of manufacturing employment leads to an increase in exports, export growth results in *reductions* in manufacturing employment. As an explanation of this finding, the author suggests that "... rising capital intensity with export production (i.e., the substitution of capital for labor) translates into fewer manufacturing jobs" (Ibid., p. 320). A similar result could occur when companies that are operating at less than full-production capacities decide to increase production for new or expanded export markets. In such instances, the short-term impact on employment could be negligible.

The work by Leichenko (2000) also corroborates earlier research by geographers, which shows that the relationship between trade and economic growth may differ across geographic space. For example, she finds that in certain regions of the United States, such as the Midwest, economic development leads the growth of exports; in contrast, in the Western region, the reverse relationship is evident. Hence, for policymakers and planners, this research underscores the importance of understanding the complex relationship between trade and economic growth in different geographic locations and under varying economic conditions so that an appropriate allocation of development funds can be made. Depending upon the region in question, it might be more prudent to postpone the allocation of scarce development resources to export promotion and, instead, direct attention to other programs designed to expand local and regional economies.

Finally, it is noteworthy that the empirical research by geographers on the relationship between trade and economic growth and development has begun to make important contributions to our understanding of the theory of international trade and development. In their criticism of economists for not engaging sufficiently in empirical analysis of actual trade relationships, the economists Davis and Weinstein (2002) argue that while economists have been very successful in advancing the theory of international trade, these efforts "... have not been matched with equally illustrious progress on the empirical side... [and] ... virtually all of the most important empirical questions remain open and at times nearly untouched" (Ibid., p. 30). It is often the case that the reverse criticism is leveled at geographers by geographers, namely, that too much attention is given to empirics, and too little effort is devoted to advancing theory (see Johnston, 1989). However, research by geographers, such as Leichenko (2000), belie such criticism. Her work is clearly designed to test the applicability of six different theories for discerning the relationship between trade and economic growth. By establishing the importance of bi-directional causality, her analysis supports the cumulative causation theory and the new international trade theory of increasing returns to scale and imperfect competition over the more traditional export-base theory, endogenous growth theory, Heckscher-Ohlin factor endowment theory, and product life cycle theory. Confirming the empirical validity of these models provides a reliable platform upon which regional and local practitioners can base specific policy recommendations.

3. ENTERPRISES, TERRITORIES, AND TRADE

Much of the literature of economic geography over the years has focused upon the relationship among firms, industry structures and systems, and the activity spaces that are created by these interrelationships (for example, see Hamilton and Linge, 1979; Hamilton and Linge, 1983; Markusen, 1996; Clark and O'Connor, 1997; and Sternberg and Arndt, 2001). The geographers Dicken and Malmberg (2001) refer to this complex set of interconnections as the "firm-territory nexus." As the authors state:

> The role of space and place in shaping the transformation of firms and industries and the impact of such transformations on the wider processes of territorial development at local, regional, national, and global scales are basic research issues in economic geography (Ibid., p. 345).

Not only are the three components of the network highly interdependent, they are also in varying ways tied to and affected by the policies, processes, and institutions of government at local, national, and international levels. For many firms, therefore, engaging in international trade activities is one of the important outcomes of the firm-territory-governance connection. Understanding the nature of these relationships is important for government officials, urban and regional planners, and executives of business establishments in formulating and implementing effective policies, programs, and business strategies.

In this section of the chapter, the authors report on a sample of research by geographers that examines the interrelationships between international trade and the firm-territory-government nexus. Two categories of studies are identified. The first examines the impacts of government trade and industry policies and locational forces on the geography of specific industry sectors and the spatial configuration of urban and regional territories. The second examines the various ways in which the changes taking place in the firm-territory-governance nexus are influencing government programs and business strategies designed to help domestic companies maintain and/or expand their international trade activities.

3.1. Trade policy, industry sectors, and territory

The research by Noponen, Graham, and Markusen (1993) is representative of efforts by geographers to assess the impact of government trade and industrial policy on firms, industry sectors, and territories. Focusing upon the impact of U.S. international trade patterns and policies on various industry sectors and regions within the nation, the authors conclude that:

> "The industry studies show that the contemporary size and location of the American steel, auto, insurance, machine tool, shipping and pharmaceutical industries are neither the product of natural comparative advantage nor of domestic cost differentials. Together, they form a powerful indictment of the free trade prescription as a national growth and regional development strategy". (Ibid., p. viii).

The basic premise is that United States government's policy of free trade has played a major role in developing and sustaining industrial leadership and international competitive advantage in many of the nation's key industry sectors. This policy has left U.S. domestic companies vulnerable to foreign competition, and, as a result, certain regions and communities, especially those around the Great Lakes, have been negatively impacted through the loss of companies and jobs.

Taking a more theoretical approach and focusing upon both trade and industry policies, Martin and Sunley (1996) critique the work of the economists Krugman (1990) and Porter (1990) regarding the usefulness of strategic trade theory and industry policy in creating national industrial competitiveness. Of particular concern is the extent to which competitiveness should depend upon government intervention and assistance, and, if deemed necessary, at what geographic level such support should be aimed. One side of the debate would favor the promotion of specialized export clusters and economic development policies in key urban areas and regions of the country. The worry, however, is that a region can become too specialized, and, in times of economic downturns, it could be susceptible to economic depression or collapse. An alternative view is to encourage industrial diversification in regions so that the local economies are dependent upon a fairly broad portfolio of industry sectors. The authors take the position that the "new industrial geography" favors flexibly specialized industrial districts, which are ". . . more adaptable to economic and technological change by virtue of the dynamism and networking of the small enterprises of which they are (invariably assumed to be) composed. However, this claim remains far from proven" (Ibid., p. 284).

The recent investigation by Britton (2002) looks at the relationship between national trade policy and the international competitiveness of advanced technology firms at a more local geographic scale, namely, the Toronto, Canada metropolitan region. The objective of the research is to determine the responses to North American trade liberalization of Canadian domestic and foreign affiliate manufacturers of telecommunications, electronics, instruments, and aerospace products. The results indicate a mix of responses, depending in part upon the companies' levels of technological and product innovation and export experience prior to economic integration, whether they were domestically owned or foreign affiliates, and their industry sector. This research is particularly valuable for policymakers in that it reveals that a "substantial minority" of small firms in the region are not responding favorably to the trade opportunities made available by integration. The author suggests that government programs to assist industrial innovations and export development are not sufficiently proactive. Moreover, it is argued that the regional economic health of the metropolitan region depends greatly upon the area's ability to retain the headquarters and production facilities of its major corporations.

A major theme in most of this research is that sub-national places and spaces continue to possess enormous power in shaping not only the geographies of the domestic space economy, but they also have an increasingly important impact on the international economic system. This is an important recognition for practitioners because the point is made that local conditions related to such factors as culture, labor markets, innovation and creativity capabilities, and economic well-being are vital to the international competitiveness of local companies. Studies by Storper (1997) and Clark and O'Connor (1997) demonstrate the growing importance of the local environment in the global economy. Storper makes the case that regions within nation-states continue to play a key role in the composition of international trade. He focuses attention on production systems that are organized to carry out continuous product innovations, such as the production of scientific instruments and aircrafts, and argues that such systems are located in highly concentrated and distinctive sub-national regions – agglomerations that he labels as "technology districts." The wealth of human and resource assets of these regions creates specific advantages that enable local businesses to compete successfully in the international arena. Similarly, in analyzing the operations of financial service systems, Clark and O'Connor conclude that the geography of specific places is very powerful because well-structured international financial systems are greatly dependent upon locally specific informational networks. Thus the geographic characteristics of local places and the informational content of financial markets that is embedded in local markets have a direct impact upon the international operations of this service sector.

Finally, Swyngedouw's recent study provides some advice for practitioners on the relationships among trade, location, and business enterprises. His focus is upon the rise to prominence of regional growth complexes and the central role of local or regional institutions in shaping processes of dynamic, innovative, and competitive economic development. He notes that:

> "As locational opportunities expand and locational capabilities increase, so does the importance of 'local' characteristics of cities and regions in maintaining or asserting their competitive advantages. Indeed, ... competitive success is indebted to specific and historically created forms of territorial and socio-institutional organization ... A host of new terms have been suggested to capture such competitive growth complexes: 'learning regions'... 'competitive cities'... 'reflexive economies' ... 'milieux innovateurs'... [and] 'associational economies'".(Ibid., pp. 546–7).

This process of what he refers to as economic 'glocalization' (global localization) is increasingly asserting itself as a prime force in articulating and shaping the space economy of the world. It is essential, therefore, for local practitioners to recall that:

> "... the existence of a close and 'hegemonic' growth coalition that weaves together public and private elites plays a foundational role in generating and maintaining competitive spaces ... a coherent and relatively homogeneous coalition of local, national, and international elites is instrumental to initiating and maintaining a 'boosterist' climate and a competitive growth trajectory" (Ibid., p. 551).

3.2. Trade, corporate strategies, and policy implications

The second category of studies in which geographers examine the interrelationships between international trade and the firm-territory-government nexus describes various strategies used by business establishments and government agencies to strengthen firm-level competitiveness in the global marketplace. For geographers, part of the motivation in undertaking these investigations is the desire to discover the nature of the relationship between firms and the territories in which they are located and in which they conduct international business. In addition, this research provides business executives and government policymakers with an assessment of how effective various assistance strategies and programs have been in different parts of the world community.

At the outset, the findings of Sternberg and Arndt (2001) regarding the expected relationship between firms and territory are instructive. They set out to determine whether the innovations of European firms are influenced more by firm-level or region-level factors. Their conclusions suggest that the research *capacity* of a region is the most significant individual determinant of a firm's innovation behavior; however, they also find that "... firm-level determinants have a greater overall influence on innovation activity than most region-level determinants" (Ibid., p. 379). In other words, the regional environment can provide support for firm-level innovations, but only those firms that already have the capacity to engage in innovative activities are likely to benefit. For regional practitioners and planners, this study provides two important policy directives to increase the innovativeness of regions. First, cooperative intra-regional and interregional networks among firms must be initiated or intensified; and second, local innovation policy must be targeted primarily to the firms with innovative capacities.

The research by Perry and Hui (1998) and Wang and Yeung (2000) in Singapore reflects the importance of understanding the relationship between firm-level strategy and the territory in which the firm is situated. The first two authors report on the success the country's Economic Development Board has had in promoting the development of cooperative linkages among various foreign multinational companies and local subcontractors. This has been accomplished by increasing the flow of information between the buyers and suppliers. As a follow up, Wang and Yeung assess the relatively recent success of the some forty subsidiaries and local suppliers of transnational chemical firms in Singapore and find that the industry has benefited from the increasing global competition within the industry and the growing market potential in developing nations. In addition, however, they attribute much of the success of the industry in that country to the ability and willingness of these local subsidiaries to develop networks among competing companies and to tap into the cluster-based advantages that are being nurtured by the country's development agency.

Geographers have also examined the learning process whereby business executives gain international marketing intelligence. This literature is part of a long-standing research agenda by people in many disciplines to understand and model the internationalization process of firms (for a review, see Bilkey, 1978 and Cornish, 1995). For example, O'Farrell, Moffat, and Wood (1995) study the internationalization process of the business service sector. Of particular interest is the process of deciding the most appropriate strategy for entering a foreign market, and how one measures cultural distance and country risk.

A similar study focuses on the export of houses from British Columbia to Japan, and on how the Canadian suppliers gain knowledge about how to penetrate the Japanese market successfully (Reiffenstein, Hayter, and Edgington, 2002). The authors are contributing to what is known in geography as the "cultural turn," which conceptualizes economic activities, such as international trade, as a cultural as well as an economic and political process. By examining the learning process of these particular manufacturers, they are confirming the point that exporting is not a discrete function, but rather the result of the ability of business people from different cultures to develop marketing and production intelligence across international borders.

These studies of the internationalization process of firms offer insights to practitioners and business executives, and they provide evidence of "best" practices in developing international competitiveness. Such studies have had a significant impact on the way certain governments are designing export-development programs and policies. For example, Scotland has created the Global Companies Strategy of Scottish Enterprise, which is designed to provide intensive and integrated international marketing assistance to particular companies (Raines and Brown, 2001). The policy instrument is based upon the assumption that companies must overcome a number of firm-specific and territorial barriers in order to be successful internationally. These include market barriers (e.g., tariff and non-tariff restrictions), information barriers, organizational deficiencies, and motivational commitments of management. To help companies overcome these barriers, the government is implementing a proactive pilot project that is providing public-sector assistance to a select group of companies. The objective is to work with managers that have already been successful in initiating some internationalization in their companies, and to assist them in shifting their existing international operations into higher value-added activities.

4. REGIONALISM AND INTERNATIONAL TRADE

In the past two decades, economic geographers have drawn attention to the success of certain industrial districts or regions in North America and Europe. As noted in section 3.1, national industrial production tends to be regionally contained (e.g. Scott's 1999 study of the music industry in Southern California), but exports of products and services from these regions are frequently globally-oriented. The observation that sector or industry-specific production is distinctively regional has not only increased understanding in the workings of agglomeration economies but also the role of place-specific structures such as local institutions, government agencies and inter-firm/inter-organizational relations in trade output.

Alongside the above sub-regional perspective, geographers have joined others in noting a parallel trend towards the international pooling of national economies and markets as a means of increasing specialization and trade among participating countries. International and trans-regionalism is not a new phenomenon as regional trade agreements have been operating since the 1960s. What is 'new' about the recent integrationist initiatives in the 1990s is that it is occurring at a time when multilateral trade liberalization is being actively pursued, when multinational enterprises are encouraging production and consumption along three continental blocs, and when regional markets are being formed between industrialized and developing countries for example the North American Free Trade Agreement or NAFTA. Under such a context, the implication for national economic growth can be significant as developing countries gain greater access to industrialized countries' markets. For geographers, interest in both forms of regionalism lies much less in their market-freeing possibilities than problems associated with regime regulation and potential regional divergence.

Two papers by Dunford and Perrons (1994) and Sanchez-Reaza and Rodriguez-Pose (2002) illustrate the above point. Dunford and Perrons showed that the market and trade liberalizing aims of European integration have not resulted in increased regional convergence. Rather, regional inequality has been reinforced between a core of advanced areas and international cities from Greater London to Northern Italy, and, peripheral areas of relative underdevelopment in the Southern Mediterranean European Union (EU) regional economies. Their paper calls for policy measures associated with greater inter-regional distribution of investment and increased long term loans from capital surplus countries to more underdeveloped regions. The impact of opening Mexico's economy to "free trade" under NAFTA is the focus of research by Sanchez-Reaza and Rodriguez-Pose. They conclude that trade liberalization policies in 1986 (accessing the GATT) and the economic union with Canada and the United States in 1994 have resulted in greater territorial polarization in Mexico, with the Mexican states that border the United States benefiting the most. This outcome of regional divergence in economic growth across the country is likely to continue under NAFTA, requiring concerted efforts on the part of the national government to design and promote economic development programs that reflect the specific needs of states located away from the borderlands.

What seems clear about understanding the complex interplay between sub and transregional development is that on one hand, international regionalism is formulated with the aim of increasing economic integration between national economies that enhances income convergence between countries. On the other hand, impact of international regulation on local and regional economies reveals a different picture with economic and social benefits being unequally shared by regions.

A corollary of concern for geographic sensibilities in trade research is that geographers are more sensitive to multiscalar development strategies. Such sensitivities are particularly needed in countries where sovereignty issues reign over economic advantages. In East and Southeast Asia for example, the first free trade area was established in 2002 in the form of AFTA or the Association for Southeast Nations (ASEAN) Free Trade Agreement. Countries here in general remain largely cautious about trans-regional integration as a trade liberalizing instrument for economic growth (Poon, 2001). In place of FTAs, countries have combined sub and trans-regional integration to create "growth triangles". Growth triangles may be found between Singapore, the southern Malaysian state of Johor and the western Indonesian province of the Riau Islands (SIJORI), the Tumen River Delta between China's northeast border with Russia and North Korea, and the Indonesia-Malaysia-Thailand Triangle. In each of the three cases, regional integration is confined to sub-regional areas between two or more geographically proximate countries. Grundy-Warr and Perry (1999) have demonstrated that growth triangles reflect the failure of larger, transnational scale regionalism that is more popular in North America and Europe. Rather, Asian policy-makers prefer a model of trade development that does not sacrifice national sovereignty and regulatory power, but that nonetheless allows transnational trade and investment, and thereby income growth and spillovers, to occur between transnationally proximate sub-national regions.

5. CONCLUSION

This chapter has reviewed the major concerns of geographers working on international trade problems. Specifically, the relationship between economic growth, development and the benefits of trade requires an understanding of differences between regions and countries. Emphasis on contextual, regional and national variability is one major reason why geographers remain largely cautious about universal policy prescriptions such as those promulgated by the World Bank on export promotion for developing countries in the 1980s. Indeed the importance of geography in trade and development is increasingly recognized in other disciplines such as Economics (e.g. Sachs, 2000; Sachs et al., 2001). But trade geographers have shown that the geography of underdevelopment is more than just "bad latitude" or being located in the wrong climatic region as Sachs has determined. Differences in culture, institutions, business and commercial practices are also important because locationally-embedded structures play a role in enabling or inhibiting trade outcomes.

For trade geographers then, the global economic landscape as constituted by international trade exchanges is not seen merely as an aggregation of sub-regional commercial flows. Rather, what might appear to be economically beneficial at a national, transnational or international level such as the international regulation of FTA spaces may sometimes mask increased sub-regional divergence as in the case of the EU or NAFTA, or, be dismissed as an instrument of economic governance as in Asia. Conversely, focusing scarce resources on a few successful regional economies in order to generate national exports may also not result in greater trans-regional regulation of wealth (Rodriguez-Pose, 1994). In practicing international trade, applied geographers have shown that a dose of geographic sensibility could help shed light on why regional and international wealth and social convergence continues to elude national and global economies despite the potential positive spillovers of international trade.

JESSIE P.H. POON
Associate Professor,
Department of Geography,
University at Buffalo-SUNY,
Buffalo, New York 14261
(email: Jesspoon@buffalo.edu)

JAMES E. McCONNELL
Professor,
Department of Geography,
University at Buffalo-SUNY,
Buffalo, New York 14261
(email: geojem@buffalo.edu)

REFERENCES

Barnes, T. (2002). Performing economic geography: two men, two books, and a cast of thousands. Environment and Planning A, 34 (3), 487–512.

Bhagwati, J. (1988). *Protectionism*. Cambridge, MA: The MIT Press.

Bilkey, W.J. (1978). An attempted integration of the literature on the export behaviour of firms. *Journal of International Business Studies*, 9, 133–146.

Britton, J.N.H. (2002). Regional implications of North American integration: A Canadian perspective on high technology manufacturing. *Regional Studies*, 36 (4), 359–374.

Clark, G.L., & O'Connor, K. (1997). The informational content of financial products and the spatial structure of the global finance industry. In K.R. Cox (ED.), *Spaces of Globalization: Reasserting the Power of the Local* (pp. 89–114). New York: Guilford.

Cornish, S.L. (1995). Marketing matters: The function of markets and marketing in the growth of firms and industries. *Progress in Human Geography*, 19, 317–337.

Davis, D.R., & Weinstein, D.E. (2002). *What role for empirics in international trade?* (Discussion Paper # 0102-05). New York: Department of Economics, Columbia University.

Dicken, P., & Malmberg, A. (2001). Firms in territories: A relational perspective. *Economic Geography*, 77 (4), 345–363.

Dunford, M. & Perrons, D. (1994). Regional inequality, regimes of accumulation and economic development in contemporary Europe. *Transactions of the Institute of British Geographers*, 19 (3), 163–82.

Edgington, D.W., & Hayter, R. (2000). Foreign direct investment and the flying geese model: Japanese electronics firms in Asia-Pacific. *Environment and Planning A*, 32 (2), 281–304.

Erickson, R. A. (1989). Export performance and state industrial growth. *Economic Geography*, 65 (4), 280–292.

Ginsburg, N. (1961). *Atlas of Economic Development*. Chicago: University of Chicago Press.

Glasmeier, A.K., & Leichenko, R.M. (1996). From free market rhetoric to free market reality: The future of the U.S. South in an era of globalization. *International Journal of Urban and Regional Research*, 20, 601–15.

Grant, R. (1993). Trading blocs or trading blows? The macroeconomic geography of US and Japanese trade policies. *Environment and Planning A*, 25, 273–291.

Gurndy-Warr, C. & Perry, M. (1999). Economic integratin or interdependence? The nation-state and the changing economic landscape of Southeast Asia. In L. Van Grunsven (ED.), *Regional Change in Industrializing Asia: regional and Local Response to Changing Competitiveness* (pp. 197–228). Aldershot: Ashgate.

Gwynne, R.N. (1996). Trade and developing countries. In P.W. Daniels and W.F. Lever (Eds.), *The Global Economy in Transition* (pp. 239–262). Essex: Longman.

Hamilton, F.E.I., & Linge, G.J.R. (Eds.). (1979). *Spatial Analysis, Industry and the Industrial Environment: Progress in Research and Applications, Vol. 1–Industrial Systems*, Chichester: John Wiley & Sons.

Hamilton, F.E.I., & Linge, G.J.R. (Eds.). (1983). *Spatial Analysis, Industry and the Industrial Environment: Progress in Research and Applications, Vol. 3–Regional Economies and Industrial Systems*, Chichester: John Wiley & Sons.

Hanink, D.M. (1994). *The International Economy: A Geographical Perspective*. New York: John Wiley & Sons.

Hoare, A. (1985). Great Britain and her exports: An exploratory regional analysis. *Tidjschrift voor economische en sociale geografie*, 76, 9–21.

Irwin, D.A. (2002). *Free Trade Under Fire*. Princeton, N.J.: Princeton University Press.

Johnston, R.J. (1976). *The World Trade System: Some Enquiries into its Spatial Structure*. New York: St. Martin's Press.

Johnston, R.J. (1989). Extending the research agenda. *Economic Geography*, 65, 338–347.

Krugman, P. (1991). *Geography and Trade*. Cambridge, MA: The MIT Press.

Krugman, P. (1990). *Rethinking International Trade*. Cambridge, MA: The MIT Press.

Leichenko, R.M. (2000). Exports, employment, and production: A causal assessment of U.S. states and regions. *Economic Geography*, 76 (4), 303–325.

Leichenko, R.M., & Coulson, N.E. (1999). Foreign industrial exports and state manufacturing performance. *Growth and Change*, 30, 479–506.

Markusen, A. (1996). Sticky places in slippery space: A typology of industrial districts. *Economic Geography*, 72 (3), 293–313.

Marshall, A. (1920). *Principles of Economics*. New York: Macmillan & Co.

MacPherson, A.D., & McConnell, J.E. (1997). Recent Canadian direct investment in the United States: An empirical perspective from Western New York. *Environment and Planning A*, 24, 121–136.

Martin, R., & Sunley, P. (1996). Paul Krugman's geographical economics and its implications for regional development theory: A critical assessment. *Economic Geography*, 72 (3), 259–292.

Meier, G.M. (1963). *International Trade and Development*. New York: Harper & Row.

Myrdal, G. (1957). *Economic Theory and Under-Developed Regions*. London: Gerald Duckworth & Co.

Noponen, H., Graham, J., & Markusen, A.R. (1993). *Trading Industries, Trading Regions: International Trade, American Industry, and Regional Economic Development*, New York: The Guilford Press.

O'Brien, K.L., & Leichenko, R.J. (2003). Winners and losers in the context of global change. *Annals of the Association of American Geographers*, 93 (1), 89–103.

O'Connor, K., & Daniels, P. (2001). The geography of international trade in services: Australia and the APEC region. *Environment and Planning A*, 33, 281–296.

O'Farrell, P.N., Moffat, L., & Wood, P.A. (1995). Internationalization by business services: A methodological critique of foreign-market entry-mode choice. *Environment and Planning A*, 27, 683–697.

Pantulu, J., & Poon, J.P.H. (2003). Foreign direct investment and international trade: evidence from US and Japan. *Journal of Economic Geography* 3 (3), 241–59.

Perry, M., & Hui, T.B. (1998). Global manufacturing and local linkage in Singapore. *Environment and Planning A*, 30, 1603–1624.

Poon, J.P.H. (1994). Export growth, economic growth, and development levels: An empirical analysis. *Geographical Analysis*, 26 (1), 37–53.

Poon, J.P.H. (2001). Regionalism in the Asia Pacific: is geography destiny? *Area* 33 (3), 252–60.

Poon, J. P.H., & Thompson, E.R. (1998). Foreign direct investment and economic growth: Evidence from Asia and Latin America. *Journal of Economic Development*, 23 (2), 141–160.

Porter, M. E. (1990). *The Competitive Advantage of Nations*. London: Macmillan.

Raines, P., & Brown, R. (2001). From 'international' to 'global': The Scottish Enterprise Global Companies strategy and new approaches to overseas expansion. *Regional Studies*, 35 (7), 657–668.

Reiffenstein, T., Hayter, R., & Edgington, D.W. (2002). Crossing cultures, learning to export: Making houses in British Columbia for consumption in Japan. *Economic Geography*, 78 (2), 195–219.

Rodriguez-Pose, A. (1994). Scoioeconomic restructuring and regional change: rethinking growth in the European Community. *Economic Geography*, 70 (4), 325–43.

Sachs, J. (2000). The *Geography of Economic Development*. Newport, Rhode Island: United States Naval War College, Jerome E. Levy Occasional Paper No. 1.

Sachs, J., Gallup, J. & Mellinger, A. (2001). *The geography of poverty*. Scientific American, March, 70–75.

Sanchez-Reaza, J., & Rodriguez-Pose, A. (2002). The impact of trade liberalization on regional disparities in Mexico. *Growth and Change*, 33 (4), 72–90.

Scott, A. (1999). The US recorded music industry: the relations between groups, location and creativity in the cultural economy. *Environment and Planning A*, 31 (11), 1965–84.

Sternberg, R., & Arndt, O. (2001). The firm or the region: What determines the innovation behavior of European firms? *Economic Geography*, 77 (4), 364–382.

Storper, M. (1997). *The Regional World: Territorial Development in a Global Economy*. New York: The Guilford Press.

Storper, M. (2000). Globalization, localization, and trade. In G.L. Clark, M.P. Feldman, and M.S. Gertler (Eds.), *The Oxford Handbook of Economic Geography* (pp. 146–165). Oxford: Oxford University Press.

Swyngedouw, E. (2000). Elite power, global forces, and the political economy of 'glocal' development. In G.L. Clark, M.P. Feldman, and M.S. Gertler (Eds.), *The Oxford Handbook of Economic Geography* (pp. 541–558). Oxford: Oxford University Press.

Thoman, R.S., & Conkling, E.C. (1967). *Geography of International Trade*. Englewood Cliffs, N.J.: Prentice-Hall, Inc.

Wang, J.H.J., & Yeung, H.W. (2000). Strategies for global competition: Transnational chemical firms and Singapore's chemical cluster. *Environment and Planning A*, 32, 847–870.

Webber, M. (1994). Enter the dragon: Lessons for Australia from Northeast Asia. *Environment and Planning A*, 26, 71–94.

Webber, M. (2000). Globalisation: Local agency, the global economy, and Australia's industrial policy. *Environment and Planning A*, 32 (7), 1163–1176.

Yeung, G. (2002). WTO accession, the changing competitiveness of foreign-financed firms and regional development in Guangdong of Southern China. *Regional Studies*, 36, 627–642.

Antoine S. Bailly

Chapter 14
Medicometry and Regional Devlopment*

Abstract

Through their role in local investment and quality of life health care services contribute to the economic development process. Regional medicometry, a rigourous technical field and an applied field concerned with the formulation of public policies, considers these policies as basic for local development. Through case studies, this paper shows the importance of the health care system in terms of investment and economic and social multipliers. Only a global view of regional health care systems can foster effective treatment of issues relating to medical infrastructures in a context of sustainable development.

From Medical Geography to Regional Medicometry

The fields of medical geography and of health economics have had long and fruitful histories. The analysis of the spatial distribution of specific illnesses and of the variations in levels of physical well-being is an ancient practice, having been initiated at the time of Hippocrates. In the 18th century, in order to better understand the antecedents of certain illnesses, physicians began investigating the relationships between the type and frequency of illnesses on the one hand, and aspects of the human and physical environment on the other hand. In the 20th century, epidemiologists and geographers have further refined the understanding of the diffusion of various illnesses and of their associated causes and consequences.

In the 20th century, in addition, an important conceptual transition has been made in the general way in which medical issues are treated; there has been a shift in emphasis from the notion of illness to that of health. Rather than emphasizing the spatial variations in illness or its causes, researchers in various fields have begun to become more interested in the areas of preventive medicine and public health. The preamble to the constitution

---

* This paper is partly based on "Regional Medicometry: Epistemological Basis and Case Studies in Switzerland", A. BAILLY and M. PERIAT, 1998, *Applied Geographic Studies,* 145–155 and "Regional Medicometry: Health Expenditures, Regional Disparities, Problems, and Policies", A. BAILLY and W. COFFEY, in D. BOYCE and others (eds), 1991, *Regional Science,* Springer-Verlag, 469–485.

A. Bailly and L. J. Gibson (eds.), Applied Geography, 273–285.
© 2004 *Kluwer Academic Publishers. Printed in the Netherlands.*

of the World Health Organization, written in 1947, exemplifies this approach, defining health as "a state of complete well-being: physical, mental and social".

In a similar manner, the investigation of the impacts of illness upon the efficiency of production systems, and of the costs to individuals and to society of providing adequate levels of health care has evolved continuously over the past two centuries; health economics has emerged as a major field within the discipline of economics (Aydalot, 1984). In its modern form, it is largely characterized by the measurement of health care expenditure and the modelling of the financial impact of various health care policies.

Both of these perspectives, in spite of their inherent utility, may be regarded as partial approaches, ones that address limited aspects of the complex issues related to spatial variations in levels of health and in the provision and cost of health care. Given the high degree of interdependency between questions involving levels of health care systems and those involving the costs of providing adequate health care, it is necessary that these issues be considered together. Further, neither the medical geography nor the health economics approach is based upon an original theoretical framework.

In this chapter, we sketch the outlines of the newly emerging field of regional medicometry, an undertaking which employs the combined viewpoints and the methods of geography, epidemiology, economics, sociology, and regional science to attempt to develop a comprehensive analysis of the multifaceted aspects of modern health care systems. We begin by identifying the theoretical underpinnings of the regional medicometric approach. We then stress the importance of adopting a broad analytical perspective. This is followed by an application to a case study from Geneva and perspectives for the future.

HEALTH SERVICES AND REGIONAL DEVELOPMENT

The set of activities referred to as health services have a great potential for stimulating the economic development not only in metropolitan areas but also in peripheral regions. Their distribution may be characterized by a spatial distribution different from other consumer services; if some health services face the basic constraint of physical proximity to their market, the patients, some specialized services do not have to follow population patterns; less populated areas or peripheral regions, having for example climatic advantages, or a rural surrounding, are able to develop health services.

These health services can constitute an important element of the economic base of a region. Often, it is not only the most rapidly growing sector of the economy, but also it can be an export-oriented activity, because it serves local population and outside patients. As such we can talk of a propulsive function, creating jobs and bringing financial injections into the local economy which, through the multiple mechanisms and the circular flow of income, stimulate local economic growth.

From this point of view, in contrast to the traditional Fischer-Clark typology of economic activity, which relegates these service activities as non-productive, a significant proportion of health services (from hospital services to pharmaceutical activities) must be regarded as basic in that they are not only exportable but also highly responsive to external demand.

Through their role in local investment and quality of life, health services contribute more and more to the economic development process. By being footloose, certain health services are able to influence the level of economic development of regions, where it is often the main employer.

If regional development is defined in terms of job creation and increased quality of life, health services are a basic activity contributing to market earned income and structural changes. In attracting skilled labour and complementary activities (social, cultural and economic) it contributes greatly to local life. Here the reasoning is that expenditure to expand the quality of health services in peripheral regions will attract other types of activities and, hence, increase the level of economic development. However, the experience, with regional policies involving health services, has been relatively slim, because these activities are still considered as expenditure and not investments. An incomplete understanding of the economics of the health sector is mainly responsible for this traditional approach.

REGIONAL MEDICOMETRY

The economics of health care is an established field that has found itself in the limelight in recent years because of concerns about rising costs and access to health care. Slower to develop are systematic analytical strategies for understanding the supply of, and demand for, medical services in a regional context. The interdisciplinary undertaking designed to do exactly this is regional medicometry.

Useful medicometric research which is able to contribute to the orientation of the medical and health care policies of our society must be based upon a multidimensional approach; it must not neglect any of the actors in the health systems. This consideration leads us to propose a definition of regional medicometry: the application of mathematical and statistical methods to the testing, the evaluation, and the prediction of medical regularities in space in such a manner as to incorporate the viewpoints of all the actors in the health system with an overall concern for issues of efficiency and equity.

The number of specific themes possible for medicometric studies is almost infinite. One which will be briefly explored here is the analysis of regional impacts of hospitals in peripheral regions. But we have first to start by a theoretical analysis. Take, for example, the impact of the establishment of a hospital complex on its region. There will be direct impact through the supply of health care which attracts patients residing within and outside the region. Further, the hospital limits the tendency of local patients to seek treatment in other regions. Direct impacts are also produced through the supply of jobs which discourage emigration of qualified local workers from other regions to the local region (Figure 1: Regional impacts).

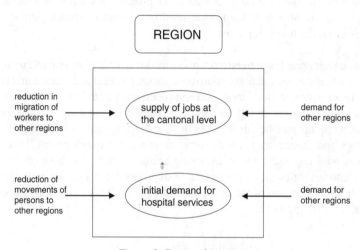

Figure 1. Regional impacts

The strengths and weaknesses of the approaches briefly reviewed above provided the foundation for the reflections made by J. Paelinck (1982) and A. Bailly (1984) who proposed to break out of the constraints of disciplinary boundaries in order to advance this new field. Through conferences on Medicometry (Bailly and Periat, 1985, Bridel and Periat, 1987), where all the actors were present, a broad analytic perspective emerged that explicitly identified the principal actors in the system, their specific objectives, and the complex interrelations between the health-care system and the social, spatial, and temporal environment. Emerging from these conferences were five principles and assumptions of medicometry (Paelinck):

P1: Models of regional medicometry should, *a priori*, be formulated in a spatially interdependent manner.
P2: P2: Spatial relationships are asymmetric so as to take into account the varying endowments of different regions in health-care facilities and pathogenic factors.
P3: P3: Health problems in one area may be explained by causal factors located in other areas.

P4:   P4: The nature of spatial interdependencies and interactions between health-care systems and other systems (economic, social ...) can be explained by spatial diffusion theories.

P5:   P5: All actions in the health-care system, including policies, investments, and diffusion processes, occur in space and time.

Based on these five principles, medicometry utilizes a global view of the health sector by integrating not only its primary function of preventing diseases and the delivery of health care, but also its multiple secondary impacts on the economic and social system.

To achieve these goals, medicometry examines health-care networks from the point of view of both time and space in order to incorporate and assess the effects of well-being on economic activity in various regions, and on its societal and demographic evolution. The pattern summarized in Figure 2 shows the complex direct and indirect economic impacts through salaries and the purchase of goods and services (Figure 2: Direct and indirect economic impacts).

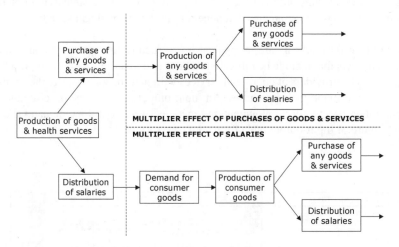

Figure 2. Direct and indirect economic impacts

THE CONCEPT OF ECONOMIC AND SOCIAL IMPACT OF HOSPITALS

The economic and social impact of a hospital is both direct on the region, and indirect on society. The direct effects are through the supply of

1. Health care to patients residing within the region, thus limiting the tendency of local patients to seek treatment in other regions.
2. Jobs that not only prevent emigration of qualified local workers, but attract workers from other regions.

Clearly, this direct impact leads to the flow of money and qualified labour into a region. The indirect effects are in the

1. economy, by the partial reinjection of financial flows into the region in the form of consumption and salaries, some of which will be spent in the region.
2. quality of life, by providing access to the high quality of health care and the proximity (accessibility) to the health infrastructures.
3. socioeconomic status, by the generation of jobs that encourage qualified persons to live in the region.
4. prestige, by attracting persons and businesses from other areas, because of the rich cultural life, prosperity, and pleasant living conditions.

Medicometric analysis provides a broader view of the place occupied by medical infrastructures in two important systems: health and the economy. It facilitates better understanding of the implications of a medical infrastructure at various social and economic levels. Medicometric analysis also enhances the understanding of the consequences of certain decisions in the field of health. Such insights enable us to implement medium- and long-term strategies that benefit both the hospital and society.

In its bid to offer medical care, a medical infrastructure must procure and combine various products that include health-care products and materials, real estate, food, and equipment. The providers of goods and services will, in turn, contribute to the economy of the region through their purchases. An input-output table can be used to calculate the multiplier effects, for employment and income, of a rise in final demand (Polese, 1994) (Figure 3: Input-output table for health care services).

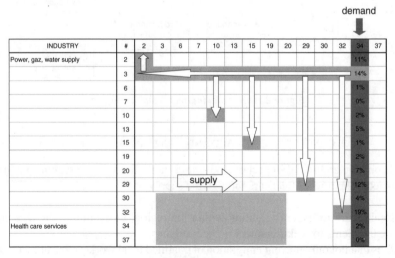

Figure 3. Input-output table for health care services

Direct impacts lead to monetary flows for the region and to indirect and induced impacts on (Figure 4: Socio-economic impacts):

- the economy-financial flows which are partially reinjected into the region in the form of consumption and salaries some of which will be spent in the region;
- the quality of life created by the high level of health care and accessibility to health infrastructure;
- the cultural life of the region which is supported by the relatively affluent and well educated workers associated with health care provision; and
- regional prestige which further attracts individuals and firms to the region because of its rich cultural life, and its pleasant living conditions.

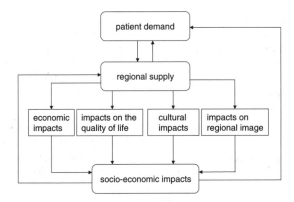

Figure 4. Socio-economic impacts

Medicometry therefore proposes a global view of the health sector, which provides jobs with a high technology component, stimulates economic activity through consumer purchases and, in so doing, participates in the improvement of the quality of life.

For example, to determine the direct economic impact of a hospital in Geneva, we calculate multiplier coefficients using matrices that integrate the apportionment of hospital expenditures and the economic structure of the region. These input-output tables give a detailed picture of the relationship of production and exchanges between the various economic actors. This study is based on the input-output table used in Swiss medicometric studies (Figure 3: Input-output table for health care services). The table consists of 42 categories that reflect some aspect of Swiss economic activity. Data for these categories are expenditures in each branch of activity in the 42 categories. The 39th column represents the allocation to the health sector. By replacing this column of expenditure by that of a given hospital, we can use the table to calculate the regional multipliers of output, value added, wage bill, and expenditures on consumption. Together with information on the hospital's value of output between intermediate consumption (purchases of goods and services) and value added, one can estimate the effects of these hospital multipliers on the canton. The breakdown of the value of output between intermediate consumption (purchases of goods and services) and value added are based on the accounts of the hospital (Table 1: Composition of hospital expenditures).

Table 1. Composition of hospital expenditures

	Francs	%
Intermediate consumption		
Within the canton of Geneva	4,394,515	8.99
Elsewhere in Switzerland	9,782,861	20.01
Abroad	67,276	0.14
	14,244,652	29.14
Value added		
Salaries paid in Geneva (canton)	11,876,138	24.29
Salaries paid outside Geneva (canton)	7,625,931	15.60
Social contributions	3,514,000	7.19
Mortgage interests	825,000	1.69
Depreciation	1,702,000	3.48
Subsidy	−9,000	0.02
Operating balance	9,110,007	18.63
	34,644,076	70.86
Value of output	48,888,728	100.00

ESTIMATION OF MULTIPLIERS

By replacing the cost structure relating to health by the cost structure of the hospital in the 42 branches of activity covered by the Swiss input-output matrix, we can calculate the following multipliers:

$$\text{Multiplier of total output: } \sum_i h_{ij} = 1.14460, \tag{1}$$

where $H = (I - AD)^{-1}$, and AD is a matrix of the coefficients of the use of local goods and services in which the jth column, health, relates to the hospital.

$$\text{Multiplier of value added: } \sum_i v_i h_{ij} = 0.7798, \tag{2}$$

where the v_is are the value-added coefficients calculated for Switzerland, with the exception of the coefficient for the health sector, which is the coefficient of the hospital, or 70.86%.

$$\text{Multiplier of salaries paid in Geneva: } \sum_i s_i h_{ij} = 0.2738, \tag{3}$$

where the s_is are the coefficients of salaries in the various branches of activity in the Swiss matrix, with the exception of the health sector, for which the salary coefficient is the coefficient relating to salaries paid to residents in the canton by the hospital, or 24.292%.

Multiplier of expenditures on intermediate consumption

$$\text{in the canton: } \sum_i s_i h_{ij} = 0.2738, \tag{4}$$

where the s_is are the coefficients of salaries in the various branches of activity in the Swiss matrix, with the exception of the health sector, for which the salary coefficient is the coefficient relating to salaries paid to residents in the canton by the hospital, or 24.292%.

Multiplier of expenditures on intermediate consumption

$$\text{in the canton: } \sum_i c_i^d h_{ij} = 0.1446, \tag{5}$$

where the c_i^ds are the coefficients of intermediate consumption of local goods and services for Switzerland, with the exception of the coefficient relating to health, which is the coefficient for the hospital's intermediate purchases in the canton, or 8.989%. This multiplier is overvalued, because the coefficient of intermediate consumption of local goods and services relates to, apart from the health sector, purchases made within the country and not within the canton. In order to adjust for this, we assumed that the ratio between the proportion of hospital purchases (in the canton) of the value of total output at the national level (0.08989), and the proportion of local purchases at the national level for the health sector (0.26094), is a constant (0.34) for the other branches of activity.

Based on this constant ratio assumption, the multiplier of expenditures on intermediate consumption in the canton was recomputed:

$$\sum_i c_i^{canton} h_{ij} = 0.10979. \tag{6}$$

According to the theory of input-output tables, these various multipliers apply to the final demand. Because we assume that hospital receipts come only from households in or outside the canton, final demand equals the value of output. The following multiplier effects are the result:

Total output of the canton: $48{,}888{,}728 \times 1.1446 = 55{,}958{,}038$.
This is an increase of 14,46% of the value of hospital output.
Value added: $48{,}888{,}728 \times 0.7798 = 38{,}123{,}430$,
or an increase of $38123430 / 34644076 = 1.1004$, which is 10.04% of hospital value added.

Cantonal wage bill: 48,888,728 × 0.2738 = 13,385,734,
or an increase of 13385734/11876138 = 1.1271, which is 12.71% of hospital
salaries disbursed.
Intermediate consumption spending in the canton: 48,888,728 × 1.10979 =
54,256,220, or an increase of 5,425,622/4394515 = 1.2346, which is 23.46% of
intermediate consumption by the hospital in the canton.

The multiplier concept has been widely used in our studies on regional hospitals in
Switzerland (Bailly and Periat, 1985). The multipliers for purchases of goods and
services vary, usually between 1 and 1.6, depending on the region and the economic
activities under consideration. A value close to 1 shows that the indirect benefits of
purchases of good and services in the region are very weak; when the multiplier is over
1.5, total expenditure on goods and services in the region is doubled, which implies a
very strong influence on the overall regional economic system.

In peripherial regions studied in different Swiss cantons, the multipliers for the purchases of goods and services have values ranging between 1.2 and 1.6. These regions
are small and offer few goods and services indispensable to the functioning of hospitals: most purchases are made outside the region. A high multiplier on the cantonal
level frequently corresponds to a weak multiplier on the local district level. Such is the
case of peripheral hospitals which have values of 1.2 in the local district and 1.5 in the
canton, or 1.2 in the local district and 1.6 in the canton.

The economic benefits of the hospital's activities are also channeled through direct and
indirect salary disbursements. Thus, the salary multiplier indicates that the activity of
the hospital is not only the source of salaries paid out by the hospital, but also of salaries
disbursed through other branches of activity. A value of 1.12, as in the Geneva case,
indicates, for example, that the increase in the wage bill is 12%. Direct and indirect
salaries trigger demand for goods and services which will be even stronger since the
"leaks", namely savings and spending outside the region, are weak.

The multipliers should be considered in terms of their effect on the economy of the
district and canton; but they must also be complemented by consideration of the frequency levels of hospital visits (visits to local or non-local hospitals) and the effect of
hospitals on jobs and the quality of life in the region. The ability to be treated in the
local language, and according to the local religion and culture is an important element
in the life of a region. Medicometry actually takes these factors into account, over and
beyond mere economic calculations. For example, we can evaluate benefits at the social
level. In the case of Geneva, because the city has a hospital with a high reputation the
canton is able not only to respond effectively to strong local demand, but also to improve the image of the canton and thereby its desirability, which is an important factor
of regional development. Moreover, the technical and professional capital and its high

level of technological development boost the level of skills in the canton, and this also influences the quality of life and image of the canton.

Hospitals are useful at the social, spatial, and temporal levels. In cases of illnesses treatable in a district hospital, patients may be hospitalized on the spot; it is a matter of security for the population because it avoids displacements toward centers outside the area (particularly for the elderly). Inversely, the absence of a hospital is regarded as a drawback, in much the same way as a lack of other service infrastructures, such as schools and postal services. If hospital facilities are taken away from peripheral regions, which are already geographically disadvantaged, these regions will witness a deterioration in their quality of life, because their inhabitants will need to look elsewhere for the amenities they cannot find locally. The presence of hospital personnel in the region enriches the economic, cultural, and social life of the region, because these workers are skilled and highly educated.

Hospitals, as a business, provide employment and offer training opportunities for young people. What is more, because of their wealth of human resources, they strengthen the level of qualification of the labor market in the district (trained and qualified personnel) and, at the same time, contribute to the maintenance and growth of the population.

High-quality infrastructure is a major advantage for the image of a region. The existence of a hospital is one of the prime incentives a region can offer to encourage economic activity and the establishment of companies, and the development of resident dependent functions. Image has come to play a crucial rôle in competition and in the search for comparative advantages.

Therefore, all investment decisions pertaining to the opening and closing of hospitals must take into account the economic position of the region, and the social and cultural features of the area served. The regional hospital is a useful economic and social infrastructure.

One goal for the future is to rationalize functions (networks of coordinated health care) so as to make them more effective and, in particular, in order to more effectively treat problems related to the aging of the population, on condition that the hospital is regarded as an economic and social investment for the peripheral regions that often have minimal infrastructures. Hospital planning can therefore only be carried out within a global perspective, integrating the spatial and local dimensions and the future envisaged for the regions concerned.

FOUR PERSPECTIVES FOR THE FUTURE

For the future, regional medicometry can be based upon four perspectives that are global in nature and that recognize the important role played by health care systems:

1. **The explicit recognition of the system's environment.** Due to the positioning of the health care system within the broader context of economic and social systems, it is necessary to conceptualize it in terms of alternative allocations of scarce resources and of highly complex interrelations. Economic, geographical, social and ethical criteria are operative at this level.

2. **A broad interpretation of efficiency.** The notion of efficiency requires one to examine questions dealing with medical investments and expenditures from the perspective of pure economic efficiency. In addition, this notion requires us to analyze the demand for health services in the framework of flexibility of choice, and to regard the supply of health care as a function of both the present and future states of the system of health care.

3. **The geographical scale of intervention.** All health-related policies must form a coherent system in which objectives and efficiency criteria must be defined in a global manner, taking into consideration the various spatial scales at which intervention can occur: a top-down approach that gives priority to broad national objectives and constraints; and a bottom-up approach that gives priority to local objectives and constraints. The allocation of expenditure can only approach an optimal form when the spatial context is specified.

4. **The temporal scale of intervention.** Similarly, the objectives and efficiency criteria of all health-related policies must be defined with reference to the temporal scales at which intervention can occur: a short-term approach that emphasizes the resolution of immediate problems; and a long-term approach that gives priority to the development and continuing support of infrastructure and comprehensive programmes, and to addressing major structural problems such as the aging of the population.

These four perspectives make it readily apparent that all narrow approaches to health care services – narrow in the spatial, temporal or professional senses – have a high probability of failure due to their inability to reflect the complexity of the system. This type of narrow approach is one of the principal factors underlying the slow progress in the search for reasonable solutions to questions of health care.

CONCLUSIONS

Empirical evidences in medicometry show that the use of the health care system can be regarded as one of the answers to regional development problems. Our research suggests that some regions benefit greatly from health service activities. By attracting a pool of skilled labour, by stimulating a local demand, by exporting services and products, regional policies devoted to health activities fulfill a basic role.

At a time when budgetary constraints are compelling cities and states to reduce public hospital expenditure, the medicometric approach leads to results that demonstrate the economic and social rôle of the health sector. These studies enable the development of future scenarios, in order to provide networks of coordinated health care, taking into account new technologies and the aging of populations. These activities clearly illustrate that, in the area of health, it is possible to reason in terms of investments and not of costs, economic multipliers, nor expenses. Only a global view of regional health systems can foster effective treatment of issues relating to medical infrastructures in a context of sustainable development within the health sector.

ANTOINE S. BAILLY
Department of Geography
University of Geneva
CH-1211 Geneva 4
Switzerland

REFERENCES

AYDALOT, P. (1985), *Economie Régionale et Urbaine*. Paris: Economica.

BAILLY, A. (1993), Health Policy in Switzerland, in Casparie, A. and others (eds), *Health Care in Europe after 1992*, London: Aldershot, 27–48.

BAILLY, A. and COFFEY, W. (1990), Regional Medicometry: Health Expenditures, Regional Disparities, Problems and Policies, in Boyce, D. and others (eds), *Regional Science*, Berlin: Springer-Verlag, 469–485.

BAILLY, A., BRIDEL, F. and PERIAT, M. (eds) (1987), *La Santé: Perspectives Médicales*, Paris: Economica.

BAILLY, A. and PERIAT, M. (eds) (1984), *Médicométrie Régionale*, Paris: Anthropos.

BAILLY, A. and PERIAT, M. (eds) (1985), Médicométrie Régionale, *Revue d'Economie Régionale et Urbaine*, Special issue.

BAILLY, A. and PERIAT, M. (1995), *Médicométrie: Une Nouvelle Approche de la Santé*, Paris: Economica.

BAILLY, A. and PERIAT, M. (1998), Regional Medicometry: Epistemological Bases and Case Studies in Switzerland, *Applied Geographic Studies*, 2, 2, 145–155.

PAELINCK, J.H.P. (1982), in K. Smith and B.V. French (eds) *Proceedings Cost-Benefit Symposium*, Rotterdam: Netherlands Economic Institute.

POLESE, M. (1994), *Economie Urbaine et Régionale. Logique Spatiale des Mutations Economiques*. Paris: Economica.

ROBERT STIMSON

CHAPTER 15
MONITORING AND BENCHMARKING REGIONAL
AND LOCAL PERFORMANCE

ABSTRACT

Measuring and benchmarking how regions and local communities perform and cope with socio-economic change is a methodological issue of interest to geographers and regional scientists. The chapter outlines two approaches used in projects in which the author has been involved. The first involved monitoring and evaluating the performance of three of Australia's metropolitan cities on a range of indicators relating to population and employment, investment in economic activities and housing markets. The second involved developing a multi-variate model to measure the socio-economic performance of local communities across Australia's cities and towns.

INTRODUCTION

Over the last decade or so a greater interest has emerged among those concerned with public policy relating to the governance, development and planning of cities and regions, in how to measure and benchmark performance. This includes, for example, performance with respect to:

*   socio-economic phenomena, such as levels of human capital skills, labor market transition, investment in economic activity, and housing markets
*   distributional issues, such as levels of income, the incidence of poverty, and the characteristics of socio-economic advantage and disadvantage among individuals and households
*   the nature and degree of social inclusion or social exclusion in communities
*   the nature and levels of social capital in communities
*   aspects of environmental quality, such as air and water pollution, waste generation and recycling, and the protection of fragile and important ecosystems.

The measurement, description and analysis of performance related to such concerns, and the evaluation of the performance of a city, a region, a community or a locality, requires access to a wide range of data at various levels of spatial scale. Often it involves the use of multi-variate spatial analytic tools employed in a GIS environment and it might

287

A. Bailly and L. J. Gibson (eds.), Applied Geography, 287–303.

involve the visualization of patterns of performance. The benchmarking of performance requires both the use of performance indicators and the setting of a standard against which the performance of a place is compared. That might involve benchmarking against a norm, such as national average, or using an index or a normalized measure, such as a location quotient or index of concentration.

Two case studies

This chapter discusses two projects in Australia in which the author has been involved in recent years. The projects involved teams of geographers (and sometimes people from other disciplines). The objective was to measure and benchmark the performance of places at various levels of scale and with respect to a number of dimensions of their socio-economic performance. The projects illustrate the application of either specifically geographic techniques of analysis or of techniques used more widely in the social sciences to analyze and measure performance of a place on specific performance indicators or to identify, through multi-variate analysis, dimensions of socio-economic performance.

The two case study projects are:

(a) Monitoring the performance of three large metropolitan cities—Sydney, Melbourne and Brisbane-South East Queensland (SEQ)—under a project funded by the Australian Housing and Urban Research Institute (AHURI) between 1996 and 1998.
(b) Identifying the dimensions and mapping the patterns of socio-economic performance across communities comprising Australia's metropolitan cities and its regional cities and towns, under a project funded by AHURI in 1998 and 1999.

Aspects of those projects are used here to illustrate the application of geographic analytic techniques in an applied context.

THE MONITORING CITIES PROJECT

Background

In the early 1990s, Kevin O'Connor, a Melbourne geographer, undertook an initiative whereby he produced two annual publications through the Centre for Population and Urban Research and the School of Geography and Environmental Science at Monash University. One was called *Monitoring Melbourne*; and the other was called *The Australian Capital Cities Report*. Later, over the period 1996 to 1998 and with funding from AHURI, O'Connor developed a collaboration with two other geographers, Robert Stimson from Brisbane and Maurice Daly from Sydney, to extend that monitoring work to include the Brisbane-SEQ Region and Sydney. Sydney, Melbourne and Brisbane respectively are the capital cities of the states of New South Wales, Victoria and Queensland, and they are Australia's three largest metropolitan regions.

Under those initiatives, a series of reports were produced using simple methodologies to monitor metropolitan capital city performance in Australia across a range of socio-economic indicators. That included benchmarking city performance. Use was made of both time series and point-in-time data available from the Australian Bureau of Statistics (ABS). The last series of Monitors produced was in 1999, based on analysis of data up to and including 1998 (Daly, 1999; O'Connor, 1999; Stimson, Shuaib, Jenkins, & Lindfield, 1999).

Context

A common approach and methodology was used in the Monitors for all three cities to analyze and evaluate their performance. This involved analyzing data and presenting output in tabular, graphic and map formats, along with brief interpretative commentary to describe and evaluate metropolitan capital city performance on a selection of socio-economic indicators. The objective was to present information in a simple, summary form that could be easily interpreted by, and be useful to, a range of people and organizations in the public and private sectors and by the general public.

Two levels of analyses were undertaken:

(a)   First, an aggregate overview of the broader state-wide perspective for the capital city. This involved analyzing performance of the state on a variety of indicators relating to: population, immigration and internal migration; production, investment and employment; consumption and housing; external linkages, including trade and passenger movements; and transport activity.

(b)   Second, analysis of performance at the metropolitan capital city level. This involved spatially disaggregated analysis across sub-regions of the metropolitan area. It involved compiling indicators relating to: population characteristics and growth; housing construction, density and prices; employment and construction activity across economic sectors including offices, factories, other business premises, shops, hotels, education facilities and health facilities; public sector employment and construction; aspects of unemployment; and social justice measures, including social welfare payments to households.

In addition, each annual series of city monitors looked at a small number of special issues or indicators, which were determined largely by release by the ABS of a particular data series or by the existence of a newsworthy planning and development issue. For example, in the Monitors for 1998 special sections analyzed aspects of housing tenure and types of households as well as aspects of the geography of jobs and patterns of travel to work at the 1996 census. In the Brisbane-SEQ Monitor an additional special section was included on the results of an analysis of industry clusters in SEQ, while the

Sydney Monitor included an additional special section on development, planning and management issues for Sydney Harbour.

By way of illustration, Table 1 reproduces the Table of Contents from the Brisbane-SEQ Monitor for 1998.

Table 1. Table of contents for the 1998 Brisbane-SEQ monitor

DATA PRESENTATION	6
AN OVERVIEW OF THE PERFORMANCE OF QUEENSLAND	
IN THE NATIONAL CONTEXT	7
Map of Population Change	8
Demography	9
Production, Investment and Employment	10
Consumption and Housing	12
External Linkages	13
Transport Activity	14
THE SEQ METRO REGION: PERFORMANCE AND PATTERNS	15
Map of the 15 Sub-Regions within the SEQ Region	16
Population	17
Housing Construction	18
Higher Density Housing	19
Median House Prices	20
Office Employment and Construction	21
Manufacturing Employment and Factory Construction	22
Wholesale Employment and Other Business Premises Construction	23
Retail Employment and Shop Construction	24
Hospitality Employment and Hotel Construction	25
Education Employment and Education Facilities Construction	26
Health and Community Services Employment and Health Facilities Construction	27
Public Sector Employment and Total Public Sector Construction	28
Aspects of Unemployment	29
Social Justice Measures	30
HOUSING TENURE AND TYPES OF HOUSEHOLDS	31
Housing Tenure and Households	33
Changing Sub-Regional Mixes of Housing Tenure Types	35
Household Family Types	38
THE GEOGRAPHY OF JOBS AND PATTERNS OF TRAVEL TO WORK	41
The Location of Jobs and the Journey-to-Work in the SEQ Region in 1996	43
THE EMERGENCE OF INDUSTRY CLUSTERS IN SEQ	47
Map of SEQ Regional Clusters	48
Industry Clusters and Their Locations	49
OVERVIEW	51

(*Source:* Stimson et al., 1999, p. 3.)

Methodology

A benchmarking approach was developed to analyze city performance on a specific socio-economic indicator. Typically that involved analyzing time series data over an 8 to 10 year period and representing it visually to demonstrate:

- the comparative performance over time of a city vis-a-vis other cities, benchmarked against each city's national share of that phenomena
- plotting variations over time in the actual volume of that phenomenon in the city under consideration.

That approach is illustrated in Figure 1. It shows the performance of Melbourne on the indicator International Air Passengers over the period 1989–1997 (O'Connor, 1999, p. 14). The approach enables the city's performance over time to be compared not only against other cities but also in terms of its changing national share of that indicator over time. Similarly, Figure 2 shows how Sydney's performance on the indicator Total Dwelling Commencements over the period 1989–1998 was benchmarked (Daly, 1999, p. 18).

To analyze variation in patterns of performance within a city, the compilation of some indicators involved time series analysis of the data (such as investment in factory construction) to monitor the changing spatial patterns of performance over time across the sub-regions of that metropolitan capital city. This included identifying changes in a sub-region's share of the metropolitan region's attraction of that activity. For example, Table 2 illustrates for Brisbane the changing patterns of the sub-regional shares in the value of factory construction over the period 1990–98 (Stimson et al. 1999, p. 22);

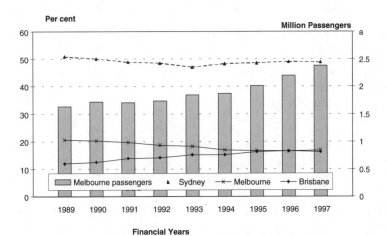

(Source: O'Connor, 1999, p. 14)

Figure 1. International air passengers, Melbourne, 1989–1997

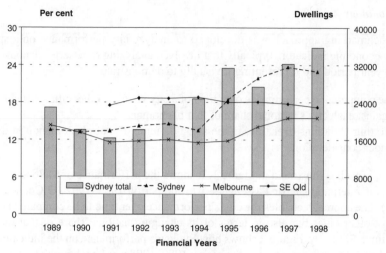

(Source: Daly, 1999, p. 15)

Figure 2. Total dwelling commencements, Sydney, 1989–1998

Table 2. Sub-regional shares of factory construction, Brisbane-SEQ, 1990–1998

	1989	1990	1991	1992	1993	1994	1995	1996	1997	1998
Brisbane City										
Core	6.0	1.6	1.2	0.2	1.1	0.1	3.4	1.2	0.5	1.5
Eastern Inner	5.5	1.1	0.9	0.8	1.5	1.7	5.4	1.0	1.1	2.4
Southern Inner	0.3	1.2	0.4	2.3	0.0	0.2	0.5	0.0	0.1	0.1
Western Inner	1.1	0.0	0.0	1.1	6.2	0.1	0.1	0.1	0.0	0.6
Northern Inner	3.6	2.9	0.6	0.8	0.0	0.3	0.3	2.7	1.8	1.8
Eastern Outer	2.1	2.2	23.3	30.3	40.5	10.9	6.2	26.2	10.6	3.2
Southern Outer	14.9	14.4	12.7	15.2	6.3	14.3	20.5	14.1	14.4	15.1
Western Outer	0.5	0.7	0.3	0.2	0.0	0.0	0.3	0.0	0.0	0.0
Northern Outer	11.9	12.1	15.5	5.3	9.7	16.7	5.9	9.9	5.3	9.0
Sub-total	45.9	36.2	54.9	56.2	65.3	44.3	42.6	55.2	33.8	33.7
Remainder of SEQ Region										
Logan-Redland	8.9	14.3	8.4	9.9	5.6	15.6	11.1	14.6	14.5	13.6
Gold Coast	31.4	32.2	19.8	18.4	14.2	15.4	13.6	10.8	21.0	27.2
Pine-Redcliffe- Caboolture	6.4	8.4	6.0	6.7	4.8	7.5	12.7	5.5	11.0	7.8
Sunshine Coast	5.0	4.8	4.5	5.2	3.0	5.7	6.7	6.2	7.1	4.9
Ipswich	2.1	4.2	5.6	2.8	3.0	7.7	12.6	6.5	3.5	11.7
Rural Fringe	0.1	0.1	0.8	0.7	4.1	3.7	0.8	1.2	9.0	1.2
Sub-Total	53.9	64.0	45.1	43.6	34.7	55.6	57.5	44.8	66.1	66.4

(*Source:* Stimson et al. 1999, p. 22)

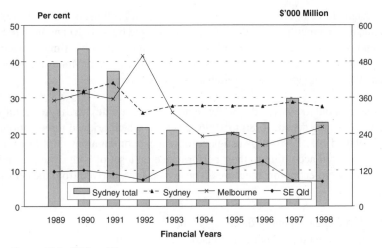

(Source: Daly, 1999, p. 22)

Figure 3. Factory construction

and Figure 3 shows in graphic form Sydney's changing share of national investment in factory construction relative to that of Melbourne and Brisbane-SEQ.

Each year the Monitor for a city concluded with a brief summary overview of trends in the performance of that city across the range of indicators analyzed. In that way the Monitor highlights the general implications of the 'report card' for that city on its performance on the range of indicators included in the Monitor.

Launch of the monitors

The Monitor for each city was launched at a public event organized in each of the three cities – Sydney, Melbourne and Brisbane – by the state branch of a national think-tank, the Committee on Economic Development for Australia (CEDA), and in the case of Brisbane that was a joint activity with the Brisbane Development Association (BDA). At those events the author(s) of the Monitors presented an overview of the findings following an introduction to the launch of the Monitor by a high profile entity such as the Lord Mayor of the city or a leading business identify. There was usually considerable press and other media coverage.

INVESTIGATING DIMENSIONS AND PATTERNS OF COMMUNITY OPPORTUNITY AND VULNERABILITY ACROSS AUSTRALIA'S CITIES AND TOWNS

Objective

In 1998–99, AHURI funded an investigation into the dimensions and patterns of community opportunity and vulnerability across Australia's cities and towns. The results were presented in a book (Baum, Stimson, O'Connor, Mullins, & Davis, 1999) which was launched at a major national conference organized by AHURI and held in Sydney at the end of November 1999. The research was undertaken by a multi-disciplinary team of five researchers at the University of Queensland and Monash University. It included three geographers, a sociologist and an economist.

The project set out to analyze the performance of communities across Australia's cities and towns over the decade 1986 to 1996, a decade which bridged three national censuses – 1986, 1991, 1996. It investigated how local communities might be classified and differentiated according to their performance on a set of both dynamic and static measures of socio-economic measures. Those measures, derived from the census data, related to:

- labor force engagement or disengagement
- the structure of employment in industries and occupations
- levels of human capital
- levels of household income
- the concentration of social advantage and disadvantage, including measures of housing financial stress.

The objective was to:

- develop a typology of community performance
- identify the dimensions on which communities are differentiated
- place communities along an opportunity-vulnerability continuum.

In that way, 'how a specific community within a metropolitan city or how a particular regional city or town has been performing over the decade 1986 to 1996 is determined with respect to its position in the wider national context' (Baum et al. 1999, p. 1). Implications of findings for both people and place-based policies were also considered in the project.

The results of the project attracted wide attention both in the media (including front page and editorial comment in *The Australian*, the national daily newspaper) and among politicians and bureaucrats, as well as in communities (including local governments) across Australia. The media interest continued over several months, and it involved researchers participating in talk-back radio, and giving many radio and television interviews.

The framework

Baum et al. (1999) developed the conceptual framework set out in Figure 4 to analyze community opportunity and vulnerability, it being derived in part from the work of

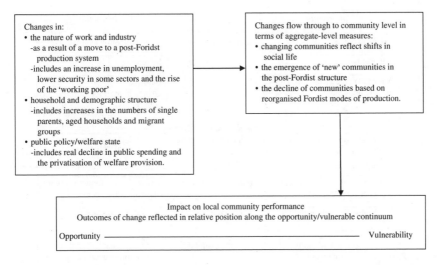

Changes in:
• the nature of work and industry
  -as a result of a move to a post-Foridst
  production system
  -includes an increase in unemployment,
  lower security in some sectors and the rise
  of the 'working poor'
• household and demographic structure
  -includes increases in the numbers of single
  parents, aged households and migrant
  groups
• public policy/welfare state
  -includes real decline in public spending and
  the privatisation of welfare provision.
	Changes flow through to community level in
	terms of aggregate-level measures:
	• changing communities reflect shifts in
	  social life
	• the emergence of 'new' communities in
	  the post-Fordist structure
	• the decline of communities based on
	  reorganised Fordist modes of production.

Impact on local community performance
Outcomes of change reflected in relative position along the opportunity/vulnerable continuum

Opportunity ———————————————————————————— Vulnerability

(Source: Baum et al. 1999, p. 7)

Figure 4. Framework for analyzing socio-economic change and community characteristics

Benassi, Kazepov and Mingione (1997). That approach looked at the interconnections between:

- changes in the economic system, including employment
- transitions in household and demographic structures
- shifts in the welfare state and public policy.

Referring to Figure 4, Baum et al. (1999, p. 7) tell how 'we move from changes taking place in social life, which tend to be most dominant at the individual or household level, to changes at the aggregate level of local communities. The key is that transitions occurring in social life are reflected in changes in a number of aggregate-level measures, as the sum of these aggregate-level measures might be seen to determine the position of a community along a continuum from opportunity to vulnerability'.

A further framework shown in Figure 5 was proposed. That summarizes what might be expected to characterize a community of opportunity and a vulnerable community.

Methodology

The study of community opportunity-vulnerability using the above and following frameworks necessitated choosing appropriate analytical statistical tools to analyze a set of variables derived from census data that were selected to both measure the dynamics of performance of communities both over the decade 1986 to 1996 and of their

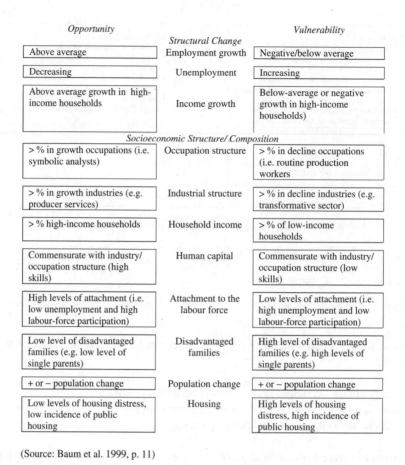

<table>
<tr><td colspan="3" align="center">Opportunity</td><td></td><td colspan="2" align="center">Vulnerability</td></tr>
</table>

	Structural Change	
Above average	Employment growth	Negative/below average
Decreasing	Unemployment	Increasing
Above average growth in high-income households	Income growth	Below-average or negative growth in high-income households)

Socioeconomic Structure/Composition

> % in growth occupations (i.e. symbolic analysts)	Occupation structure	> % in decline occupations (i.e. routine production workers
> % in growth industries (e.g. producer services)	Industrial structure	> % in decline industries (e.g. transformative sector)
> % high-income households	Household income	> % of low-income households
Commensurate with industry/ occupation structure (high skills)	Human capital	Commensurate with industry/ occupation structure (low skills)
High levels of attachment (i.e. low unemployment and high labour-force participation)	Attachment to the labour force	Low levels of attachment (i.e. high unemployment and low labour-force participation)
Low level of disadvantaged families (e.g. low level of single parents)	Disadvantaged families	High level of disadvantaged families (e.g. high levels of single parents)
+ or − population change	Population change	+ or − population change
Low levels of housing distress, low incidence of public housing	Housing	High levels of housing distress, high incidence of public housing

(Source: Baum et al. 1999, p. 11)

Figure 5. The conceptual framework for describing community opportunity and vulnerability

performance at one point in time at the 1996 census. The methodology chosen involved a two-stage process used previously by Hill, Brennan and Wolman (1998) and Coulton, Chow, Wang and Su (1996). It involved first using a hierarchical cluster analysis technique to group localities on their socio-economic characteristics. Following this, stepwise multiple discriminant analysis was used to identify the main factors differentiating between the groups or clusters. Baum et al. (1999, pp. 14–15) explain how the output from those analyses was used in several ways. The discriminant analysis produced correlations between the individual functions and the independent variables and are reported in the structure matrix in the software program SPSS, and are used to identify the properties of each function. Once identified, those are then used in the interpretation of the clusters (the dependent variables). The analysis produced for each observation (community) a series of discriminant scores. Those were then used in two ways:

- the centroids of the clusters (representing a cluster's mean on each function) were used to identify key differences between the clusters, thus providing a general structure for identifying the way the characteristics of each cluster differs from those of the other clusters
- the discriminant scores were also used to compose a summary discriminant score, this being derived by weighting each observation's score on a given function by the percentage of variation explained by that function, and then summing the individual scores for each observation (i.e. community).

An additional output produced was a 'hit ratio' or 'classification score', which identified the proportion of the observations that were correctly classified. Once the discriminant analysis was run and the require statistics computed, Baum et al. (1999) then analyzed the clusters of community opportunity and vulnerability by reference to their associations with the discriminant functions, their summary discriminant score, and a set of mean scores (cluster profiles) which related to the socio-economic variables used (independent variables). The summary discriminant scores were then used to position each community along the community opportunity-vulnerability continuum.

The above analyses were conducted separately for three sets of communities, using as their spatial units Statistical Local Areas (SLAs), which predominantly equate with local government authority areas or sub-areas within them. Thus separate analyses were conducted for:

- 240 SLAs across seven mega metro regions including and surrounding the capital city metropolitan areas
- 122 large regional cities and towns with populations of 10,000 and above at the 1996 census
- 136 small regional towns with populations of 4,000 to 10,000 at the 1996 census.

The clusters of community opportunity and vulnerability

The above frameworks and analytical tools were applied to identify a range of clusters of community opportunity and vulnerability across Australia's cities and towns. The typology of community clusters and their position on the community opportunity-vulnerability continuum are shown in Figure 6:

(a)   The 240 SLAs comprising the metro city regions were allocated into nine clusters. Fifty per cent of the population of these metro regions lived in four clusters of community opportunity, and 31 per cent lived in four clusters of community vulnerability, while the remaining 29 per cent lived in a marginal opportunity-vulnerability cluster.

(b)   The 122 large regional cities and towns split evenly into four opportunity and three vulnerable clusters.

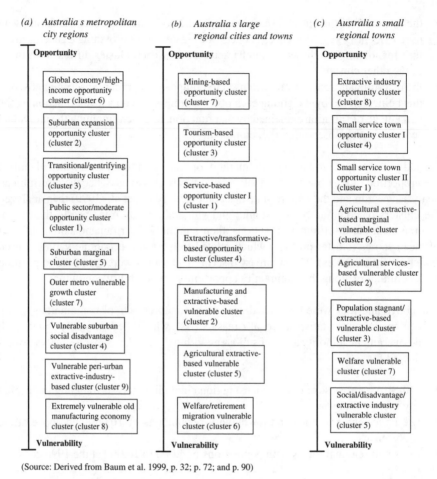

(Source: Derived from Baum et al. 1999, p. 32; p. 72; and p. 90)

Figure 6. Clusters in the continuum of community opportunity and vulnerability

(c)   The classification of the 136 smaller regional towns resulted in three-quarters of
      them falling into five vulnerable clusters while one-quarter were classified into
      three opportunity clusters.

For a full definition and description of the characteristics and spatial patterns of those
three sets of clusters, the reader is referred to Baum et al. (1999, pp. 31–43; pp. 72–77;
and pp. 90–99).

The spatial distribution of those clusters of community opportunity and vulnerability
were mapped both within all seven of the metro city regions and across the large and the
smaller regional towns. In addition, rank order lists were produced placing each SLA
on a community opportunity-vulnerability continuum. By way of illustration Figure 7

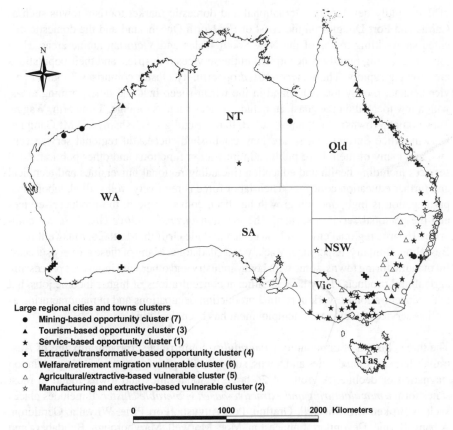

Figure 7. Patterns of community opportunity and vulnerability across Australia's large regional cities and towns

reproduces the result for the clusters of opportunity and vulnerability across Australia's 122 large regional cities and towns with populations greater than 10,000 in 1996.

Description of opportunity and vulnerability clusters in the large regional cities and towns

The four *opportunity* clusters across Australia's large regional cities and towns are associated with a small number of the successful mining and tourism towns and as well with many of the traditional service towns in the wheat-sheep belt and with some of the regional towns that retain manufacturing as well as service functions. The *mining-based* opportunity cluster comprises five towns located mostly in research areas of Queensland and Western Australia (such as Mt Isa, Kalgoorlie and Port Hedland) and are characterized by high-income jobs concentrated in that extractive industry, with very low levels of unemployment. The *tourism-based* cluster comprises a small group of five rapidly developing international and domestic market tourism towns such as Cairns and Port Douglas on the coast of Far North Queensland and the domestic oriented snow-skiing towns of the New South Wales and Victorian alpine areas. Here jobs are growing rapidly concentrated in the services industries, and their populations are growing rapidly. The *services-based* opportunity cluster comprises 36 larger service centers, mostly located inland in the wheat-sheep belt and other farming areas, with a few located on the coast. It includes places such as Orange, Tamworth, Wagga, Toowoomba, Townsville, Wangaratta, Ballarat, Bendigo, Horsham, and Mt Gambier, Darwin, Alice Springs. These are long established successful regional service centers, and many of them have public administration functions and other public funded services including health and education (including regional universities and technical and further education colleges). Their labor force is relatively well skilled, labor force participation is high, and job growth has been considerable in the producer services and the personal services sectors. The *extractive/transformative-based cluster* comprises 16 large regional cities and towns that include Griffith, Maitland, Muswellbrook, Busselton, Bunbury, Esperance, Mackay and Gladstone. Many of these are very successful manufacturing towns, some with heavy industry and other with industries processing agricultural products. They have significant concentrations of higher income jobs, but with concentration of work in routine production occupations and in manufacturing as well as services industries. Unemployment has been declining.

The three clusters of community *vulnerability* comprise a surprising mixture of Australia's larger regional cities and towns, some rapidly growing and others in population stagnation or decline. A group of 24 large regional cities and towns found in most states form a *manufacturing and extractive-based vulnerable cluster*. It includes places such as Broken Hill, Inverell, Grafton, Port Augusta, Port Pirie, Whyalla, Geraldton, Albany, Bernie, Devonport, Launceston, Moe, Morwell, Maryborough, Bundaberg and Rockhampton. Many of these were once prosperous mining and manufacturing towns with industries ranging from textiles to abattoirs, but they have suffered from tariff reduction and economic restructuring. They have increasing proportions of low-income households, unemployment is high, and housing cost stress is evident. An *agricultural-extractive-based cluster* of 24 large regional cities and towns occurs in all states except Western Australia, and the Northern Territory, and it includes places like Bega, Cowra, Forbes, Gunnedah, Young, Taree, Kyabram, Kerang, Port Fairy, Koroit, Leongatha, Murray Bridge, Atherton, Mareeba, Bowen and Innisfail. These range from towns in cane growing areas to those in fruit production areas, dairy towns, and some places in the wheat-sheep belt. Typically they once had prosperous agricultural service and manufacturing functions that have been in decline for some time. Thus, out-migration is often evident, unemployment is relatively high, and job growth and employment opportunities are restricted. Finally, perhaps the most unexpected group of vulnerable large towns comprises 14 rapidly growing (in terms of population) urban centers along the east coast ('sunbelt' areas) that form a *welfare/retirement migration cluster*. It includes places such as Byron Bay, Coffs Harbour, Bellingen, Foster-Tuncurry,

Kempsey, Lismore, Nambucca, Yeppoon and Hervey Bay. While attracting high levels of in-migration, these towns are characterized by extremely high concentrations of welfare dependency, they have very high levels of unemployment, and the only growth industries are in retailing, personal services and tourism, all low-wage sectors. They have significant concentrations of both older people and single-parent families.

For a detailed discussion of the policy implications of this opportunity-vulnerability differentiation across the larger towns of regional and rural Australia, see Stimson et al. (2001) and Stimson, Baum and O'Connor (2003).

Discriminant functions differentiating the clusters

The step-wise multiple determinant analyses conducted (Table 3) revealed that the important discriminators of community opportunity and vulnerability in Australia relate predominantly to:

- levels of human capital and income
- labour market engagement in industries and occupations relating to the new economy, and to rates of unemployment
- overall levels of labour market engagement.

Table 3. The functions and their contribution to the total variance in discriminating between communities of opportunity and vulnerability in Australia

Metropolitan city regions	*Large regional cities and towns*	*Small regional towns*
1. Human capital/affluence (43.1%)	1. Economic/employment advantage (41.0%)	1. Income advantage (49.0%)
2. Employment expansion (23.5%)	2. Industry structure (25.8%)	2. Industry-employment (21.0%)
3. Industry employment structure (13.1%)	3. Employment/population growth (17.2%)	3. Human capital (13.3%)
4. Structural employment decline (9.4%)	4. Non-identified function (12.3%)	4. Population change (10.8%)
5. Older persons (6.5%)	5. Occupation structure (3.5%)	5. Occupation and social disadvantage (3.3%)
6. Income change (2.2%)	6. Disadvantage (0.2%)	6. Youth unemployment (2.4%)
7. Non-identified function (1.5%)		7. Mortgage hardship (0.2%)
8. Low human capital (0.7%)		

(*Source:* Baum et al. 1999, p. 106)

Population size per se, and measures of socio-economic disadvantage relating to housing cost stress, were not found to be important discriminators. This was especially the case in the metro city areas and the large regional cities and towns. If anything, high rates of population growth seemed to be more related to vulnerability among the larger cities and towns, it being particularly associated with the 'sun belt' non-metropolitan urbanization phenomenon. However, it was in the small regional towns where population growth per se was identified as a significant factor discriminating between clusters of opportunity and vulnerability, and here usually population decline was related to vulnerability.

From a consideration of the literature on social polarization, social inequity and social exclusion, Baum et al. (1999) note how it might have been expected that differentiation in community opportunity and vulnerability across Australia's cities and towns would be explained – at least to some degree of significance – by phenomena such as the incidence of concentrations of disadvantaged groups, including the aged, single-parent households, and public housing tenants, as well as by the incidence of spatial concentration of housing-related financial stress. However, as Table 3 shows, the discriminant functions identified as being associated with such measures did not play a particularly important, and certainly not a decisive role in discriminating between clusters of community opportunity and vulnerability.

CONCLUSION

The two case studies discussed in this chapter have been presented to demonstrate how both simple and more complex geographic analytic techniques may be applied to measure and benchmark regional and local socio-economic performance. The results from projects such as the two discussed here serve to illustrate how geographical analysis can produce outputs that are of interest both in a policy context and in the context of capturing attention in the media and among business and public sector agencies, at the national, regional and local levels.

ACKNOWLEDGMENTS

The author acknowledges the contribution of colleagues Kevin O'Connor, Maurice Daly, Scott Baum, Patrick Mullins, Rex Davis, Michael Lindfield, Olivia Jenkins and Fikreth Shuaib with whom he collaborated in the original projects on which this chapter draws.

ROBERT STIMSON
Professor of Geographical Sciences and Planning, and
Director of the Centre for Research into Sustainable Urban and Regional Futures
University of Queensland
Brisbane, Queensland, Australia

REFERENCES

Baum, S., Stimson, R., O'Connor, K., Mullins, P., & Davis, R. (1999). *Community opportunity and vulnerability in Australia's cities and towns: Characteristics, patterns and implications.* Brisbane: Australian Housing and Urban Research Institute, University of Queensland Press.

Benassi, D., Kazepov, Y., & Mingione, E. (1997). Socio-economic restructuring and urban poverty under different welfare regimes. In F. Moulaert & A. Scott (Eds.), *Cities, enterprises and society on the eve of the 21st century* (pp. 174–215). London: Pinter.

Coulton, C., Chow, J., Wang, E., & Su, M. (1996). Geographic concentration of affluence and poverty in 100 metropolitan areas, 1990. *Urban Affairs Review, 32*(4), 186–216.

Daly, M. (1999) *Monitoring Sydney 1998: Evaluating the performance of the Sydney metropolitan region.* Brisbane: Australian Housing and Urban Research Institute, University of Queensland Press.

Hill, E., Brennan, J., & Wolman, H. (1998). What is a central city in the United States? Applying a statistical technique for developing taxonomies. *Urban Studies, 35*(11), 1935–1969.

O'Connor, K. (1999). *Monitoring Melbourne 1998: Evaluating the performance of the Melbourne metropolitan region.* Brisbane: Department of Geography and Environmental Science and Centre for Population and Urban Research, Monash University and Australian Housing and Urban Research Institute, University of Queensland Press.

Stimson, R., Baum, S., Mullins, P., & O'Connor, K. (2001). Australia's regional cities and towns: modeling community opportunity and vulnerability, *Australasian Journal of Regional Studies, 7*(1), 23–62.

Stimson, R., Baum, S. & O'Connor, K. (2003). The social and economic performance of Australia's large regional cities and towns: implications for rural and regional policy, *Australasian Geographical Studies, 41*(2): 131–147.

Stimson, R., Shuaib, F., Jenkins, O., & Lindfield, M. (1999). *Monitoring Brisbane and the south east Queensland Region 1998: Evaluating performance of the south east Queensland metropolitan region.* Brisbane: Australian Housing and Urban Research Institute, University of Queensland Press.

ANTOINE BAILLY AND LAY GIBSON*

CHAPTER 16
APPLIED GEOGRAPHY FOR THE FUTURE

All academic fields evolve and geography is no exception. Geographers are broadening their perspectives, borrowing from related disciplines from other sciences including social behavioral and natural sciences. New examples of applied research have been explored in this book ranging from G.I.S. to medicometry, in different parts of the world. This is a chance to consider the future geography in light of increasing opportunities in the world of application.

APPLIED GEOGRAPHY POTENTIALS

Applied geography defined by Pacione (2001) "as the application of geographical knowledge and skills to the resolution of social, economic and environmental problems" give us the possibility of altering the direction of our field. Geographers previously dealt with problems such as those of local planning and regional applied problems. Now geography provides foundations for a variety of contributions for the management of spatial, environmental and geopolitical problems. But geographers still fail to commute with governemental enterprises. Politicians and business people often get this aknowledgements from everyday sources and not from experts. As Enyedi writes in his chapter, "employers today are not looking for geographers since they generally do not know what geographers actually do."

We do not have enough high profile applied projects with NGO's or world renown institutions to be able to show our true capabilities to the international media. Geographers often get caught up with local and regional details and don't get to the big picture. Even the word "Economic geography" belongs now to the economists if we believe P. Krugman! Many fields are more successful than us in developing their image in the media.

What should we do to develop applied geography potential? Of course this book is one step as other books recently published in the field by M. Pacione and M. Phliponneau. But this is not enough. We should create more "Festivals of Geography", similar to the

---

* The authors want to thank Reginald Golledge and Grady Meehan for their useful comments.

A. Bailly and L. J. Gibson (eds.), Applied Geography, 305–307.

Saint-Dié-Des-Vosges one, a success story for more than 10 years, involving politicians, entrepreneurs, the media, teachers and researchers in geography. We should be more involved in a range of popular magazines from the the National Geographic, the Harvard Business Review. We should learn to write for these audiences. We need also to do more people products and services to private and public firms to change the awareness level of geography. Geographers elected to the Royal Society in the U.K., or in the USA to the National Academy of Science are signs of recognized achievement. Changes are going on in each country where geographers are honored and get high public positions, but moving at a pace that leaves the field open for predators.

NEW INITIATIVES

History, archeology, earth sciences and zoology are being propulsed through television programming. But often the geographical contribution is not clearly identified as geography. We need to get geography in its present form more widely recognized or we will not survive.

New initiatives will be needed to heighten our profile in the outside world, to get more support for graduate research and teaching. We need more and better channels of communication. The business of managing geography could be done through our own research institutes, but also through IGU and its Applied Geography Commission. After four years our Commission has produced a new book, has set up a network of friends of geography (the Advisory Committee) among public and private decision makers. It has also participated in the Festival of Geography and other international events. The next step will be to create a popular book to diffuse the knowledge. First, we need to make every effort to create a new academic culture. Second we need to create a new partenership with the world outside geography and to encourage a public recognition of our potential and ability in solving spatial, environmental event and societal problems. Third we need to make the public aware of our potential.

LESSONS FROM THE BOOK

Applied geography's future lies in revealing its different cognitive capacities for problem solving. In Portugal, in Central Europe, as in the U.K., the USA, France and Australia... applied geography has made its way in spatial, environmental and societal planning. The area of interest involve space and gender, socio-spatial inequalities, disabilities, business and environmental problems. New techniques that include G.I.S. and more efficient models for geographic analysis encourage more world spread application.

The context is favorable for geography to be active outside the traditional academic base. It still remains well linked with the necessary fundamental research. Many of our present day geography students utilize their knowledge to work in government and business. They are the foundation of an active network of applied-professional geographers around the world. Our geography departments should be more open to the professional environment.

Through the history of applied geography, case studies and examples, this book shows new paths and the necessity to develop applied geography in our curriculum. They should be integrated with the B.A. and M.A. programs. Fields trips and applied seminars have always been popular with our students. We should reinforce these activities and the part they play in our student's training. We need to complement the traditional academic education with new programmes for practitioners. As all authors in this book, we propose a future for applied geography, a geography devoted to a better life for people and society.

CHAPTER 17
BIOGRAPHIES AND FIELDS

ANTOINE BAILLY

Biography

Antoine Bailly (b. 1944, France) is Professor of Economic and Urban Geography at the University of Geneva (Switzerland). He is President of the Swiss National Geographic Committee, President of the European Club for Health and President of the Regional Science Association International.

Field

Since my university studies in geography and regional science, I have always been involved in urban and regional planning: in France at the DATAR (French Ministry for Planning), in Québec at the National Institute for Urban Research, in Switzerland with a variety of local and federal authorities. After developing applied researches on the service sector and regional development, I created, with two colleagues, regional medicometry, a field devoted to the location of health care infrastructures. This field is now used in different countries for planning purposes and is tought at the University of Geneva (Switzerland).

GYORGY ENYEDI

Biography

Professor Gyorgy Enyedi (b. 1930, Hungary) graduated from Economics (1953) and Geography (Ph.D., 1957) at the Budapest University of Economics. He was elected as member of the Hungarian Academy of Sciences (1982) and of the Academia Europaea (London, 1990). He served as Vice President of the IGU (1984–1992). As an applied geographer, he has worked for national government agencies and international inter-governmental agencies on the field of regional development, environmental policies and settlement development, as consultant or project leader. At present he is the chair-person of the advisory commission on regional development for the Hungarian Prime Minister office.

A. Bailly and L. J. Gibson (eds.), Applied Geography, 309–318.

Field

My field within applied geography may be labeled as analysing, forecasting of regional development processes and planning for intervening corrections to these processes. This field (regional planning and policy) has a great number of practitioner organizations: central, regional and local government agencies (e.g. DATAR in France), urban and regional planning enterprises (e.g. VATI in Hungary), international organization of the (e.g. European Commission).

JOHN W. FRAZIER

Biography

John W. Frazier is a Professor of Urban Geography and Co-Chair of the Department of Geography, as well as Director of the GIS Core Facility, Binghamton University, Binghamton, N.Y. He has authored four books and numerous articles on the applied aspects of geography. The recipient of more three-quarters of a million dollars in grants and contracts, Professor Frazier has been funded by NSF, EPA, and other agencies and worked as a consultant for HUD's Fair Housing and Equal Opportunity Division, 1994–96.

Professor Frazier has received a number of distinguishing awards, including the New York State University Professors' Service Award (1994), a national AGSG Applied Geography Project Award (1994), a KSU Distinguished Geographer Award (1994), a Distinguished Service Award of the national Applied Geography Conferences (1995), and the James R. Anderson Medal of Applied Geography (1996), the highest honor bestowed by the Association of American Geographers for Applied Geography.

Field

John W. Frazier has worked as a consultant for approximately 25 years, serving all levels of government in the United States and many private sector firms. He established his own company in 1988 and served as its president until its sale in 1999. During his tenure as professional consultant and faculty member at Binghamton University, Frazier consulted with locla, regional, state and the federal governments, advising them on various aspects of planning applications related to human geography. In the 1990s, he advised HUD's Fair Housing and Equal Opportunity Division on the applications of geographic concepts and GIS technology to fair housing issues.

JORGE GASPAR

Biography

Academic education in Lisbon (1960–1965) and Lund (1967–1968) Universities; 1968–1972); training as an applied geographer in private companies and some public departments. OECD and European Union Commission's consultant. Participated in some of the most important urban and regional planning projects in Portugal, along the last 30 years. Recently, co-ordinated the Portuguese Focal Point in the ESDP: European Spatial Development Perspective. In January 2003, appointed by the Portuguese Government to co-ordinate the National Spatial Planning Programme.

Field

As a professional geographer and urban planner I have worked in various domains: prospective, regional planning, urban planning, infrastructures and equipments locali- sation studies, strategic planning, urban rehabilitation and renovation, integrated devel- opment, sector and regional plans ex-post and ex-ante evaluation. This work has been carried out mainly in Portugal and in the European Union, with a certain number of projects in urban planning in Macao and in Angola.

In Portugal and in most of the European countries, the major employers are the State, central administration, local and regional authorities. The European Commission has also revealed an interesting demand, especially through DG Regio. The capacity of employment in the private sector, varies from country to country – in my specific area, the consultation societies and large engineering, architecture and urban planning offices stand out. With the private sector, there is often a competition, but not always a fair one, from the universities, through its departments and study centres or associated company like structures.

The European Commission is undoubtedly the main employer, but, through different projects in Africa, various Portuguese geographers have worked in the Portuguese speaking countries – GIS, environmental studies, tourism and training, with the financial support of the World Bank.

ARTHUR GETIS

Biography

Arthur Getis is the Birch Chair at San Diego State University. With Manfred Fischer he edits the Journal of Geographical Systems. Supported by NIH, he is doing applied geographic research on the transmission of dengue fever. He has received Distinguished Scholarship honors from the North American RSA and the AAG.

Field

Dengue fever is a rapidly spreading viral disease of the tropics. Children and the poor are most susceptible. Especially in its hemorraghic and shock syndrome forms, dengue is deadly. His current research is in the application of GIS and spatial statistics to the study of dengue transmission in Peru and Thailand. This work is supported by the

National Institutes of Health. Those in this field work for such agencies as the Centers for Disease Control (CDC) in the US and the Epidemic Intelligence Service of CDC in Europe. In addition, universities and pharmaceutical firms are increasingly employing spatially oriented researchers.

LAY JAMES GIBSON

Biography

Lay James Gibson is Professor, Geography and Regional Development and Director, Economic Development Research Program, University of Arizona, Tucson, USA. His Ph.D. is in Geography from the University of California, Los Angeles. Gibson is a long-time member of the Association of American Geographers. He is the recipient of that Association's James Anderson Medal for his work in applied economic geography. He is the Past President of the Regional Science Association International (RSAI). He serves on the Board of the Pacific Regional Science Conference Organization; he is a Past President of this organization. Gibson is also a Past President, Life Fellow, and Board Member of the Western Regional Science Association. Gibson served on the Arizona State Land Board for three terms; he was appointed to this Board by Governor Bruce Babbitt. He is a member of the International Economic Development Council (IEDC). The American Economic Development Council (now IEDC) has awarded him both the Richard Preston and Howard Roepke prizes for his contributions to the development profession.

Field

Lay Gibson is an active researcher and consultant in the fields of economic development and regional planning. Applied geographers in these fields work on projects which are designed to inform public and private sector decision makers who are seeking efficient locations for facilities or who are concerned with implementing strategies to more effectively manage regional systems. Typical assignments deal with regional growth including analyses of economic impacts of project development, facility feasibility studies, project development and implementation strategies, target industry and cluster studies, research using economic base theory and central place theory, studies of regional comparative advantage and strategic plans. The economic development specialist's client list might include corporations, local development organizations, government agencies, chambers of commerce, Indian Tribes and non-governmental organizations.

REGINALD G. GOLLEDGE

Biography

Reginald Golledge is a Professor of Geography at the University of California Santa Barbara. His BA (Hons) and MA were obtained from the University of New England

(Australia) and his PhD from The University of Iowa (USA). Past interests focused on modeling transportation flows in freight and human behavioral environments, shopping center location, and consumer behavior. After becoming legally blind in 1984, he shifted his research emphasis to the practicalities of navigation and wayfinding without the benefit of sight. Since then, he has focused much attention on research and activities designed to use geographic knowledge to improve the quality of life for disabled people – a new direction for Applied Geographers.

Field

Although isolated researchers have worked on disability-related topics for decades (particularly related to ethnic groups and poverty), the greatest emphasis has occurred in the last 10–15 years. Both ablest and disabled geographers have emphasized the need for barrier-free environments and socially-responsible treatment of disabled people, and both have pointed to the economically debilitating effects of the technological revolution and the Digital Divide. Specialty Groups of geographers interested in disability exist in the USA (AAG Disability Specialty Group), the UK and Ireland (where research and social activism has impacted government policies towards disability and emphasized barrier-free environments), Australia and New Zealand (again, where social activism is common), Japan (where researchers focus on creating barrier-free environments and uses of e-technology to improve accessibility to public transport), and many of the EEC countries (particularly France, Germany, Denmark, Norway, Sweden, Italy, and Switzerland where geographers interested in applied disability studies have flourished. Other European countries (east and west) have emerging interests in this field, particularly in making environments accessible to travelers. Practitioner organizations include National Societies for the Blind (e.g., Royal Blind Society of the UK), universities, and federal and local governments. Organizations such as the Braille Institute often employ geographers to make tactual maps and embossed signs (e.g., as in Japan) or help design and build optimal barrier-free residential environments (as in Denmark). Perhaps the greatest potential employers now are Public Transit and Mass Transportation Systems where technology is used to create Smart Environments that make movement more accessible for disabled people. More information can be gained from the electronic organization found at http://isc.temple.edu/neighbor/research.

KINGSLEY E. HAYNES

Biography

Dr. Kingsley E. Haynes is University Professor and Dean of the School of Public Policy at George Mason University, Fairfax, Virginia 22030. He holds his Ph.D. in Geography and Environmental Engineering from the Johns Hopkins University. Haynes was the director of McGill's Urban Studies Center, Texas' Joint Center for West Texas Environmental Studies and Indiana University's Regional Economic Development

Institute. He worked overseas on air transport projects in Brazil, regional development programming in Malaysia, water resource and environmental management projects for the Ford Foundation in Egypt, Sudan, and Jordan and investment evaluation programs in Kuwait and Saudi Arabia. In North America he has worked on coastal zone management projects in Texas and Indiana, environment and energy projects in the Yellowstone and Ohio River Basins, and transportation projects in the U.S. Northeast Corridor, Quebec and at the National Center for Intelligent Transport Systems.

Field

Policy analysis is the applied sub field of political geography and the geography of public administration. This field addresses questions of immediate concern to policy makers which often contain issues of spatial distribution of equity and impact. The ability to both respond in a timely manner and to project or predict possible outcomes or policy decisions in their spatial context is of central concern to this application field. In some ways it harness back to the work of geographers in the field of planning and retail analysis but the focus is usually on issues of public service, management of public investments and the consequence of public decisions.

My work in applied geography is on regional economic development policy including the general field of public administration, public management and public policy. Given my interests in regional economic analysis and regional science I have focused on public infrastructure investments in the information, transportation and communication (ITC) sector particularly as they relate to public private partnership activity.

The major contributions in geography to relevant policy analysis has come from the outputs of the IGU's Commission on Geography and Public Administration which followed the outstanding leadership of its early chair Professor Rebert Bennett and his early work on fiscal geography. It is too bad that those outstanding contributions of the Commission have not found themselves in print more often and have not generated a wider following in geography. The quality of their output deserves both.

Possibly one of the the limiting dimensions to the full appreciation of this outstanding policy analysis work in geography is the lack of building a curriculum in the field of geography that fully utilizes this output and trains students to become active participants in the policy process in government directly and in quasi-government and private sector participants in government decision making.

Lei Ding

Lei Ding is presently pursuing a Ph.D. in Public Policy at School of Public Policy, George Mason University. His research interest focuses on regional economic development and the relationship between telecommunications and economic development. He received a MS degree in Management Science and Engineering and a Bachelor degree in Mechanical Engineering from Tsinghua University, China.

QINGSHU XIE

Dr. Qingshu Xie is Research Analyst at MacroSys Research and Technology, Washington, D.C., USA. His research interests include urban economic development, urban land economics, regional economic growth, environmental policy, and public policy analysis.

MICHAEL PACIONE

Biography

Michael Pacione has held academic positions in Queen's University, Belfast; the University of Guelph, Ontario; and the University of Vienna, and currently occupies the Chair of Geography at the University of Strathclyde in Glasgow. In 2002 he was awarded the Higher Doctorate degree of DSc in recognition of his 'original and distinguished contribution to learning in the field of Urban Geography'.

Field

His principal research interest is in the field of applied urban geography, with particular reference to urban problems, urban planning, and urban policy. He has published twenty-five books and more than 100 research papers in an international range of academic and professional journals, and has served as academic advisor to the Scottish Executive on matters relating to city development.

MICHEL PHLIPPONNEAU

Biography

Born in 1921, Michel Phlipponneau has followed, at the Sorbonne, the classical cursus of the classical academic geographer. D. Phil in 1955, he has achieved his entire academic career at the department of Geography of the University of Haute-Bretagne, Rennes, from 1949 to 1984.

His engagement as an applied geographer is mostly linked to his actions as a regional planner for the CELIB (One of the earliest and most active regional development lobbies).

This naturally led him to a further engagement as a political actor: departmental councillor, regional councillor for Brittany, deputy-mayor and president of the urban district of Rennes from 1977 to 1989.

In the mean time, M. Phlipponneau was chairman of the commission on applied geography of the IGU, from 1968 to 1980.

Field

Since the fifties, Michel Phlipponneau has been one of the leaders of applied geography in France. He himself showed, by his actions, that all the branches of geography led to applications.

As an academic geographer, he was himself a consultant, and has worked under contact for private firms, local authorities, administrations, foreign governments (Quebec Ministry of Industry and Commerce), international organizations (UN, for the development of Haïti and Burundi; OECD, for the development of Eastern Thrace). An it is also as an academic geographer that he has trained non teaching-professional geographers, working full time for different types of users.

If in France, in 1961, less than 10 geographers per year were becoming professionals, today geography departments deliver more than one thousand professional diplomas every year, thus training more professional than academic geographers.

JESSIE P.H. POON

Biography

Jessie P.H. Poon is associate professor at the University at Buffalo-SUNY (UB). Her research focuses on trade and development as well as the impact of regional agreements on economic activities in Asia. She is director of UB's Asia Pacific Economic Cooperation (APEC) study center and a member of the Canada-US Trade center.

JAMES MCCONNELL

James McConnell was trained as a geographer at the Ohio State University. He is currently a professor at the University at Buffalo, with interests in international business, industrial geography, and regional development. His research on export decision-making has been used by government agencies and businesses to promote and expand trade.

Field

Geographers in international trade are employed in U.S. Federal Reserve Banks; Export Assistance Centers of the U.S. Department of Commerce; state, county, and metropolitan economic development agencies; the U.S. Census Bureau; logistics and customs-broker companies; import and export trading and consulting companies; and private-sector manufacturing and service firms. Practitioner organizations that are interested in the field include World Trade Centers in the U.S. and foreign countries, Asia

Pacific Economic Cooperation centers, the Federation of International Trade Associations (www.fita.org), and dozens of international trade consulting organizations (e.g., Trade Compass: http://www.tradecompass.com)".

ROBERT STIMSON

Biography

Robert Stimson is Professor of Geographical Sciences and Planning and Director of the Centre for Research into Sustainable Urban and Regional Futures at the University of Queensland. He trained as a geographer at the University of New England in the 1960s, and in 1979 received his doctorate at Flinders University.

Stimson is a life member of the Regional Science Association International (Australia and New Zealand Section), is a former Director of the Australian Institute of Urban Studies in Canberra, and has been director of the Australian Housing and Urban Research Institute at the University of Queensland and Queensland University of Technology. He is a member of the Brisbane Development Association. He has conducted research and consultancies for the Commonwealth and various State Governments, local councils, the private sector and national bodies in Australia, and for AusAid and the World Bank in South East Asia.

Field

Throughout his career Robert Stimson has sought to demonstrate the application of geographical analysis to public policy issues. That includes investigating the patterns of immigrant settlement in Australian cities, the impacts of globalisation, economic restructuring and population shifts on the performance of cities, towns and regions; the social implications of locational disadvantage in cities; the provision of retirement housing, social housing policy and practice; and regional development and planning strategies for the Brisbane-SEQ region.

In 1990–91 Stimson was Project Director of The Brisbane Plan for the Brisbane City Council.

Stimson is a regular media commentator on urban and regional development and planning issues.

HARRY J.P. TIMMERMANS

Biography

Harry Timmermans (1952) is chaired professor of urban planning and Director of the European Institute of Retailing and Services Studies (EIRASS) at the Eindhoven

University of Technology, P.O. Box 513, 5600 MB Eindhoven, The Netherlands, eirass@bwk.tue.nl.

Between 1991 and 1996 he also was chaired professor of retail marketing at the University of Alberta, Edmonton, Canada. He holds research interests in modelling spatial choice behavior and spatial decision support systems in a variety of application contexts, including retailing, tourism, recreation, transportation and housing. He is editor of the Journal of Retailing and Consumer Services.

Field

Specialist in retail, tourism and transportation geography are hired by planning authorities, national and international consulting firms such as Rand Corporation and urban developers. Examples of practitioner organisations are the Transportation Research Board in the USA, the Travel and Tourism Association with chapters in the States and Europe and the International Council of Shopping Centers, also with a European and American link.

INDEX

AAG, 202–204, 206, 208, 209
Academic geography, 53, 65
Accessibility, 242, 243, 248–251
Applicable geography, 190, 191, 194, 202
Applied geography, 47, 48, 53, 55, 56, 58–63, 65, 66, 187–195, 197–199, 201–209, 231, 305, 306
Applied Geography Conferences, 202, 203, 208–210
Applied research, 28–31, 33, 37, 40–42

Benchmarking, 287–289, 291
Best practice, 113
Blindness, 213, 217, 232
Border changes, 174

Cartography and visualization, 98
Central Europe, 169, 170, 173, 175, 180–182, 185
Chicago School, 200
Clusters, 113, 131
Computational geography, 98
Conceptualization of space, 97
Confirmatory spatial data analysis, 98
Core-periphery, 253
Cross-commuting, 118
Cross-shopping, 117, 118

Design aspects of GIScience and technology, 97
Developed nations, 258
Developing countries, 253, 257, 258, 267, 268, 270
Direct investment, 269, 271
Disability, 214–216, 223, 225, 230–232
Disadvantage, 213, 215
Discrete choice, 136, 140
Discrimination, 213, 223, 225

Economic base theory, 113
Economic development, 113

Education, 151, 155, 162, 163, 165, 166
Employment, 278, 283
Employment multipliers, 113, 130
Employment of applied geographers, 183
Endogenous growth theory, 261
Environmental legibility, 238, 240
European Union, 152, 166, 168
Explorations, 47–51
Exploratory spatial data analysis, 98
Export promotion, 257, 261, 268
Exports, 253, 254, 257–261, 266, 268, 270

Feasibility, 136, 141
Firm-territory nexus, 262

Geographic information science, 95
Geographical societies, 53, 56
Geography and politics, 153, 154
Geostatistics, 102
GIScience and criminal justice, 107
GIScience and demographic research, 105
GIScience and disaster response, 104
GIScience and economic research, 106
GIScience and epidemiology, 104
GIScience and political science, 107
GIScience and public health, 105
GIScience and sociological research, 106
GIScience and telecommunications, 105
GIScience and the environment, 107
GIScience and urban and transportation research, 106
GIScience, government, and public policy, 107
Globalization, 253, 255, 256, 258, 269
Government, 10, 12, 21, 306

Health care, 273–279, 283–285

Imperfect competition, 261
In-commuters, 116
Income convergence, 69

319

Institutions, 257, 262, 264, 266, 268
International competitiveness, 263, 266
International trade, 253–258, 261, 262, 264, 266, 268, 269, 271

Landmarks, 236–240, 245, 247, 249
Local, 273–277, 280–284

Marxist, 27, 31, 36, 40, 42, 43
Media, 19–21, 305, 306
Medical geography, 273, 274
Medicometry, 273, 276, 279, 282, 285
Model, 9, 14
Models, 133, 136–143
Multiplier effects, 130

New geography, 152, 157

Out-commuters, 123

Pattern analysis, 100
Planning, 283
Policy analysis, 69, 73
Policymakers, 254, 256, 260, 261, 263, 265
Political geography, 69, 89, 92, 153
Postmodernism, 27
Practice, 23, 24, 32, 33, 39, 40, 42, 43
Preference, 139, 140
Principles, 24, 27, 28, 40, 42, 43
Problem-oriented approach, 29
Product life cycle theory, 261
Professional geographer, 47
Professional, social, legal aspects of GIScience and technology, 98
Public policy, 32, 41–43

Quality of life, 31, 39

Region, 274, 276–279, 282, 283
Regional planning, 171, 172, 175–177, 181, 183, 185
Regional science, 15, 16, 20, 22
Relevance, 23, 25, 28, 33, 38–44
Retailing, 143–147

Shopping centers, 135, 138–142
Signage, 237, 239, 242, 246, 248, 249, 251
Smart environments, 244
Social justice, 27
Social studies, 155, 165
Social usefulness, 42
Space, 11, 17–19, 306
Spatial association, 101
Spatial choice, 137, 139
Spatial data acquisition, sources, and standards, 98
Spatial data manipulation, 98
Spatial data models and data structures, 97
Spatial econometrics, 103, 109
Spatial theory, 187, 195, 196, 199
Spatial thinking, 195, 196, 200
Stores, 133, 135, 138, 140–142

Trade policy, 257, 259, 263
Transfer payments, 122
Transportation networks, 234, 241
Travel behavior, 235, 238, 246, 248

Useful knowledge, 24, 25, 29, 34, 190, 207

Values in research, 28, 30

War-related activities, 192
Wayfinding, 233, 234, 241, 247–250, 252
World economy, 29